The Cloud in IoT-enabled Spaces

The Cloud in IoT-enabled Spaces

Fadi Al-Turjman

CRC Press is an imprint of the
Taylor & Francis Group, an **informa** business

CRC Press
Taylor & Francis Group
6000 Broken Sound Parkway NW, Suite 300
Boca Raton, FL 33487-2742

Printed on acid-free paper

International Standard Book Number-13: 978-0-367-27814-4 (Hardback)

Visit the Taylor & Francis Web site at
http://www.taylorandfrancis.com

and the CRC Press Web site at
http://www.crcpress.com

"Good, better, best. Never let it rest. 'Til your good is better and your better is best.."

St. Jerome

This book is dedicated to my wonderful family, my little stars, and my dearest students..

Do your best to be remembered as someone who did the best with the time and talent you got. And remember: Above the Cloud are the stars with their beauty.

Fadi Al-Turjman

Contents

Preface and Outline

Smart spaces are solutions where the Internet of Things (IoT)-enabling technologies have been employed towards further advances in the lifestyle. It tightly integrates with the existing Cloud infrastructure to impact several fields in the academia and/or industry. IoT-enabled spaces have made revolutionary advances in the Cloud that we know so far. Cloud is used not only in computing applications but also in our daily activities towards more intelligent health, traffic, and even farming systems. Cloud enabled with IoT features is expected to integrate smoothly with the emerging fifth-generation (5G) networks and beyond in the next-generation telecommunication paradigms. This opens the door for other exciting research and application opportunities. In this book, we are considering significant research topics towards realizing the future vision of Cloud in IoT-enabled spaces.

Cloud computing (CC) and IoT-based techniques have been merged with wireless networks (WNs) for the purpose of storage and data access anytime and anywhere. In the Cloud, users can find the set of required hardware devices, network connections, storage spaces, data services, and application interfaces that are easily accessed over the Internet. As the CC paradigm will be shortly integrated with intelligent IoT spaces, user authentication, task scheduling, medium access, and the forthcoming smart applications (indoor and outdoor) will all be hot research topics and open issues to be further investigated and validated.

This book overviews these issues and proposes the most up-to-date alternatives and solutions. The objective is to pave the way for IoT-enabled spaces in the next-generation CC paradigm and open the door for further innovative ideas. Accordingly, our main contributions in this book can be summarized as follows:

- We start by overviewing the importance of IoT-enabled spaces in CC paradigms and the necessity to integrate numerous tasks, which can compete for the same set of resources. We propose an optimized task scheduling approach that can help in releasing the Cloud stress towards more reliable and fault-tolerant performance, especially in smart survivability applications.
- We introduce a new scheduling and routing approach to enhance the end user experience while better utilizing the existing Cloud/IoT network

resources by providing improved transmission speed for emerging big data applications.

- We also target the wireless medium access problem under rapid mobility conditions in smart city spaces. We take the advantage of CC and put forward an agile vehicular cloud and mobile IoT communication framework.
- Furthermore, we consider the big data storage issue and propose a novel cache replacement approach for smart applications enabled by the emerging software-defined networks (SDNs) concept at the edge of the Cloud. This replacement approach provides a significant availability for the most valuable and difficult-to-sense data in IoT spaces.
- As for the security issue, a seamless key agreement approach for the extension of user authentication over the Cloud is proposed. Using an elliptic curve, the proposed approach offers proper mutual authentication and session key agreement between the mobile user and the base station.
- Moreover, we propose a framework that addresses data delivery issues in IoT-enabled Cloud by considering the uncertainty factors in this paradigm. We compare the performance of this framework in both a centralized and a distributed work fashion. The results show a 60% speedup in the distributed approach compared to its centralized version when the uncertainty entries increase.
- Finally, this book provides a comprehensive overview of the efforts that have been done so far in realizing smart spaces in homes, cities, and even farms. It highlights the developments done in the field of these smart spaces and the challenges the researchers are trying to resolve under the umbrella of CC. It simply says "Cloud Computing is grid computing". It enables us to use existing luxury infrastructures without the need to buy it. ***Shortly there will be no need for powerful PCs (Personal Computers). We will be in need for connected PEDs (Personal Edge Devices).***

Fadi Al-Turjman

Author

Prof. Fadi Al-Turjman received his Ph.D. degree in computer science from Queen's University, Canada, in 2011. He is a Professor with Antalya Bilim University, Turkey. He is a leading authority in the areas of smart/cognitive, wireless and mobile networks' architectures, protocols, deployments, and performance evaluation. His record spans over 200 publications in journals, conferences, patents, books, and book chapters, in addition to numerous keynotes and plenary talks at flagship venues. He has authored/edited more than 20 published books about cognition, security, and wireless sensor networks' deployments in smart environments with Taylor & Francis, and the Springer (Top tier publishers in the area). He was a recipient of several recognitions and best papers' awards at top international conferences. He also received the prestigious *Best Research Paper Award* from Elsevier *Computer Communications* (*COMCOM*) Journal for the last three years prior to 2018, in addition to the *Top Researcher Award* for 2018 at Antalya Bilim University, Turkey. He led a number of international symposia and workshops in flagship IEEE ComSoc conferences. He is serving as the Lead Guest Editor in several journals, including the *IET Wireless Sensor Systems*, Springer *EURASIP Journal on Wireless Communications and Networking*, MDPI *Sensors*, Wiley & Hindawi *Wireless Communications and Mobile Computing*, and the Elsevier *COMCOM* and *Internet of Things*.

Contributors

Hussain Al-Rizzo
Systems Engineering Department
Donaghey College of Engineering & Information Technology, University of Arkansas at Little Rock
Little Rock, Arkansas

Fadi Al-Turjman
Department of Computer Engineering
Antalya Bilim University
Antalya, Turkey

Mohammed Zaki Hasan
Systems Engineering Department
Donaghey College of Engineering & Information Technology, University of Arkansas at Little Rock
Little Rock, Arkansas

Aissa Houdjedj
Department of Computer Engineering
Antalya Bilim University
Antalya, Turkey

Arman Malekloo
Department of Computer Engineering
Middle East Technical University
Northern Cyprus Campus, Turkey

Mohamad Sanwal
Department of Computer Engineering
Antalya Bilim University
Antalya, Turkey

Hadi Zahmatkesh
Department of Computer Engineering
Middle East Technical University
Ankara, Turkey

Chapter 1

Task Scheduling in Cloud-Based Survivability Applications Using Swarm Optimization in IoT[1]

Fadi Al-Turjman
Antalya Bilim University
Mohammed Zaki Hasan and Hussain Al-Rizzo
University of Arkansas at Little Rock

Contents

[1] **F. Al-Turjman**, M. Z. Hasan, and H. Al-Rizzo, "Task Scheduling in Cloud-based Survivability Applications Using Swarm Optimization in IoT", *Transactions on Emerging Telecommunications*, 2019. DOI. 10.1002/ett.3539.

1.1 Introduction

Survivability services are often the most demanding in urgent applications, and various types of disasters. Newfangled devices, systems, applications, and interfaces are expected to become key interconnected enabling technologies in Internet of Things (IoT) [1]. Existing communication schemes for IoT devices run over Wireless Area Network (WAN), 3G and 4G networks. Therefore, IoT allows achieving multiple urgent tasks and gather useful and actionable information [2]. Indeed, it will allow performing several tasks over the cloud including beforehand warning, communicating, remote controlling, etc. [3]. And thus, the IoT paradigm is in need for a large-scale distributed system such as the cloud, where numerous and heterogeneous resources with respect to hardware and software are managed. For example, Figure 1.1 depicts applications of IoT based real-time scenario for

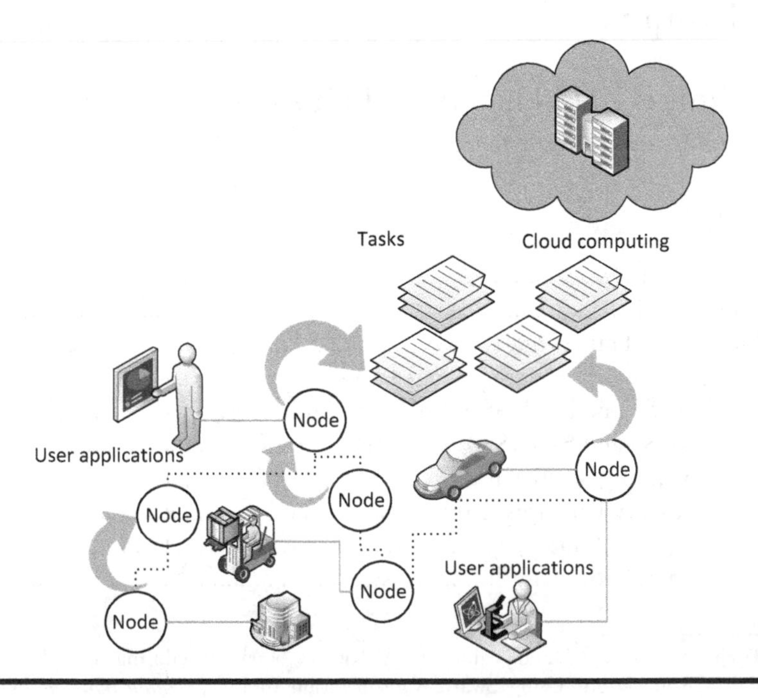

Figure 1.1 Types of IoT requests in cloud environment.

remote monitoring system in a smart city, whereas nodes continuously monitor critical parameters depends on the application. Therefore, the critical nature of these applications in smart environment compels to have a reliable and stable connections towards the cloud computing.

Stability and the survivability of the connection are important in these types of applications. Moreover, IoT needs to guarantee certain quality properties as the minimum requirement to meet service providers' and end users' requirements. Quality of Service (QoS) is thus a very important aspect to be considered in IoT applications, because it assures that services will be delivered to end-users while ensuring their QoS requirements via modeling their human-type traffic which is usually done by the means of Poisson process (PP). However, PP is not well suited for modeling coordinated data traffic, which is expected to play a large role in IoT applications [4]. Such IoT applications necessitates QoS requirements levels including the efficient utilization of network resources while delivering quality services such as competitive response time, high availability and best resources allocation. Mapping user level requirements into the network level resource requirements as well as mapping QoS requirements into pricing signals depend on specific network properties, specific implementation of the user requirements, and definition of the survivability applications. Therefore, this mapping process faces several challenges that impede their adaptation in IoT. Such challenges include scalability, resource allocation, energy efficient congestion, network bandwidth, and theft protection service for cloud data and devices. To address these challenges which can negatively affect the performance of the IoT applications, we recommend the mapping process of the IoT applications at the network edges in order to provide better QoS monitoring in survivability applications. But how many of these applications can work simultaneously and consistently? This is the key point.

Nowadays, several IoT devices and services might connect and interact with each other in a single home or a small business at any time. This creates countless, unpredictable demands for the available services and resources over the Cloud. It causes a complicated issue for users who do not own sufficient computing resources [5]. Additionally, the required QoS for a user to access the Cloud services dramatically is increasing. In addition to resource heterogeneity, scalability and context awareness factors can make the resource management a very challenging task over the Cloud [6]. We claim that the use of IoT system as an infrastructure can make large difference against the existing traditional practices in cloud computing paradigms. Knowing that resource allocation mechanisms in cloud computing paradigm necessitate sophisticated strategies in order to better allocate resources while satisfying users' demands and requirements [5]. Therefore, integrating cloud computing and IoT is not an easy task and it might be difficult to establish, distribute and maintain various applications while meeting the peak demands of users with preserved QoS requirements.

In IoT, various applications for different devices attempt to exploit the underutilized different classes of data traffic [7]. Because of the varying capabilities of the available heterogeneous servers and devices, multiple tasks scheduling process is required to meet different users' QoS requirements. This process depends on the allocation policy type, flexibility and priority of the data traffic classes [8]. In fact, there are different classes for the experienced data traffic in IoT paradigms and it necessitates different QoS constraints that need to be addressed while maximizing the available Cloud resources' utilization. Therefore, it is essential to propose an appropriate scheduling algorithm to provide an optimal queue management for multi-classes data traffic in order to ensure the maximum possible utilization of the IoT resources allocation among the heterogeneous systems of servers and personal devices. There might be cases where typical scheduling algorithms in Cloud computing paradigm cannot satisfy the QoS constraints. And such inefficient scheduling can lead to undesired lengthy delay and low network throughput.

In this chapter we propose an efficient scheduling approach in order to optimize the management queue while guaranteeing the QoS levels. We develop our scheduling algorithm based on a meta-heuristics approach while optimizing the profit of the IoT resource allocation. The proposed algorithm provides an integration between IoT and Cloud computing allowing optimal scheduling for applications to the users in several real-world domains. A robust Particle Swarm Optimization (PSO) for Dynamic Dedicated Server Scheduling (DDSS) and heterogeneous Dynamic Dedicated Server Scheduling (h-DDSS) is proposed to handle massive incoming tasks. PSO has the advantages of simply searching and converging to optimal solution. We employ the PSO in serving the users' requests by updating the velocity of each particle corresponding to a different device/server type. The value of this velocity takes into consideration the resource allocation policy and the QoS constraints to better quantify the difference between the high and low data traffic priorities. We emphasize that each task has a strict QoS constraint while scheduling the multi-incoming tasks to assure better resource allocation as has been done in Dead-line Constrained Task Scheduling (DCTS) problem [9].

The rest of the chapter is organized as follows. Section 1.2 presents previous works related to task scheduling in Cloud computing and IoT. Section 1.3 describes the framework architecture of IoT in cloud computing environments. Section 1.4 presents the architecture of two scheduling algorithms for homogeneous and heterogeneous objects/devices. Section 1.5 presents the two optimization algorithms. Section 1.6 presents the performance of the two algorithms with two task scheduling. Section 1.7 concludes this work.

1.2 Related Work

Infrastructure as a Service (IaaS) is considered an essential part of cloud computing that becomes as the foundation for high level Platform as a Service (PaaS) and Software as a Service (SaaS) models. IaaS providers allow users to allocate the resources in the form of different Virtual Machines (VMs) characterized by QoS parameters, machine configurations and pricing service model [10]. The service provision follows several infrastructure approaches such as centralized, distributed and hybrid for applications' running, and resource allocation mechanisms in IoT services supported by the cloud computing paradigm. Moreover, users' applications requests are aperiodic (i.e. uses's demands cannot be static and constantly change with time) and the number and types of requested VM instances are hard to predict dynamically provisioning of resources. Therefore, IaaS service model can experience massive requests which cannot be satisfied while guaranteeing the QoS parameters.

Typically, a resource allocation problem is considered as an optimization problem and there are several attempts to solve it using heuristic algorithms under different assumptions and constraints as in [9], [11], and [12]. Authors in [9] have proposed a strategy based on PSO to process a task scheduling approach for IaaS cloud. The authors tries to minimize the execution time and cost by considering the deadline (delay) constraints. Authors in [13] focus on minimizing the total execution cost and transferring time in order to reduce the executing time. Therefore, they use PSO to achieve the computing intensive task scheduling by transforming updated velocity vector values from the continuous to discrete permutations. A new variant of the continuous PSO algorithm, named Integer-PSO, is proposed in [14] to solve the bi-objective task scheduling problem in pubic or in private cloud for independent tasks. The authors perform the smallest position value (SPV) rule based PSO technique which is achieved through the weighted sum approach. There are several studies based on multi-objective functions for task scheduling using many optimization techniques but they have not been addressed in the case of private or public cloud with independent tasks to be scheduled. They have been only proposed for work flow scheduling in a hybrid cloud. Our approach proposes a new variant of PSO algorithm taking into consideration QoS parameters requested by the end user and committed by the the task arrival rate and service rate which are highly unpredictable and dynamic in nature. Authors in [15] propose solving such problem to improve the utilization of services and shorten the task execution time in case of the dynamic task arrival. The strategy of selecting tasks is based on the current optimal services state according to the real-time status of the candidate services that can shorten the task execution time.

The authors in [11] surveyed heuristic algorithms in the literature in order to schedule autonomous tasks in homogeneous and heterogeneous cloud environments. It is worth mentioning that Cloud resources can be either provided by physical or virtualized approaches [3]. On the other hand, authors in [1] examine the infrastructure of different IoT protocols in order to identify the key technical challenges in implementing virtualized services in Cloud environments. A straightforward solution can be represented by a distributed infrastructure which supports key features such as scalable services, on-demand usage, and over-purchase resources in advance. Nevertheless, a few attempts in the literature have taken into consideration the usage of heterogeneous servers in centralized infrastructures.

Centralized infrastructure brings many well-known challenges such as resources over provisioning, unlike distributed clouds which hires infrastructure on demand, and acquires dedicated connectivity and resources from communication providers. The authors in [16] introduce "embarrassingly distributed applications" which do not require massive internal communication among server pools, It is created out of a small distributed data centers. The limitations of this work is that different classes of users' requests are not considered. Meanwhile, authors in [17] choose "distribute voluntary resources" to form the building of Cloud as "nebulas" that is more dispersed and has low deployment cost. However, it suffers the central management major issue with respect to reliability and state maintenance in case of failures. An alternative solution could be achieved by increasing profits for the service providers offering cloud services through a kind of federation agreement. This approach is derived from management agreement policy. However, applying this approach in practice is neither simple and nor easy [8]. Therefore, a scheduling algorithm which considers the different priorities for different classes of data traffic must be designed.

In this chapter, we provide a task scheduling approach by considering the priority of IoT users' requests. The application is decomposed into a sequence of distributed tasks taken into account these tasks represent performing a survival operation such as computing the average of the temperature in a given geographical area, measuring the light intensity in a room, video surveillance of a specific geographical area, or a combination of these. Therefore, the determination of level of priority for a specific application request is used to decide the next request to be scheduled and served on an iterative and asynchronous robust optimization of the task allocations among neighboring nodes in IoT. This determination of level of priority for a specific application revealed survivability services in which at each instant of time, each node in the network has some positive probability to interact with one of its neighbors. A similar approach is reported in [18] where task scheduling adaptable to homogeneous and heterogeneous environments is performed with taking into account various aspects related to system performance metrics, such as drop rate, throughput, and utilization in IoT. However, the major difference between the reported in this chapter and [18] is that our distributed algorithm is based on robust particle swarm optimization in which tasks scheduling between devices of the IoT and Cloud computing resources is able to increase the capabilities of the IoT system and to achieve greater overall performance in application execution.

1.3 Cloud-Based Framework

The framework of resource allocation is described in Figure 1.2. It offers a public interface component for received users' applications requests/tasks. In this framework, each user requests an application from different Application Entry Point (AEP). These entry-points might be physical devices, such as smart phones which

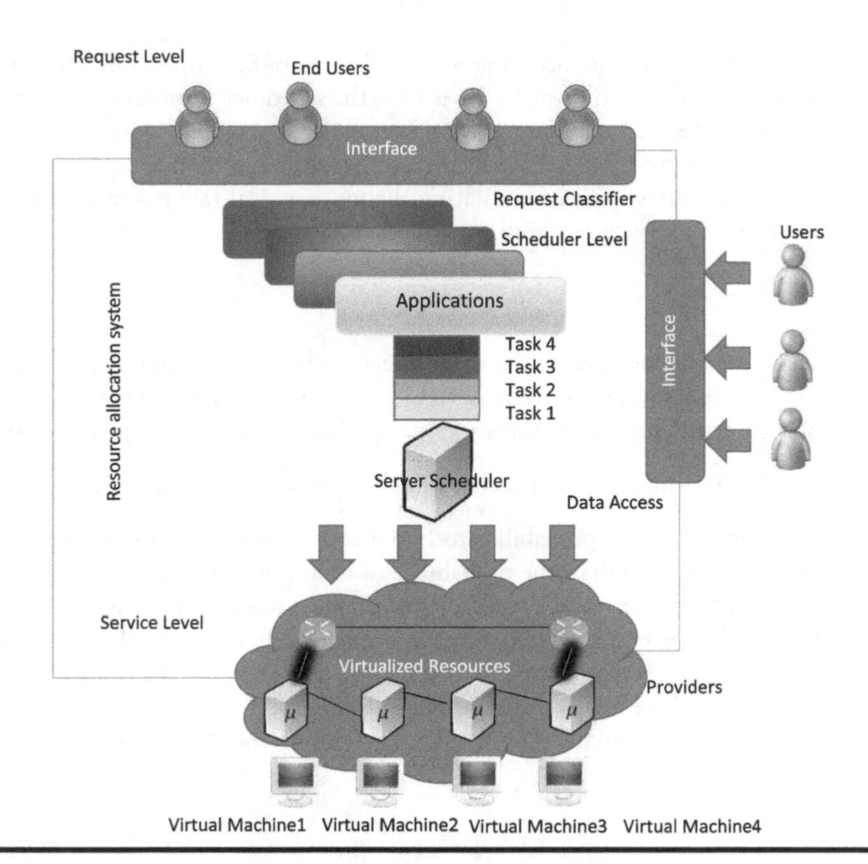

Figure 1.2 Homogeneous and heterogeneous dedicated server groups for different applications types of IoT requests in cloud environment.

are part of IoT infrastructure or they might be personal computers running web portals such as a user interface, virtual machine (VM), or even a gateway. We claim that each VM is orthogonal to four most upper and lower layers which are defined as user interface, request manager, scheduler, service level and external cloud interface. All tasks are received and classified using classifier component that is solely responsible for classifying user's tasks according to priority level of the data traffic class. We use C_i to determine the class number of each task that needs to be scheduled for being served, where $i = 1,2,\ldots,m$. Each data traffic has arrival rate denoted by λ_i, therefore λ_1, and λ_2 will be used for C_1, and C_2 respectively and so on. Suppose that $CP = CP_1, CP_2,\ldots, CP_n$ is a set of cloud providers. Assume CP_1 is the private cloud and $CP_2,\ldots, CP_n$ are ECs. $VM = VM_1, VM_2,\ldots, VM_I$ is a set of VM types and $A = a_1, a_2,\ldots, a_w$ is a set of applications. Each application $a_\epsilon \left(\epsilon \in \{1,2,\ldots,w\}\right)$ consists of a number of parallel and independent tasks $Task_j = \left\{task_{j1}, task_{j1},\ldots, task_{jTask_j}\right\}$ that have strict QoS parameters in terms of delay and throughput. Each task is queued via a queue manager who is responsible for collecting and queuing the incoming user's tasks in a buffer. Thus, a new arriving user's task will be dropped if the buffer is full. The scheduler component is used to schedule and serve the incoming tasks with a service rate μ. It allocates resources under specific constraints. According to equilibrium equations in queuing system [19], we are interested in formulating the mean arrival rate per task which is equal to the mean service rate, and we should expect

$$\mu_1 \le \mu_2 \le \cdots \le \mu_m \tag{1.1}$$

whereas μ_1 is the service rate of the slowest server when the length queue is small. Meanwhile μ_2 is associated with the second lowest server, and so on till μ_m, which is associated with the fastest server that experiences the largest queue length. Therefore $\lambda_i \to 0$ and $\mu_i \to 0$ in such a way that $\frac{\lambda_i}{\mu_i}$ is constant. We assume heavy data traffic flow, and the probability to find a task to serve (i.e., to be scheduled in the system) is higher than the probability to find it in any other server. That is because the task is expected to stay longer in the slow server in comparison to the fastest one. In order to know which server is more likely to be the slowest or the fastest and when all server m are busy or not, we consider that the probability of being the last one depends on the knowledge of which server started which service. This can be learned through the shared traffic among the servers. And thus, we assume

$$\lambda = \lambda_i \quad \text{for} \quad i = 1, 2, \ldots, m \tag{1.2}$$

and μ_m is variable and defined as

$$\mu_m = \begin{cases} \mu_1, & \text{if} \quad i = 1 \\ \mu_2, & \text{if} \quad i = 2 \\ \vdots & \vdots \\ \mu_1 + \mu_2 + \cdots + \mu_m, & \text{if} \quad i \ge m \end{cases} \tag{1.3}$$

Basically, we need to compute an upper bound for how long a $Task_i$ can stay in the slowest server. Thus, Eqs. 1.2 and 1.4 can be reformulated in three ways: (i) when the queue system contains fewer *Tasks*, and μ_i is varying, (ii) when more *Tasks* in the queue system and μ_i is constant, and (iii) when the queue system contains fewer *Tasks*, and the utilization ratio ρ is varying. Therefore, we need to define the available number of servers in order to facilitate the development and visualization of the utilized resources.

$$m_i = \sum\nolimits_{j=1}^{i} \mu_j \quad \text{if} \quad i \le m \tag{1.4}$$

and

$$m_i = m_l = \sum\nolimits_{j=1}^{l} \mu_l \quad \text{if} \quad i \geq m \tag{1.5}$$

Therefore, when $i = 1$, the $m = \mu_1$. This means that there exists one $task_i$ in the queue and there is no server (i.e., high probability to be the slowest server). Suppose, if there are two requests in the queue system, then the service rate is $\mu_{t2} = \mu_1 + \mu_2$. Therefore, the service rate will continually increase until all server are utilized. We discuss in next section 1.4 the upper bound of the used queue system as an approximation for the framework performance in terms of QoS parameters such as throughput and delay for DDSS and h-DDSS. The notations used in the rest of the chapter are listed in Table 1.1.

Table 1.1 Notation

Symbol	*Quantity*
DDSS	Dynamic Dedicated Server Scheduling
$h-DDSS$	Heterogeneous Dynamic Dedicated Server Scheduling
VM	Virtual Machines
C	Class of incoming data or task need to server
λ	Arrival rate for each class of incoming data traffic
μ	Service rate for each class of incoming data traffic
μ_i	Initial service rate of dedicated i servers for incoming data traffic
μ_{ti}	Service rate of dedicated servers for incoming data traffic until i^{th} during specific period time t
AEP	Application entry points
CP	Cloud provider
EC	External Cloud
A	Set of Applications
m, k	Number of dedicated servers for incoming data traffic
l	The total number of available server in queue system
r	The number of classes data traffic

(Continued)

Table 1.1 (*Continued*) Notation

Symbol	*Quantity*
p_i	Probability of i number of applications requests in the queue system
χ	The queue length process
n	The number of state of Marhov chain
ρ	The system utilization
N	The size of queue
p_0	The probability of steady state of queue system
Drop	Drop probability of request from the queue system
Throughput	The throughput of the queue system
η	Average occupancy in the queue system
Delay	Average delay of the queue system
ξ	The constriction coefficient
v_i	Velocity toward selecting optimal solution
P_i	Personal best position for $Task_i$
G_i	Global best position for $Task_i$
p_{best}	Personal-best
g_{best}	Global-best
ϕ_1	Personal best coefficient
ϕ_2	Neighbor best coefficient
ξ	Constriction coefficient

1.4 Homogeneous and Heterogeneous DDSS

The architecture of Homogeneous and Heterogeneous Dynamic Dedicated Server Scheduling are depicted in Figures 1.3 and 1.4. DDSS and h-DDSS frequently update the number of dedicated CPs with a VM for each class data traffic according to the priority level of the requested application as well as the arrival rates of C_1, and C_2. In contrast to homogeneous DDSS, service rates of CP can vary in h-DDSS. And the capabilities of CP in h-DDSS can also vary.

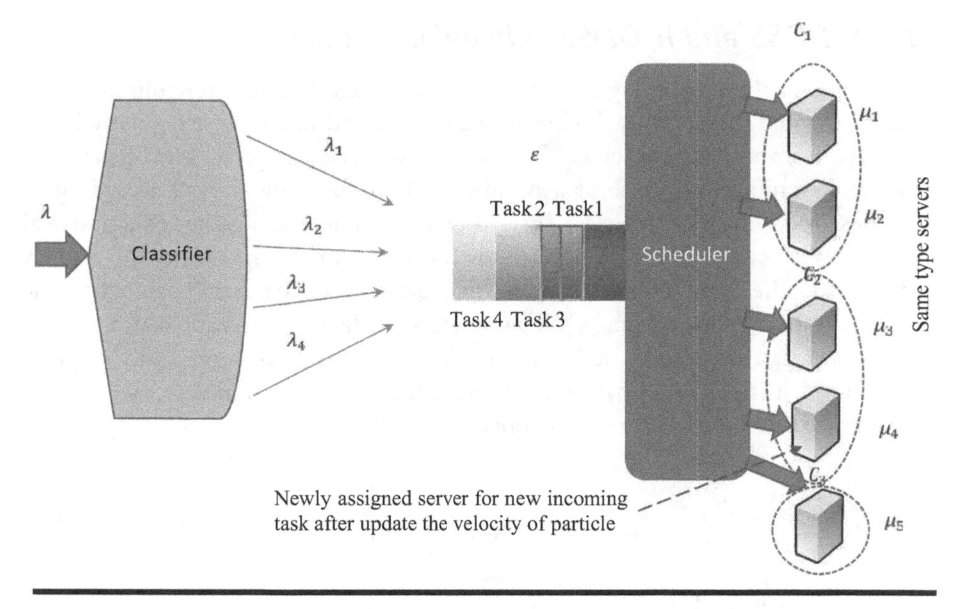

Figure 1.3 Homogeneous dynamic dedicated server scheduling DDSS.

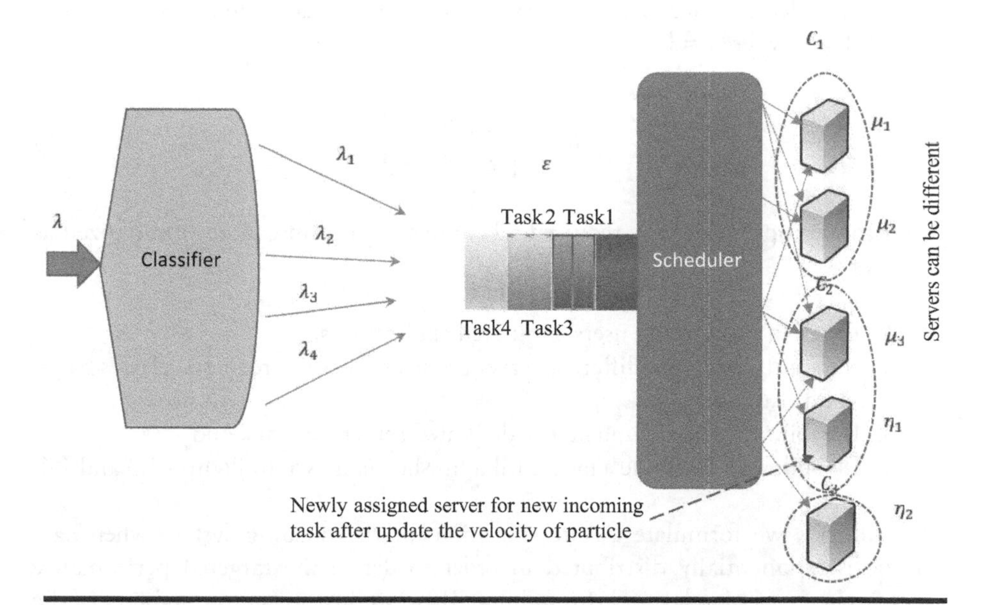

Figure 1.4 Heterogeneous dynamic dedicated server scheduling h-DDSS.

1.4.1 DDSS and h-DDSS Scheduling Model

The duties of the scheduling algorithm is summarized in characterizing the number of assigned servers to each class of data traffic and updating it regularly based on Eqs. 1.6 and 1.8. The classifier classifies all arrived requests and forwards it to the corresponding queue of a number of *VMs* for being served according to the designed allocation policies. This process continues during the request arrival period. We focus on three parameters in this period, the arrival rate, the priority of each CP, the number of servers and the required service rate to enable dynamic tasks scheduling with optimal QoS parameters for the requested application. These parameters can be used in deriving the allocated number of servers in the queue system which is denoted by m. Thus, these parameters are assigned to each class of data traffic and for each requested application as follows:

$$m = \left\lfloor \frac{l\Psi_i\lambda_i}{\sum_{i=1}^{n} \Psi_i\lambda_i} \right\rfloor \tag{1.6}$$

The dedicated number of servers is calculated as follows [19]:

$$k = l - m \tag{1.7}$$

Where k and m define the number of dedicated servers for the incoming data. Eq. 1.6 can be extended to process a multi-class data traffic with r data traffic types by assuming $\Psi_i\lambda_i \neq 0$ where $i = \{1,\ldots,r\}$ as:

$$\mu_{tm} = \left\lfloor \frac{\mu_{total}\Psi_i\lambda_i}{\Sigma_{i=1}^{r}\Psi_i\lambda_i} \right\rfloor \tag{1.8}$$

And thus, properties of the required scheduling algorithm can be summarized as follows:

1. The ability to classify users' requested applications,
2. The ability to assign different server groups to distinct requests' classes based on Eq. 1.6,
3. The ability regularly update the dedicated servers' counts, and
4. The ability to continue a task until it finishes as shown in Figures 1.3 and 1.4.

Accordingly, we formulate a state transition for the queuing system when data traffic is exponentially distributed in order to derive the targeted performance metrics in terms of delay and throughput. Then, we investigate the stability of the state transition for a discrete-time queuing system in order to assign the arrived tasks/requests to a specific application over the Cloud.

We remark that it is possible to perform multiple requests scheduling in a specific time period based on the developed equilibrium equations. However, the scheduling algorithm shall consider the dependency of service rates on the incoming request type/class in order to avoid the high dropping rate. This means that the service rate increment shall continue until all servers are utilized.

In the following, we develop the equilibrium equations for the above described state transition diagrams, which are shown in Figures 1.5 and 1.6 for DDSS and h-DDSS, respectively. The queue length $\{\chi_n \mid n = 0,1,2,\ldots\}$ is defined as a Markov chain with a finite state space and it satisfies the stability condition $\rho < 0$ to have a unique stationary probability distribution which can be written as follows:

$$\mu_1 p_1 = p_0 \lambda_1 \Leftrightarrow p_1 = \frac{\lambda_1}{\mu_1} p_0 = \rho_1 p_0 \tag{1.9}$$

$$\mu_2 p_2 = p_1 \lambda_2 \Leftrightarrow p_2 = \frac{\lambda_2}{\mu_2} p_1 = \rho_2 p_1 \tag{1.10}$$

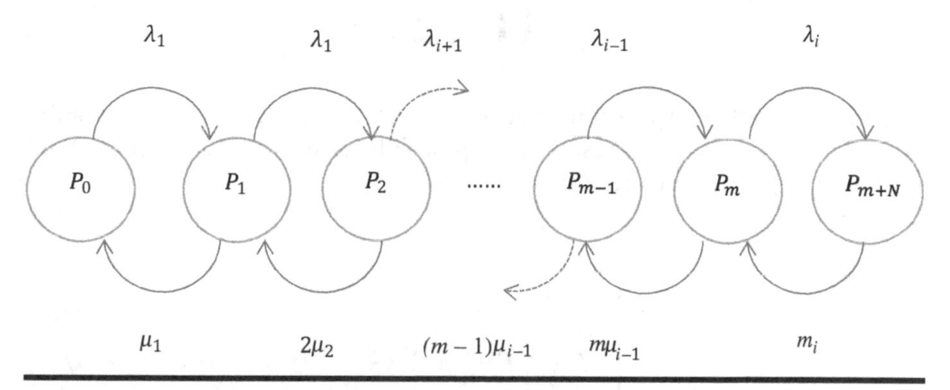

Figure 1.5 State transition diagram for homogeneous DDSS.

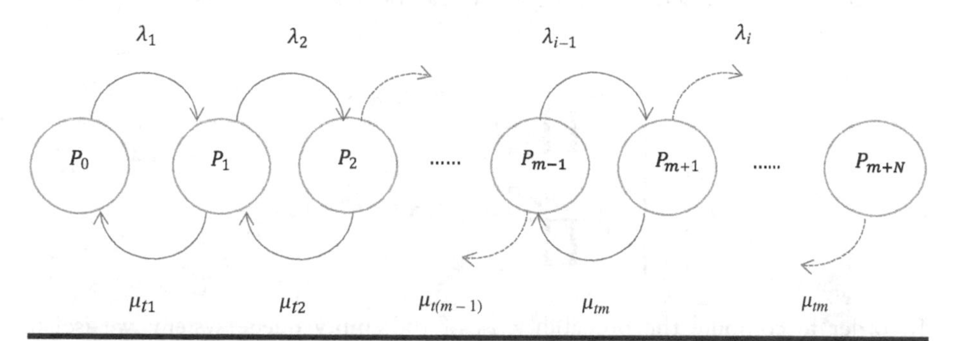

Figure 1.6 State transition diagram for heterogeneous dynamic dedicated server scheduling (h-DDSS).

$$\mu_3 p_3 = p_2 \lambda_3 \Leftrightarrow p_3 = \frac{\lambda_3}{\mu_3} p_2 = \rho_3 p_2 \tag{1.11}$$

$$\vdots \tag{1.12}$$

This will be repeated i times. Therefore, the last term can be written as:

$$p_i = p_0 \frac{\lambda^i}{\sum_{j=1}^{k} \mu_j} \tag{1.13}$$

To develop the equilibrium equation, the probability p_i of processing a task in the queue system shall be written as follows:

$$p_i = p_0 \frac{\lambda^i}{\prod_{k=1}^{i} \left(\sum_{j=1}^{k} \mu_j \right)} \tag{1.14}$$

In contrary to DDSS, we assume that $\mu_1 \leq \mu_2 \ldots \leq \mu_m$ in scheduling the incoming tasks for an h-DDSS. And thus, the state probabilities for the homogeneous DDSS scheduling algorithm can be written as follows [19]:

$$p_i = \begin{cases} p_0 \dfrac{\rho^i}{i!} & 1 \leq i \leq m \\ p_0 \dfrac{m^m}{m!} \rho_2^i & m \leq i \leq m+N \end{cases} \tag{1.15}$$

And for the h-DDSS scheduling algorithm can be written as follows:

$$p_i = \begin{cases} p_0 \dfrac{\lambda_1^i}{\prod_{j=1}^{i} \mu_{tj}} & i \leq m \\ p_0 \dfrac{\mu_{tm}^m \rho^i}{\prod_{j=1}^{m} \mu_{tj}} & m \leq i \leq m+N \end{cases} \tag{1.16}$$

In order to compute the probability p_0 of the empty queue system, we isolate Eqs. 1.15 and 1.16 using the fact that the probability p_i must obey the following relationship [18]:

$$\sum_{i=0}^{n} p_i = 1, \tag{1.17}$$

An expression for p_0 can be formulated for DDSS as follows [18]:

$$p_0^{-1} = \begin{cases} 1+\sum_{i=1}^{m} \frac{\rho^i}{i!} + \frac{m^m}{m!} \sum_{i=m+1}^{m+N} \rho_2^i & \rho_2 \neq 1 \\ 1+\sum_{i=1}^{m} \frac{\rho^i}{i!} + N \frac{m^m}{m!} & \rho_2 = 1 \end{cases} \tag{1.18}$$

Meanwhile, it can be stated as follows for the h-DDSS:

$$p_0^{-1} = \begin{cases} 1+\sum_{i=1}^{m} \frac{\lambda^i}{\prod_{j=1}^{i} \mu_{tj}} + \frac{m_{tm}^m}{\prod_{j=1}^{m}} \sum_{i=m+1}^{m+N} \rho^i & \rho \neq 1 \\ 1+\sum_{i=1}^{m} \frac{\lambda^i}{\prod_{j=1}^{i} \mu_{tj}} + N \frac{m_{tm}^m \mu_{tj}}{\prod_{j=1}^{m} \mu_{tj}} & \rho = 1 \end{cases} \tag{1.19}$$

In order to maintain stability, the queue system utilization must be as follows [19]:

$$\rho = \frac{\lambda}{\prod_{j=1}^{c} \mu_j} = \frac{\lambda}{m_c} \tag{1.20}$$

And

$$\rho_2 = \frac{\lambda}{m\mu} \tag{1.21}$$

Whereas for the h-DDSS, it should be as follows [19]:

$$\rho = \frac{\lambda}{\mu_{tm}} \tag{1.22}$$

1.4.2 DDSS and h-DDSS Performance Metrics

It is possible to formulate the performance metrics of the discussed queuing system from p_i and p_0. The average system throughput can be derived through the packet drop rate, which is represented by the final state probability p_{m+N}. Therefore, the packet drop rate and throughput for DDSS scheduling algorithms can be obtained as follows [17]:

$$Drop = p_0 \frac{m^m}{m!} \rho_2^{m+N} \tag{1.23}$$

Meanwhile, for the h-DDSS scheduling algorithm, it can be obtained as follows [18]:

$$Drop = p_0 \frac{\mu_{tm}^{m+N} \rho^{m+N}}{\prod_{j=1}^{m} \mu_{tj}} \tag{1.24}$$

$$Throughput = \lambda(1 - Drop) \tag{1.25}$$

The average data traffic occupancy can be obtained for both cases using the state transition probabilities as follows:

$$\eta = \sum_{j=1}^{N} j p_j \tag{1.26}$$

However, we assume a queue system with m servers, and a queue size equivalent to summation of a finite number of state probabilities as referred in Eq. 1.15. Thus, we can rewrite Eq. 1.23 as

$$\eta = \sum_{j=m+1}^{m+N} (j - m) p_j \tag{1.27}$$

where

$$\eta = \begin{cases} p_0 \rho_2 \dfrac{m^m}{m!} \left(\dfrac{1-(N+1)\rho_2^N + N\rho_2^{N+1}}{(1-\rho_2)^2} \right) & \rho_2 \neq 1 \\ p_0 \dfrac{m^m}{m!} \left(\dfrac{N(N+1)}{2} \right) & \rho_2 = 1 \end{cases} \tag{1.28}$$

And as follows for n of the h-DDSS case:

$$\eta = \begin{cases} p_0 \dfrac{\mu_{tm}^m}{\prod_{i=1}^m \mu_{ti}} \rho^{m+1} \left(\dfrac{1-(N+1)\rho^N + N\rho^{N+1}}{(1-\rho)^2} \right) & \rho \neq 1 \\ p_0 \dfrac{\mu_{tm}^m}{\prod_{i=1}^m \mu_{ti}} \left(\dfrac{N(N+1)}{2} \right) & \rho_2 = 1 \end{cases} \tag{1.29}$$

Using Little's law and Eqs. 1.25, 1.28, and 1.29, the average delay can be obtained as follows:

$$Delay = \frac{\eta}{Throughput} \tag{1.30}$$

And thus, the targeted optimization problem can be formulated as follows:

$$\max Throughput - Delay \tag{1.31}$$

Subject to

$$m_i = \sum_{j=1}^{i} \mu_j \quad \text{for} \quad \imath \leq m \tag{1.32a}$$

$$\sum_{Task_j}^{n} task_{jk\in} = 1 \tag{1.32b}$$

$$m_i = \sum_{j=1}^{i} \mu_j \quad \text{for} \quad i \geq m \tag{1.32c}$$

$$\lambda = \lambda_i \quad \text{for} \quad i = 1, 2, \ldots, m \tag{1.32d}$$

$$\forall_j \in \{1, 2, \ldots, task_j\} \quad \forall_\in \in \{1, 2, \ldots, w\} \tag{1.32e}$$

1.5 PSO Scheduling Approach

PSO is a relatively new stochastic optimization approach inspired by the social behavior of animal and swarm theory [9]. PSO can be applied to various complex and robust optimization problems. It has several advantages, including quick convergence, high precision and relatively easy implementation. Therefore, we use a canonical PSO (CPSO) algorithm and a fully-informed PSO (FIPS) algorithm in our scheduling approach. In CPSO and FIPS, each particle represents the set of priorities for all tasks $Task_j = \{task_{j1}, task_{j1}, \ldots, task_{jTask_j}\}$. Each task flies in the search space with dynamically adjusted velocity according to its own and other companions' flying experience. To improve the search capability, the particle velocity is updated based on the desired QoS parameters which adaptively change during the swarm evolution. A swarm of particles, which are represented as tasks in a *D*-dimensional search space, is represented by three tuples $\{X_i, V_i, P_i\}$. $X_i = \{x_{taskj1}, task_{j2}, \ldots, task_{jTask_j}\}$ and $V_i = \{v_{taskj1}, task_{j2}, \ldots, task_{jTask_j}\}$ denote the position and velocity of a particle $Task_j$. $P_i = \{p_{taskj1}, task_{j2}, \ldots, task_{jTask_j}\}$ represent the personal best p_{Task_j}. $G_i = \{g_{taskj1}, task_{j2}, \ldots, task_{jTask_j}\}$ represent the global best g_{Task_j} that are tracked by the entire swarm. The value of each element in the vector V_i can be clamped to the range of $[Delay_{min}, Delay_{max}]$ and $[Throughput_{min}, Throughput_{max}]$ to control the excessive roaming of particles outside the search space.

We aim in this section to improve the PSO with a better search capability to effectively solve the DCTS problem which is considered to be NP-hard problem in the literature. In order to increase the robustness of the proposed scheduling algorithm, we propose a few modifications on the classical PSO approach. These modifications are based on two perspectives; theoretical one that takes into consideration the convergence to a global optimal solution and the experimental one which depends on defining the relationship among the particles, neighborhood topology and the velocity updates. This relationship may define the word "convergence" in term of having converged global solution to the swarm's best known position $\overrightarrow{g_i}$. Some recent works provide guidelines about the PSO parameters which can be lead to the convergence, divergence, or oscillation of the swarm's particles which may or may not be the optimum [20]. Other studies, [21]-[23] classify the limitations related to convergence in PSO into four groups: convergence to a point, patterns of movement, convergence to local and global optimum solution. Since the convergence in our proposed algorithm rely on updating of the structure of the objective function, thus we are able to find the global optimum solution. We determine the logical communication among swarm particles in order to know whether a particle's best known position $\overrightarrow{p_i}$ and the swarm's best known position $\overrightarrow{g_i}$ is a neighbor of another particle or not. Each particle connects with its neighbors to exchange information with each other so that they converge towards the optimal solution. Accordingly, we study the convergence of the proposed algorithm by using stochastic optimization approach in order to enhance the performance and select QoS parameters. In other words, we modified the velocity updating strategy in order to improve the global search capability and robustness. The updating strategy is used in the comprehensive learning particle swarm optimizer (CLPSO) [8], in which each particle learns from its p_{best} a tournament selection. In this study, we assume that:

1. $\overrightarrow{p_i}$ and $\overrightarrow{g_i}$ are always within finite domain.
2. P^* representing the global optimal solution in the search space, $\| P^* \| < \infty$, where $\lim_{p_i \to P^*}$ and $\lim_{g_i} \to P^*$. In other word, a stochastic optimization approach is said to converge to point $\overrightarrow{X_{p_i}}$ in the search space with probability if

$$\forall \xi > 0, \lim_{i \to \infty} P(\| \overrightarrow{x_{g_i}} - \overrightarrow{X_{p_i}} \| < \xi) \tag{1.33}$$

where P is the probability measure, x_{g_i} is generated solution by PSO at each iteration, and $\overrightarrow{X_{p_i}}$ is any point selected in the search space including local solution referring to objective function.

Using these assumptions we can prove the convergence of the proposed algorithm in the sense of the velocity of a particle is accelerating according to the system utilization ratio, arrival rates and service rates in order to increase the performance of a data traffic class in the system. Three modified velocity updating strategies to introduce three coefficients ξ, ϕ_1, and ϕ_2 are used to improve the convergence of global search robustness and capabilities of our proposed architectures with dynamic environments and analyzed the swarm behavior with different parameters values. The pseudo-code of the proposed algorithm in terms of CPSO and FIPS is depicted in Algorithm 1.1.

1.5.1 Velocity Updating Strategies

1.5.1.1 Strategy I

The first strategy accelerates the associated particle's velocity when the ratio of incoming arrival rates λ_i increases to reach the value of the total service rates μ_i of all the allocated servers. Accelerating the velocity updates can significantly affect the delay metric. And thus, *CP* will select a particle's p_{best} for all incoming data traffic that optimizes the delay. Furthermore, Eq. 1.34 will be used to obtain the information exchanged between the selected two particles. The velocity is updated as follows [2]:

$$\vec{v_v} := \xi(v_v \tag{1.34a}$$

$$+\vec{\lambda_i}(0,\phi_1)+\left(\vec{p_i}-\vec{x_i}\right) \tag{1.34b}$$

$$+\vec{\lambda_i}(0,\phi_2)+\left(\vec{g_i}-\vec{x_i}\right) \tag{1.34c}$$

where $\vec{\lambda_i}$ and $\vec{\mu_i}$ represent the distribution of the arrival rate and the service rate of the incoming data traffic, respectively. ξ is a constriction coefficient, which helps in providing a balanced global exploration versus the local one. It is defined as

$$\xi = \frac{2}{\phi+\sqrt{\phi^2-4\phi}}, \ with\ \phi = \phi_1+\phi_2 > 4 \tag{1.35}$$

1.5.1.2 Strategy II

The second strategy accelerates the velocity of the particles when the ratio of the service rates μ_i increases to the total incoming class arrival rates λ_i for all the allocated servers handling the multi-class data traffic. Accordingly, this will significantly affect the throughput metric. And hence, *CP* selects a particle's p_{best} for all incoming data traffic classes while optimizing the throughput. In the following, we show how the velocity is updated [2]:

$$\vec{v_v} := \xi(v_v \tag{1.36a}$$

$$+\overrightarrow{\mu_i}\left(0,\phi_1\right)+\left(\overrightarrow{p_i}-\overrightarrow{x_i}\right) \tag{1.36b}$$

$$+\overrightarrow{\mu_i}\left(0,\phi_2\right)+\left(\overrightarrow{g_i}-\overrightarrow{x_i}\right) \tag{1.36c}$$

1.5.1.3 Strategy III

In the third strategy, the acceleration of the velocity updates increases when the utilization ratio calculated by the ratio of the incoming arrival rates λ_i to the total service rate μ_i is increased. And thus, accelerating the velocity updates can significantly affect the utilization values for DDSS and h-DDSS approaches. In the following, we show how this velocity can be updated [2]:

$$\overrightarrow{V_v} := \xi(V_v \tag{1.37a}$$

$$+\overrightarrow{\rho_i}\left(0,\phi_1\right)+\left(\overrightarrow{p_i}-\overrightarrow{x_i}\right) \tag{1.37b}$$

$$+\overrightarrow{\rho_i}\left(0,\phi_2\right)+\left(\overrightarrow{g_i}-\overrightarrow{x_i}\right) \tag{1.37c}$$

During the evolution process, Eqs. 1.34a, 1.36a and 1.37a are referred to as the momentum that represents the particle's current selection direction. Meanwhile, Eqs. 1.34b, 1.36b and 1.37b are referred to as the social components, which represent the force of attraction towards the best solution evaluated so far by the neighbors. Eqs. 1.34c, 1.36c and 1.37c represent the cognitive component, which attracts the optimal solution towards the previous solutions. And hence, the main difference between the CPSO algorithm and the FISP algorithm is the velocity update function. It describes not only the best position of a node, but also all of its neighbors positions. Therefore, the velocity update function is defined as [2]

$$\overrightarrow{v_i} = \xi\left(\overrightarrow{v_i} + \frac{1}{n}\sum_{v=1}^{n+m}\overbrace{\vec{\lambda}\left|\vec{\mu}\right|\overrightarrow{\rho_i}}\left(0,\phi_1\right)+\left(\left(\overrightarrow{p_v}-\overrightarrow{x_v}\right)\right)\right) \tag{1.38}$$

Taking the advantage of the exchanged information about all personal-best p_{best} messages can significantly strengthen the particle's ability to learn from other particles' experiences and guide its selection direction.

1.5.2 Convergence Strategies

Several limitation of PSO have been identified so far. These limitations are concerned with convergence and transformation invariance. Basically, we depend on the classification in [21] that related to convergence in PSO to investigate the convergence of the proposed algorithm to a point, i.e., particle or pattern of movements of sub group of particles that is usually conducted for iterative stochastic optimization, evaluation of the particles' current positions, and consequently initialization of the personal bests and global best. Therefore, we understand whether the sequence of generated solutions after following a random initialization of positions and velocities vectors of the objective functions produced by our proposed algorithm is convergent to local or global solution so far. This persuade us to realize that the final solution founded by the algorithm is at least a local or global optimum, since the current velocity component is left undisturbed, while the personal and global best components are each scaled by a random scalar drawn uniformly from $[0,2]$.

Actually, we initiate a set of parameters values of the algorithm such a constriction coefficient, number of particles and the communication topology of particles in order to locate higher quality solutions which are related to the exploration/exploitation ability of the algorithm. Therefore, we restrict the velocity vector to the boundaries of the search space and prevent effective search. Moreover, we need to define the boundaries for the inertia weight and acceleration coefficients of particle's velocity in such a way that the positions of the particles converge to local solution or global in the search space. The use of the inertia weight ξ improved the performance of the algorithm by performing self organized adaptive positions and velocities vectors of the objective functions. It is used to limit the impact of the previous velocity updating of the particles, providing a balance between exploration and exploitation. Furthermore, the inertia weight value is reduced in each iteration to help the particle swarm to attain more diversity and prevent taken into account another solutions when particles are being too close to another. It should be noted that the behavior of particles should be investigated while they search for a new personal best solution, since the importance of exploration and exploitation cannot be stressed enough. Therefore, the validity of the convergence boundaries is found when the social interaction between global and personal best solutions are steady with specific topology in terms of bus, ring, or mesh. Finally, we simplify the formulation of update rules in Eq. 1.35 to prove that particle positions converge to a local stable solution point if

$$\xi = \frac{2k}{\left|2 - c - \sqrt{c^2 - 4c}\right|}, \textit{ with} \tag{1.39}$$

$$c = c_1 + c_2 > 4, c_1 = \frac{\phi_1}{\xi}, \textit{ and } c_1 = \frac{\phi_2}{\xi} \tag{1.40}$$

where ξ is equal to inertia weight, and k is a value in the interval $(0,1]$. With these settings, the value of ξ and k will control the slower convergence to fixed point in terms of local or global as will be discussed in the following section.

Algorithm 1.1: Canonical Particle Swarm Scheduling Approach

1. input: the tasks $Task_{\#} = \{task_{\#1}, task_{\#1}, \ldots, task_{\#Task\#}\}$;
2. Compute λ for each Task $\lambda_{Task_j} = \left\{task_{j1}, task_{j1}, \ldots, task_{jTask_j}\right\}$;
3. Compute μ for each Task $\mu_{Task_j} = \left\{task_{j1}, task_{j1}, \ldots, task_{jTask_j}\right\}$;
4. Compute ρ for each Task $\rho_{Task_j} = \left\{task_{j1}, task_{j1}, \ldots, task_{jTask_j}\right\}$;
5. input: Objective functions $\vec{Z}$
6. $\vec{V} := \left\{Task_j = \left\{task_{j1}, task_{j1}, \ldots, \text{task}_{jTask_j}\right\}\right\} := InitNode\left(Delay_{min}, Delay_{max}\right)$ $\forall_{\upsilon} \in \{1, \ldots, task_{jTask_j} : \vec{i_{\upsilon}} := \vec{U}\left(Delay_{min}, Delay_{max}\right)\}\}$
7. $\vec{V} := \left\{Task_j = \left\{task_{j1}, task_{j1}, \ldots, task_{jTask_j}\right\}\right\} :=$ $InitNode\left(Throughput_{min}, Throughput_{max}\right)$, $\forall_{\upsilon} \in \left\{1, \ldots, task_{jTask_j} : \vec{i_{\upsilon}} := \vec{U}\left(Throughput_{min}, Throughput_{max}\right)\right\}$
8. $\vec{V} := \left\{\vec{\upsilon}_{j1}, \ldots, \vec{\upsilon}_{jTask_j}\right\} := InitParticleVelocities\left(Delay_{min}, Delay_{max}\right)$, $\forall_{\upsilon} \in \{1, \ldots, i\} : \vec{\upsilon}_{1,\ldots,i} := \left(Delay_{max} - Delay_{min}\right) \otimes \vec{U}(0,1) - \frac{1}{2}\left(Delay_{max} - Delay_{min}\right)$
9. $\vec{V} := \left\{\vec{\upsilon}_{j1}, \ldots, \vec{\upsilon}_{jTask_j}\right\} := InitParticleVelocities\left(Throughput_{min}, Throughput_{max}\right)$, $\forall_{\upsilon} \in \{1, \ldots, i\} : \vec{\upsilon}_{1,\ldots,i} := \left(Throughput_{max} - Throughput_{min}\right) \otimes \vec{U}(0,1) - \frac{1}{2}\left(Throughput_{max} - Throughput_{min}\right)$
10. $\Upsilon = \left\{Task_j = \left\{task_{j1}, task_{j1}, \ldots, task_{jTask_j}\right\}\right\} := EvaluateObjectfunction(\vec{Z}), \forall_{\upsilon}$ $\left\{task_{j1}, task_{j1}, \ldots, task_{jTask_j}\right\} = \Upsilon_{\upsilon} := f\left(\vec{Z}_{\upsilon}\right)$
11. $P = \left\{\vec{p}_{Task1}, \ldots, \vec{p}_{jTaskj}\right\} := Initllocallyoptimal(Z) \rightarrow Z$
12. $P = \left\{p^{Z}_{Task1}, \ldots, \vec{p}^{Z}_{jTaskj}\right\} := InitObjectivefunction(\Upsilon) \rightarrow \Upsilon$
13. $G = \left\{\vec{g}_{Task1}, \ldots, \vec{g}_{jTaskj}\right\} := Initgloballyoptimal\left(P, T_p\right) \rightarrow P$
14. $P = \left\{g^{Z}_{Task1}, \ldots, g^{Z}_{jTaskj}\right\} := Initgloballyoptimal\left(P^{Z}, T_p\right) \rightarrow P^{Z}$
15. **while** termination condition nor met **do**
16. **for** each Task υ of # **do**
17. $\vec{\upsilon}_{\upsilon} = \chi \cdot \left(\vec{\upsilon}_{\upsilon} + \vec{Z}(0, \varphi_1) \otimes \left(\vec{p_{\upsilon}} - \vec{\upsilon}_{i}\right) + \vec{Z}(0, \varphi_1) \otimes \left(\vec{g_{\upsilon}} - \vec{\upsilon}_{\upsilon}\right)\right)$

18. $\vec{v}_p := \vec{Z}_p + \vec{v}_p$
19. **end for**
20. $\Upsilon := EvaluateObjectivefunction\left(\vec{Z}\right)$
21. $P, P^Z := Updatelocallyoptimal\left(\vec{Z}\right) \rightarrow \forall_p \in \{task_{j1}, \ldots, task_{jTaskj} : \vec{p}_p, p_p^Z :=$

$$\begin{Bmatrix} \vec{Z}_p, y_i & \text{if } \Upsilon_v \text{better than } p_p^f \\ \vec{p}_p, p_p^f & \text{otherwise} \end{Bmatrix}$$

22. $G, G^f := Updategloballyoptimal\left(P, P^Z, T_p\right) \rightarrow \forall_p \in$

$$\left\{task_{j1}, \ldots, task_j Task_j : \overrightarrow{g_p}, g_p^Z := best\left(P_{T_p}, P_{T_p}^Z\right),\right.$$

$$\left. where\ T_p\ are\ the\ neighbors\ of\ p\right\}$$

23. **end while**
24. best solution found

1.6 Performance Evaluation

To validate the proposed algorithm, we assume a queuing system under different traffic flows depending on the QoS requirements when the requests for an IoT application is placed. We assume a Poisson probability density function distribution for all these requests, and a service rate for each application that is exponentially distributed. The type of the queue discipline used in this simulation is FIFO and the service rates for all servers are equal in case of a homogeneous system and vary in case of heterogeneous system. Each queue/buffer has a capacity to hold up to 30 tasks. We ran the simulation for 1000 iterations and 50 trials with different arrival rates and service rates. Table 1.2 depicts the simulation parameters used to implement the proposed algorithm. The values in Table 1.2 are selected based on a real-life scenario as it behaves in a natural Cloud environment. We choose the

Table 1.2 SIMULATION PARAMETERS

Parameter	*Value*
The number of servers	30
Probability of arrival rates for incoming	$P_x(t) = ((\lambda.t)^x / x!)\exp^{-\lambda.t}$
Queue size	5
Users' request priority level	1

highest priority level for C_i while tracking the arrival rates relationship. Arrival rates of C_i depend on i to track the relation of the arrivals counts in a specific time period as stated by $P_x(t) = \left((\lambda.t)^x / x!\right) \exp^{-\lambda.t}$, where t is used to define the time interval 0 to t, x is the total number of arrivals in this interval, and λ is the average arrival rate in arrivals/second.

1.6.1 Performance of DDSS and h-DDSS

We show the difference between DDSS and h-DDSS by displaying the performance of CPSO and FIPS algorithms while optimizing the QoS parameters in terms of throughput and delay as referred in Eqs. 1.31 and 1.36, respectively. Figures 1.7 and 1.8 show the evaluation of DDSS and h-DDSS while investigating the performance of CPSO and FIPS in scheduling the arrived requests/tasks when throughput of users' requests is optimized. Observed throughput using FIPS algorithm in DDSS is higher than the CPSO as the assigned number of servers is enough to serve the related class of data traffic. On the other hand, there is a small difference between the performance of CPSO and FIPS in throughput of h-DDSS since CPSO is prone to converge at a local optimal solution. And thus, it cannot converge to the optimum task scheduling. That is because the h-DDSS scheduling algorithm dynamically arranges the number of assigned servers according to

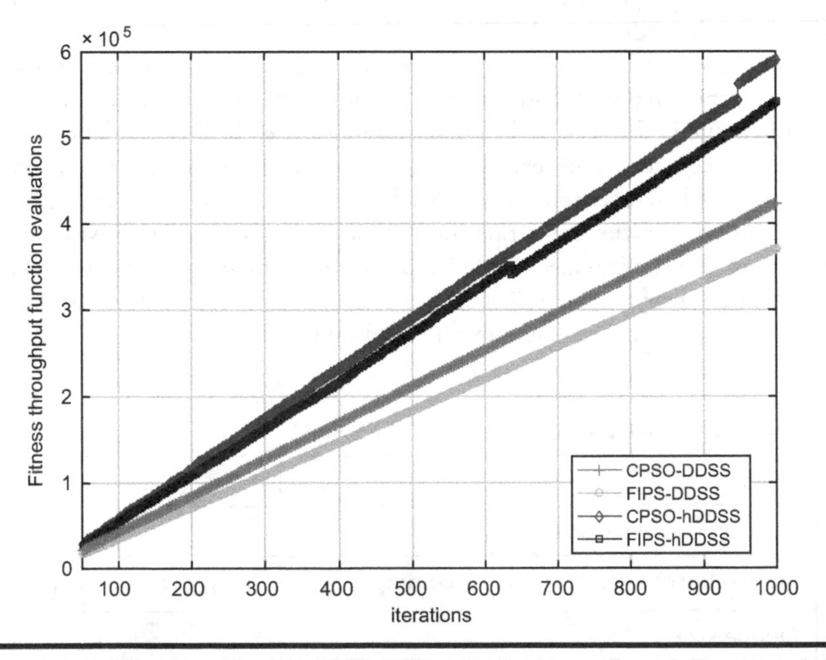

Figure 1.7 Throughput fitness function for DDSS and h-DDSS using CPSO and FIPS algorithms.

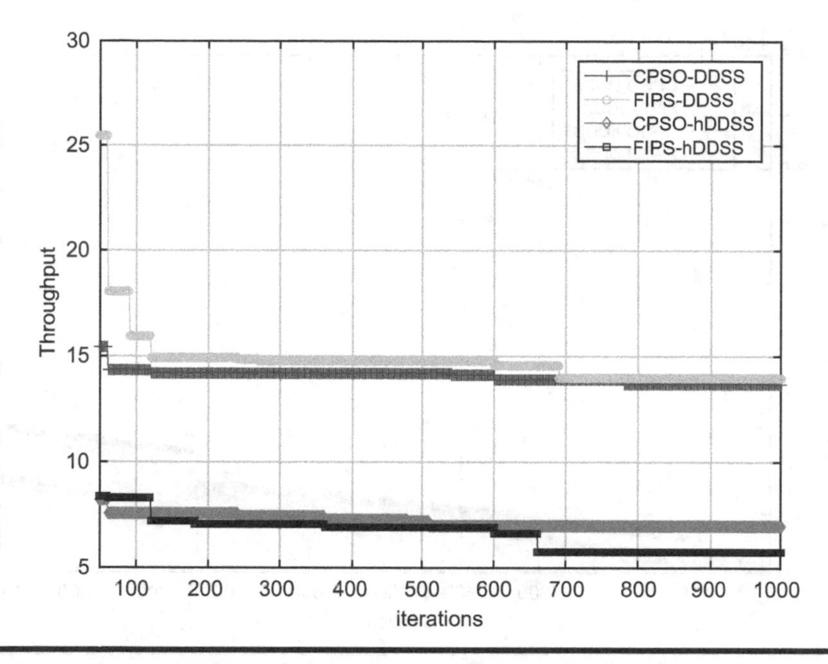

Figure 1.8 Optimized throughput fitness function for DDSS and h-DDSS using CPSO and FIPS algorithms.

FIPS mechanism in addition to adapting the velocity updates which prevents new allocated servers to drop any incoming users' requests.

Figures 1.9 and 1.10 show the delay fitness for DDSS and h-DDSS by investigating the performance of CPSO and FIPS in obtaining the optimal task scheduling. It seems that the optimized delay using FIPS in DDSS is lower that CPSO since the assigned number of servers is enough to serve related class data traffic. The delay results of all scheduling algorithms have the same pattern because of the dynamic scheduling for each incoming uses's request. Therefore, we can conclude that there is no significant difference between the performance of CPSO and FIPS in the delay metric. That is because the particle behavior in FIPS avoids to stick at a local optimal solution while increasing the number of assigned servers. In other words, the convergence boundaries analysis for FIPS is depends on the number of particles and the type of communication topology among particles which are different from CPSO. In addition, FIPS algorithm does not just take into consideration the p_i of the best neighbor solution, but also all the connected neighbors. This is the key real strength of the FIPS which uses a stochastically weighted average of the information available in p_i and g_i. If two particles are close to one another, therefore, they cycle around the region of solution defined by them, eventually converging on the center of that region. Meanwhile, if the particle's previous best is distant from the neighborhood's

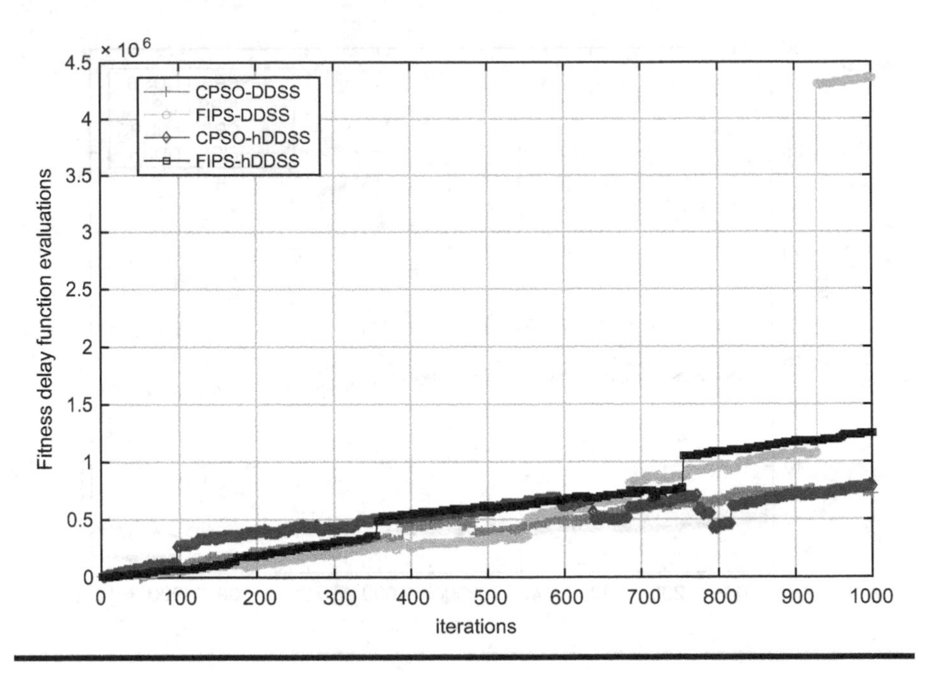

Figure 1.9 Delay fitness function for DDSS and h-DDSS using CPSO and FIPS algorithms.

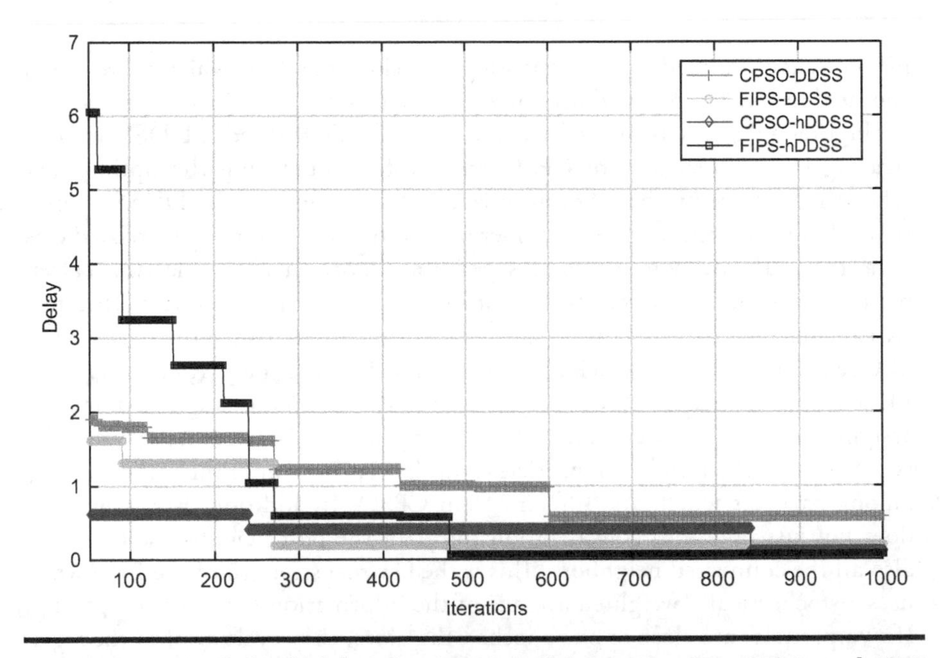

Figure 1.10 Optimized delay for DDSS and h-DDSS using CPSO and FIPS algorithms.

previous best, then the particle's trajectory is much wider, enabling it to explore that broad region between both positions. Therefore, it can converge to a good task scheduling.

Figures 1.11 and 1.12 show the utilization of DDSS and h-DDSS with different priority levels for C_i. Figure 1.11 shows the optimized utilization for an h-DDSS system using FIPS and CPSO, which is considerably higher than the DDSS system. That is because of the equally increasing arrival and service rates. Furthermore, the performance of FIPS in h-DDSS and DDSS is better than CPSO because the FIPS provides an optimal task scheduling through the utilized velocity update function, which describes not just the best position particle taken into account but all of its neighbors as referred to in Eq. 1.35. Furthermore, FIPS depends on selecting values of coefficients within the convergence boundaries which might prevent a particle from moving unboundedly and guaranteeing that the particle converges to global best solution. Therefore, the assigned number of servers is enough to serve the related different priority levels of the incoming C_i. Meanwhile, CPSO presents a considerably worst performance compared to FIPS in h-DDSS and DDSS since the velocity update function describes only the best position particle without taking into consideration all of its neighbors as referred in Eqs. 1.31, 1.34, and 1.36. Therefore, the assigned number of servers is not enough to serve the related different priority level of incoming C_i.

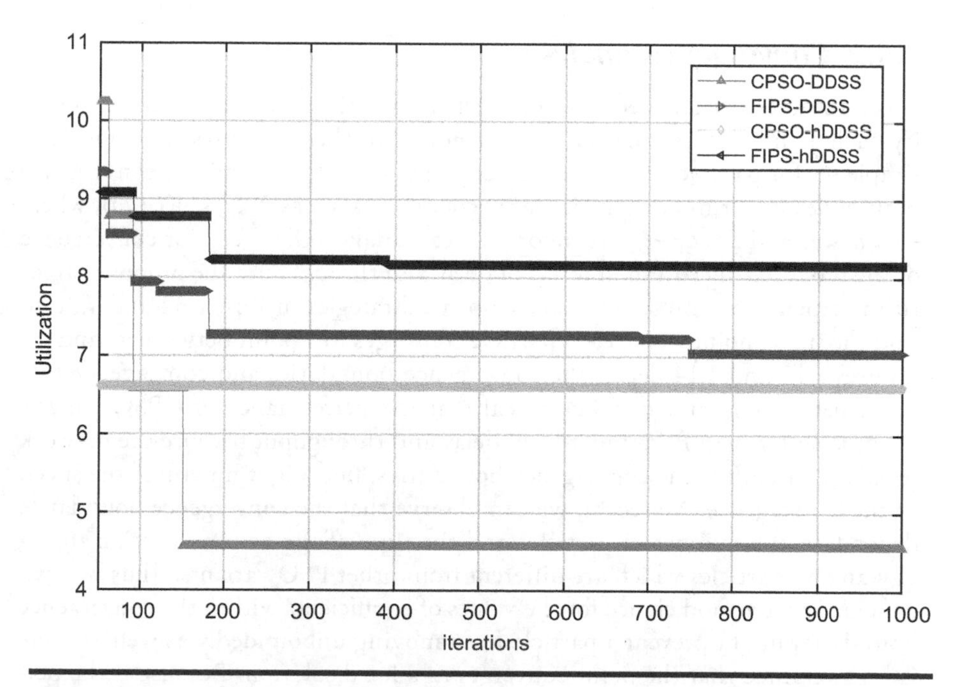

Figure 1.11 Optimized utilization fitness function for DDSS and h-DDSS.

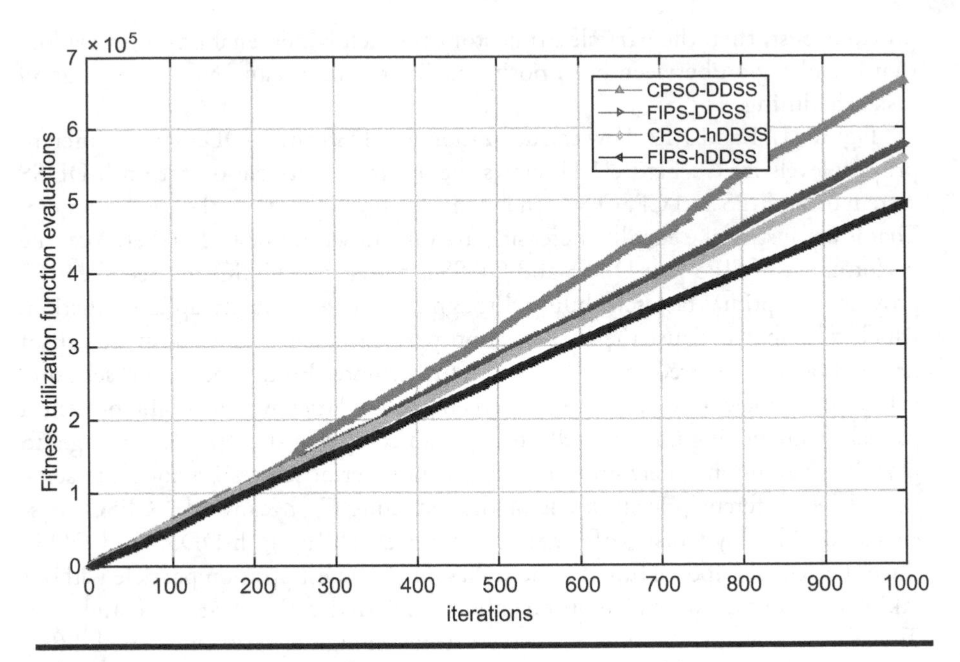

Figure 1.12 Fitness function for utilization of DDSS and h-DDSS.

1.6.2 Convergence Analysis

Several studies investigated the convergence analysis in different terms of stability such as first and second-order. In contrast to these analyses, we consider a simple set for parameters of each velocity update vector to find the convergence boundaries under more realistic traffic condition such as $0 < \xi < 4(1+\omega)$, where $-1 < \omega < 1$, and $\xi = \xi_1 + \xi_2$. Therefore, we can analyze the speed for convergence of the proposed algorithm in terms of delay and throughput. The proposed algorithm depends on using the velocity update strategies in Eqs. 1.34 and 1.36 to find the stable point where each particle converges to a point between $\overrightarrow{p_i}$ and $\overrightarrow{g_i}$. Figures 1.13 and 1.14 depict the convergence boundaries and convergence to a point between $\overrightarrow{p_i}$ and $\overrightarrow{g_i}$. They reveal that the performance of CPSO convergence is better than FIPS in terms of delay and throughput because the network topology can affect the convergence boundaries, but might not affect the speed of the convergence. Moreover, we can observe that the convergence boundaries depend on the number of particles and the type of the communication topology among particles which are different from other PSO variants. Thus, we can conclude that a good choice for the values of coefficients within the convergence boundaries might prevent a particle from moving unboundedly as well as there is no guarantee that the point converges to a local solution. Because if the personal best of a particle is guaranteed to converge to a local optimum, then it is

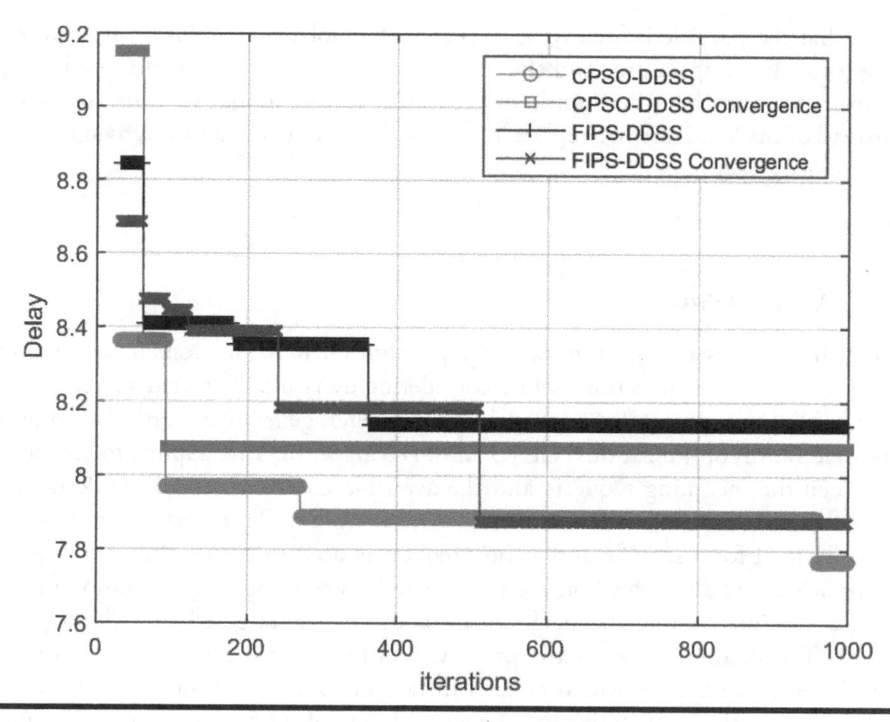

Figure 1.13 Fitness function for utilization of DDSS and h-DDSS.

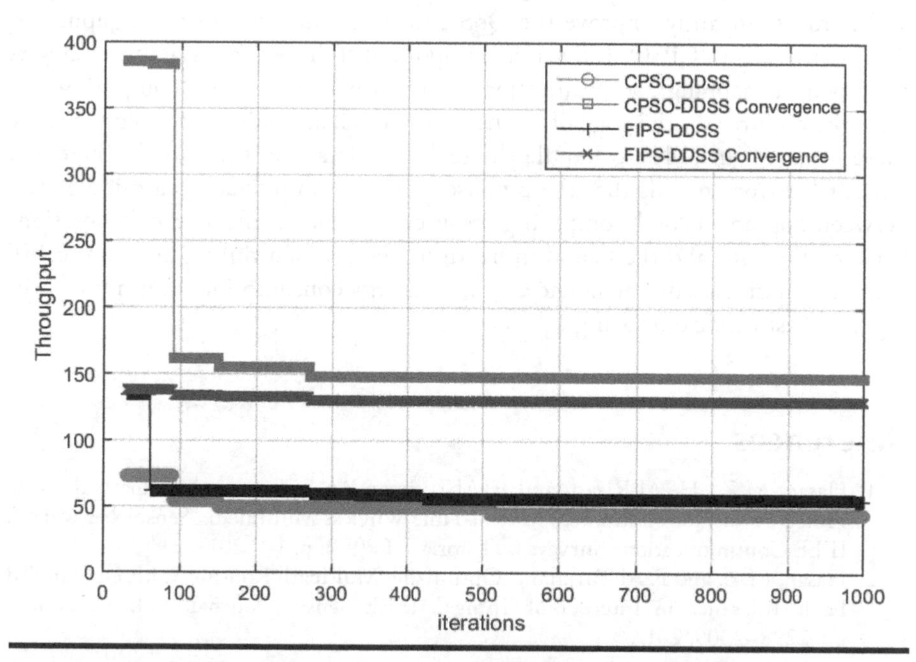

Figure 1.14 Fitness function for utilization of DDSS and h-DDSS.

said that the particle is locally convergent. The global best solution is converging to the local optimum, and thus the proposed algorithm is said to be locally convergent. On the other hand, and to converge to a global solution optimum instead of the local one, we replace the global/personal best solution by updating the velocity vector.

1.7 Conclusion

This chapter presents an optimization algorithm for users' requests scheduling in Cloud-based IoT applications while considering dynamic dedicated server scheduling (DDSS) approach for homogeneous and heterogeneous systems. Two robust particle swarm optimizations, CPSO and FIPS algorithms are applied to prioritize between the incoming requests and the available Cloud resources, while taking into consideration the different classes of data traffic. The determination of the priority level for a specific application request is used to decide the next request to be scheduled and served on an iterative and asynchronous optimization of the task allocations among the neighboring nodes in the targeted IoT/Cloud paradigm. This determination of the priority level for a specific application revealed survivability services in which at each instant of time each node in the network has some positive probability to interact with one of its neighbors. Results show that the performance of DDSS and h-DDSS architectures based on FIPS have the ability to significantly improve the QoS parameters in terms of throughput and delay compared to CPSO. Indeed, it is important to investigate suitable strategies to provide a mapping communication resources with the Cloud computing-side for more resources at the edge of the network as well as a suitable strategy to share these resources in order to handle the real-time data over the Cloud. Therefore, our further work reveals that the proposed scheme can interact as a middle ware between Fog and Cloud computing resources, to enable effective collaborations between the edge and the Cloud in handling the task scheduling problem in IoT with minimum closed-loop latency, optimal energy consumption, minimum caching, and best traffic offloading.

References

1. Hasan, M.Z., H. Al-Rizzo, and F. Al-Turjman, "A Survey on Multipath Routing Protocols for QoS Assurances in Real-Time Wireless Multimedia Sensor Networks". IEEE Communications Surveys & Tutorials, PP(99): p. 1-1, 2017.
2. Hasan, M.Z. and F. Al-Turjman, "Optimizing Multipath Routing With Guaranteed Fault Tolerance in Internet of Things". IEEE Sensors Journal, Vol. 17, Issue:7, pp. 2298–2309, 2017.

3. Hasan, M.Z., H. Al-Rizzo, and M. Gunay, "Lifetime maximization by partitioning approach in wireless sensor networks". EURASIP Journal on Wireless Communications and Networking, Vol. 2017, Issue: 1, pp. 15, 2017.
4. Hasan, M.Z., F. Al-Turjman, and H. Al-Rizzo, "Analysis of Cross-Layer Design of Quality-of-Service Forward Geographic Wireless Sensor Network Routing Strategies in Green Internet of Things". IEEE Access, Vol. 2018, issue 6, pp. 20371–20389, 2018.
5. Dey, N., et al., "Internet of Things and Big Data Analytics Toward Next-Generation Intelligence". 2017: Springer International Publishing.
6. Delicato, F.C., P.F. Pires, and T. Batista, "Resource Management for Internet of Things", 2017: Springer International Publishing.
7. P. T. Endo et al., "Resource allocation for distributed cloud: concepts and research challenges," in IEEE Network, Vol. 25, no. 4, pp. 42–46, July-August 2011.
8. G. Colistra, V. Pilloni, and L. Atzori, "Task allocation in group of nodes in the IoT: A consensus approach," in IEEE International Conference on Communications (ICC), Vol. 2, issue. 1, pp. 3848–3853, 2014.
9. X. Zuo, G. Zhang, and W. Tan, Self-adaptive learning PSO-based dead-line constrained task scheduling for hybrid IaaS cloud, IEEE Trans. Autom. Sci. Eng., Vol. 11, no. 2, pp. 564–573, Apr. 2014.
10. Buyya, R., J. Broberg, and A.M. Goscinski, "Cloud Computing Principles and Paradigms". 2011: Wiley Publishing. 664.
11. Madni, S.H.H., et al., "Performance comparison of heuristic algorithms for task scheduling in IaaS cloud computing environment". PLOS ONE, Vol. 12, issue 5, pp. e0176321, 2017.
12. Farahmandpour, Z., et al., "Service virtualisation of internet-of-things devices: techniques and challenges", in Proceedings of the 3rd International Workshop on Rapid Continuous Software Engineering. IEEE Press: Buenos Aires, Argentina. pp. 32–35, 2017.
13. Guo, L., et al., Task scheduling optimization in cloud computing based on heuristic algorithm. JNW, Vol. 7, issue 3, pp. 547–553, 2012.
14. Beegom A.S.A., Rajasree M.S. A Particle Swarm Optimization Based Pareto Optimal Task Scheduling in Cloud Computing. In: Tan Y., Shi Y., Coello C.A.C. (eds) Advances in Swarm Intelligence. ICSI 2014. Lecture Notes in Computer Science, vol 8795. Springer, Cham, 2014.
15. Zhou, L., et al., An event-triggered dynamic scheduling method for randomly arriving tasks in cloud manufacturing. International Journal of Computer Integrated Manufacturing, 2018. 31(3): p. 318–333.
16. K. Church, A. Greenberg, and J. Hamilton, "On Delivering Embarrassingly Distributed Cloud Services", in Proceedings of Seventh ACM Workshop on Hot Topics in Networks (HotNets-VII), (Calgary, Alberta, Canada), Oct. 2008.
17. Chandra, A. and J. Weissman, "Nebulas: using distributed voluntary resources to build clouds", in Proceedings of the conference on Hot topics in cloud computing. USENIX Association: San Diego, California, 2009.
18. Narman, H.S., et al., "Scheduling internet of things applications in cloud computing". Annals of Telecommunications, 2017. Vol. 72, issue 1, pp. 79–93, 2017.
19. Alves, F.S.Q., et al., "Upper Bounds on Performance Measures of Heterogeneous M/M/c Queues". Mathematical Problems in Engineering, 2011.

20. Hasan, M.Z. and F. Al-Turjman, SWARM-based data delivery in Social Internet of Things. Future Generation Computer Systems, 2017.
21. Yuan, Q. and G. Yin, "Analyzing convergence and rates of convergence of particle swarm optimization algorithms using stochastic approximation methods". IEEE Transactions on Automatic Control, Vol. 60, issue 7, pp. 1760–1773, 2015.
22. ping Tian, D., A review of convergence analysis of particle swarm optimization. International Journal of Grid and Distributed Computing, 2013. 6(6): p. 117–128.
23. Cordero, C.G., Parameter adaptation and criticality in particle swarm optimization. arXiv preprint arXiv:1705.06966, 2017.

Chapter 2

Scheduling and Routing for Sensors' Data in Cloud-Based IoT[1]

Fadi Al-Turjman

Antalya Bilim University

Contents

[1] **F. Al-Turjman**, "Cognitive Routing Protocol for Disaster-inspired Internet of Things", *Elsevier Future Generation Computer Systems*, Volume 92, March 2019, Pages 1103-1115.

2.1 Introduction

Sensing technology has played a significant role in detection and containment of disasters in numerous disciplines. The most notable among these applications are those functioning in harsh environments, such as pollution and flood detection, forestry fire prevention and earthquake monitoring applications [1][2]. For instance, a network of sensors can be used to monitor the motion of toxic gasses over vast areas [3]. In [4], redwood trees in risk have been monitored via a wireless sensor network. Furthermore, the existence of a disinfectant that works better and more efficient than conventional traditional ones, by providing long lasting antiviral effect against major viruses has proved the importance of sensing technology in disaster management [5]. The above mentioned examples are just few of the many areas where sensing technology has made massive improvements. However, this technology is still suffering extreme limitations in terms of energy and connectivity while collaborating in wireless network-based systems.

The connectivity and links between sensing-devices (nodes) distributed to monitor a specific phenomenon have led to the idea of WSNs followed by the Internet of Things (IoT) proposal. In IoT, things are connected to the internet via local gateways. Integrating IoT with other heterogeneous network systems such as WiFi, LiFi, LTE, etc. will expand the array of services that can be provided to public users as well as decision makers. Several IoT's design aspects that stem from its unique features in terms of limited-energy constraints, short communication range between geo-located objects, and low processing power need to be incorporated into its routing protocols in order to realize the IoT paradigm. For example, there have been several attempts to design lightweight solutions for IoT devices, and operating systems in order to reserve energy [6]-[8]. However, challenges against routing protocols design in terms of energy are still being investigated with no currently fully developed solutions. WSNs in IoT consume energy in almost all processes [9]. They consume energy while making data transmission, data sensing and data processing. There have been a few attempts towards achieving energy efficiency in such networks via wireless multi-hop networking and have proposed adequate schemes, e.g. [10]-[12]. However, such schemes either assume static network topology, which render these schemes impractical for real-life network implementation, because IoT-based networks exhibit random topology due to the mobility of nodes, or are restricted to two hop from source to the sink routing schemes. Due to the fact that IoT-networks' sensors are usually limited in their processing power, communication range and energy capabilities, design and implementation of routing algorithms are considered a nontrivial task.

Moreover, accommodating various levels of harshness in surroundings in terms of temperature, dust, humidity, etc., dictates a resilience requirement to an expanded set of failure possibilities that includes partial or complete failure of the IoT sensor nodes, and reduced levels of activity or accuracy undergone as batteries deplete, which forms serious threats for losing critical in-network data before being utilized. This requirement stresses the integrity of cognition elements in IoT routing protocols. Moreover, an IoT sensor network needs to sustain different levels of mobility on disaster situations [13]. Since the IoT connected things rely on each other to gather and process data, mobility may be temporarily or permanently detrimental to the network operation by breaking some functional communication links that affects the in-network data retrieval. Hence, nodes and links are prone to several risks leading to high probabilities of failures and several nodes in the network may become disconnected/failed. As a reason of that, we are characterizing such circumstances by the Probability of Node Failure (PNF). Consequently, for a successful and reliable operation of the IoT paradigm in disaster-inspired scenarios, a cognitive energy-efficient routing approach shall be applied.

Recently, there have been significant attempts in the direction of building a cognitive sensor network, where researchers have made use of artificial neural networks, genetic algorithms, game theory and even software agents to implement distributed and intelligent decision making in sensor networks [2][18][19]. However, there is no single framework that can be used to implement cognition in sensor networks supporting the IoT paradigm in a way that is domain and application independent.

In this chapter, we propose a cognitive data delivery approach that addresses the challenges of data delivery in IoT networks comprised of energy-constrained IoT sensors. Two key elements in machine learning are utilized in our approach in order to implement cognition; *reasoning* and *learning* elements. Reasoning helps in prioritizing the attributes of a given traffic flow, and choosing the next hop along the data delivery path to the destination. While reasoning realizes short-term objectives and makes decisions based on the current network status, learning helps in achieving long-term goals of the network, such as improving its lifetime. Feedback obtained from the network's history aids the learning process, and helps in planning proactive responses. This model will help us to specify which path to follow in order to determine the optimal usage of the available resources for a wide range of IoT applications in safety and security. Where the proposed approach is energy-efficient and designed to optimize the current network status for guaranteed Quality of Service (QoS) via machine learning [18]. It provides efficient and self-healing data delivery while choosing and selecting reliable communication links [19]. Moreover, the proposed approach caters for the grid-based distribution of the employed IoT sensors on the monitored object to efficiently and effectively cope with the dynamicity of IoT-network topology [20].

The rest of the chapter is organized as follows: Section 2.2 reviews previous related studies. Section 2.3 discusses our system models. Section 2.4 describes our proposed routing approach for IoT paradigm. Section 2.5 provides performance evaluation for the proposed approach. Finally, Section 2.6 provides the conclusions and future directions.

2.2 Related Work

Connectivity in IoT involves finding reliable routes from IoT "things" to the IoT gateway. Utilizing duty-cycling, a routing algorithm can be designed to significant balance the network load and optimize the energy-consumption, especially in energy-constrained IoT networks. Duty-cycling also has the advantage of allowing the network to circumvent elements and zones rendered temporarily unavailable due to energy unavailability, medium variations, or mobility. In mission critical IoT networks, it is also important, when designing the routing algorithms, to facilitate prioritization between different traffic types. Another critical problem to overcome is the uneven energy consumption across the network where elements near the gateway would deplete their energy faster than those at a distance. Feedback from the MAC and physical layers, in addition to information about the residual energy and the current load of IoT nodes, can be utilized to identify and bypass critical links, effectively prolonging the network lifetime and increasing network throughput through load balancing. Hence, there are considerable advantages in coupling an IoT routing protocol with the underlying MAC layer protocol through cross-layer design. Reducing the ratio of lost packets during channel impairments is an important reliability objective, as well [21]. Where, energy saving via cognitive radio can be applied at the MAC layer of the network stack [22][23].

The MAC layer is usually expected to adapt the number of retransmissions depending on channel quality. Current MAC protocols, however, typically limit the number of back offs and retransmissions. Due to the uniqueness of the short-range communication, which enables enormous bandwidth and negligible transmission delay, the probability of collision is considerably less in IoT than that of traditional wireless networks. Hence, we can safely assume that packet loss is not a major issue for IoT networks. Similarly, the unique characteristics of short-range communication band and specific challenges of IoT mainly energy and processing constraints and random network topology exclude direct implementation (without adaptation) of traditional wireless sensor networks routing schemes. In the following, we discuss traditional sensor network protocols with preferable features to be considered in IoT routing protocol in order to reduce energy consumption.

Multipath versus Single-path: Recent studies show that multipath routing protocols for sensor networks are better than single-path routing protocols in terms of QoS parameters satisfaction. Multipath routing protocols provide higher probability to achieve packet reachability utilizing redundant paths to the sink [14]. The Reliable Information Forwarding using multiple paths (ReInForM) protocol as described in [15] employs probabilistic flooding to deliver information awareness packets with desired priority levels of reliability at proportional costs for sensor networks. The routing mechanism is based on local knowledge of network conditions, such as channel error, hops-to-sink counting, and out-degree. Unfortunately, this protocol is not designed specifically for real-time or multimedia traffic; therefore, it does not consider delay deadlines of packets when selecting the multiple paths.

A chosen path might not be able to meet the delay requirements, yet it will be used to propagate duplicates potentially consuming valuable energy and unduly occupying useful channel bandwidth without improving the system performance. The work in [29] presents a data flooding dissemination scheme. The proposed scheme assumes a square grid network architecture where sensor-nodes are distributed densely at the vertices of the grid. Utilizing the uniform nodes' patterns and lattice algebra, the scheme dismisses node addressing requirements and employs a simple flooding scheme for data dissemination. The scheme relies on classifying each node as either an infrastructure or single-user node, depending on its reception quality. Only infrastructure nodes are allowed to forward any received packets to their neighbors. However, this classification is dynamic and adaptable. While the proposed scheme simplifies the communication model, it overlooks nevertheless the cost of classification and real-time signal processing of each packet. Additionally, it assumes a fixed structure and a static node deployment. Sensor-nodes typically display random behavior. IoT-networks can move around us for certain health applications [27], and therefore, may need to be associated with different neighbors and thus may not always follow a fixed structure. The authors of [28] propose a sound routing scheme for energy harvesting in IoT-networks. The routing scheme assumes a hierarchical cluster-based architecture. Packet transmission from the source to the cluster head can be direct or multihop based on the probability of saving energy through transmission, optimizing throughput and minimizing sensors' load. Still, however, an additional challenge comes from the very limited computational resources of sensing-devices.

Geographic-based Routing Protocols: These algorithms take advantage of the node position information to make routing techniques more efficient. For example, the Geographic Adaptive Fidelity (GAF) protocol makes it possible to optimize the performance of sensor networks by specifying the equivalent nodes in a given WSN [16]. Firstly, two nodes are designated as equivalent nodes. These two nodes are considered to be equivalent when they maintain the same set of neighbor nodes and so they can belong to the same communication routes. To specify these two equivalent nodes, it is a crucial step in this approach to precisely identify nodes' positions. Additionally, a virtual grid must be created at this step. This grid is divided into cells. The set of nodes that belong to the same cell are considered equivalent in terms of relaying/forwarding packets to another cell member. Another routing protocol, GEAR [17] (Geographic and Energy-Aware Routing) aims to improve energy efficiency by forwarding queries to aimed regions. In this routing protocol, sensors need to have localization hardware such as a GPS unit or a localization system. The work presented in [25] proposes a geographic routing protocol; the nodes of the IoT-network is assumed to comprise two types of anchor nodes, which have higher communication and processing capabilities than user nodes. User nodes are required to localize their position with reference to these anchor nodes. The authors assume that the network topology is square with four anchors located at the vertices of the square corners. The authors claim that two anchors are sufficient for each node to measure its location. The forwarding scheme operates at two phases; setup phase and operation phase. The setup phase is designed to assist user nodes in measuring their distances from the anchors. In the operation phase, a source node selects the anchor nodes and incorporates this information in a packet header proposed by the authors. A receiving node checks its location, the destination location and the source location to decide on forwarding or dropping the packet. However, the proposed scheme is topology-dependent and assumes a fixed topology, which may not be applicable for IoT networks. Additionally, the scheme requires addressing for all nodes, which forms a significant challenge in IoT networks with large-scale applications. Moreover, the localization techniques used can be inaccurate and lead to dramatic degradation in energy consumption.

Shortest Path Routing: In the Nearest Neighbor Algorithm (NNA), when a packet is transmitted from a node to another, it follows the shortest path based on the available common control channels. NNA assumes that if a packet always follows the shortest path, it will use it until it reaches destination node. In short, this algorithm uses a 4-direction transmission (left, right, up, down) only so it actually does not consider the shortest path, it considers the shortest neighbor relay node in order to send data. As a result of this, hop count unnecessarily increases and therefore energy consumption is negatively affected. Meanwhile, in the Shortest Path Algorithm (SPA), when a packet is transmitted from a node, the algorithm calculates the shortest path from recent node to destination instead of node-to-node, and the packet follows this path until it reaches to destination. SPA, uses 8-directions (up, upper-left, upper-right, down, down-left, down-right, right and left) and considers the shortest path to destination rather than the shortest neighbor of the relay node. Thus, In SPA hop count decreases and energy consumption of the nodes decreases in comparison with NNA. Nevertheless, SPA is the simplest routing protocol that takes in to consideration the path length as a unique design factor affecting the network energy. In practice, this assumption is not accurate due several other design factors such as the communication link condition and reliability.

In this research, we propose a Cognitive Energy-Efficient Algorithm (CEEA) as routing protocol. CEEA assumes a multitier IoT-network, and cluster/tier-wide synchronization. It is a topology-independent protocol which copes with the randomness nature in IoT-networks. CEEA determines the path from the Routing Node (RN) to the destination node in view of each node's remaining energy. The remaining energy of neighbors of recent RN's is controlled each time before a packet is sent from the RN. If one of the neighbor RNs' energy is below half of its initial value, a new path will be determined for the packet to follow. In addition, when all neighbors' remaining energy is below half of the initial energy, the system uses the same strategy. As a result of that, even if hop count increases in comparison with SPA, energy efficiency is improved for RNs and so is network lifetime.

2.3 System Model

The main objective of IoT in smart environments is to monitor physical and/or chemical changes and pass the information to a data center for processing [30]. IoT nodes may have varying sensing capabilities. Due to energy constraints, nodes do not communicate with each other but rather pass the sensed data to routing nodes (RN). RNs take collected data to a Gateway (GCN) that is usually connected to the internet for remote collection/processing. This communication type continues till the IoT network death. IoT networks have to overcome several challenges. Energy consumption for communication is the most significant one. Energy-efficient routing protocols can prolong IoT systems' lifetime dramatically. In this section we list the assumed system model for the proposed CEEA routing protocol.

2.3.1 Network Architecture

The typical communication range in IoT is expected to be between 1 cm and 150 m [9]. This means that the transmission range is still limited, making multi-hop routing particularly important for IoT networks. Furthermore, when IoT nodes are mobile, the direction of a communication route is not deterministic and is dependent on the drift velocity of sensory machines, which may lead to communication delay. This necessitates efficient schemes for multi-hop path creation and management. IoT networks can be divided into three categories: in-object, on-object and off-objects IoT. An overview of the structure of IoT network under such circumstances can be summarized as:

- IoT Sensor-Nodes (SNs): These are assumed to be small and simple IoT sensor-devices. Due to their limited energy, limited memory and reduced communication capabilities, they can only perform simple computation task and can transmit over very short distances. The nodes could be composed of sensor and communication units.
- Relay-Nodes (RNs): These are the Relay-devices with slightly larger computational resources than SNs and can aggregate information from a limited number of SNs and also can control the behavior of SNs by sending simple instructions (such as on/off, sleep, read value, etc.). These added capabilities would increase their size; thus, their deployment would be more invasive.
- Cognitive Relay Nodes (CRNs): They are used to aggregate the information forwarded by RNs and send the information to other CRN devices. At the same time, they can send the information from short-range-scale to large-scale. In this chapter we identify these nodes as cognitive nodes (CRNs).
- Gateway (GCN): It enables to control or monitor the entire IoT system remotely over the Internet.

It's worth pointing out here that IEEE 802.15.4 protocol is considered at the CRN to specify a sub-layer for Medium Access Control (MAC) and a physical layer (PHY) for low-rate wireless private area networks (LR-WPAN) because of some desired features such as low power consumption, low data rate, low cost, and high message throughput [9]. Thus, the IEEE 802.15.4 based CSMA access method can be considered at the MAC layer. This inherently reflects the communication channel reliability. Based on [26], this channel reliability can be characterized by a reliability design factor as follows:

$$C_R = \left(\left(1 - P_{blocking}\right) * \left(1 - P_{c-fail}\right) * \left(1 - P_{p-discard}\right)\right) \tag{2.1}$$

Where $P_{blocking}$ represents the blocking probability due to a buffer-full condition; P_{c-fail} is the common channel access failure probability due to channel condition (i.e. SNR) and $P_{p-discard}$ is the probability that a packet is discarded on reaching the maximum number of retries limit. This reliability factor is responsible to make a decision when equivalent energy levels' at RNs are faced. And it reflects the probability that a frame is not blocked, lost due to common channel access failure, or discarded as a result of reaching the maximum number of retries limit.

2.3.2 Lifetime of IoT Network

IoT Network Lifetime is defined as the time or number of transmission rounds beyond which the network can no longer deliver useful information to the outside end-user. This is reflected by the network's inability to find a data delivery path with satisfactory values for quality of information (QoI) attributes such as delay, reliability and throughput, as determined by the end-user [28]. This definition not only provides information to satisfying the application requirements, but also considers the status of the network and sensing resources in defining the network lifetime. It also justifies the fact that if the network does not have the necessary resources to send packets, it cannot satisfy the end-user, and so it should be considered as a dead IoT network. The IoT network lifetime can therefore be evaluated in three ways;

2.3.2.1 Lifetime Based on Number of Alive SNs

Several variants do exist with this model. The simple model identifies the time until the death of the first SN in the network as the lifetime of the network. Another variant evaluates lifetime until the death of 'k' out of 'n' SNs in the network, where k <n. The lifetime is the range between the death of '*k*' nodes from '*n*' nodes in non-critical ones [31].

2.3.2.2 *Lifetime Based on SN Coverage*

This model defines the lifetime of the network in terms of the coverage of region of interest. If it is used to ensure that all points inside a region of interest are covered, it is denoted by volume coverage. When an identified number of target points are to be covered, it is denoted as target coverage.

2.3.2.3 *Lifetime Based on Coverage and Alive SNs*

This type of metrics is mostly found in Ad-hoc IoT networks. In this option, lifetime is defined as the period during which most of the nodes are connected with each other. Because in IoT each node has to communicate with a gateway node, this metric cannot be used as is. Another issue with this metrics is that the lifetime is based on the total number of packets transmitted to the gateway. Nevertheless, in most of the related works this metrics become useless [32].

2.3.3 *Energy Conservation & Dead Node Issue*

Energy conservation is one of the most important issues in IoT design. SNs are restricted in carrying out the network layer functions, their main task is to flood the data to their one hop routers. Hence, the multi-hop forwarding between source and gateway is normally performed by RNs which have relatively higher capabilities than SNs. And thus, we define the energy consumed at a RN by $\boldsymbol{E_{RN} = C\,(T * (E_{TX}) + R * (E_{RX}))}$. Most of the energy consumption at the RN is due to data communication, indicated by E_{TX} for energy consumed during transmission and E_{RX} for energy consumed during data reception. C represents the cost function of the energy consumed T represents the number of transmitted packets and R represents number of received packets. As discussed above CRN main function is data aggregation and routing of traffic received from the RNs via cognition elements. The capabilities of the CRN are higher than those of RN, hence our assumption of the cognitive decision process to be performed by the CRNs, which is expected to consume additional energy compared to regular RNs [33]. Additional energy consumption is divided into two parts: one is protocol overhead incurring during cognitive data delivery due to feedback from the IoT-network during the learning process and the exchange of values of QoI attributes such as delay, reliability and throughput while making routing decisions and the other one is the increased transmit power for increasing the communication range of CRNs.

$$\boldsymbol{E_{CRN} = C\left(T * (E_{TX}) + R * (E_{RX})\right) + C\left(Ag * \left(E_{ag}\right)\right) + C\left(P * \left(E_{cog} - E_{pro}\right)\right)} \quad (2.2)$$

In Eq. 2.2 T, R, Ag, and P, represents the total number of packets that are transmitted, received, aggregated and processed by the cognitive elements respectively, in each transmission round. $(T * (E_{TX}) + R * (E_{RX}))$ is the energy cost incurred during data transmission and reception, $(Ag * (E_{ag}))$ represents the energy cost incurred during data aggregation and $C(P * (E_{cog} - E_{pro}))$ indicates the energy cost due to protocol and processing overhead during the cognitive processes. Forming Eq. 2.2 in terms of the energy cost of RNs we get:

$$E_{CRN} \geq E_{RN} + \left(Ag * \left(E_{ag}\right)\right) + C\left(E_{cog} - E_{pro}\right) \tag{2.3}$$

If the RN and CRNs use the same transmit power, the equality sign becomes positive in Eq. 2.3. In order to ensure that the energy cost of *CRNs* does not offset the advantages it offers in terms of adapting to traffic flow dynamics and network topology alterations, the cost can be optimized by maximizing the number of RNs and minimizing the number of *CRNs* in the deployment [28].

In this work, we refer to one-hop neighbors' communication as the first tier of nodes. Since no other node can reach the monitoring station directly, traffic from every other node will have to be forwarded, in the last hop, by one of these first tier nodes. Similarly, the two-hop neighbors of the monitoring station will forward data for all nodes except the one-hop neighbors and themselves, etc. If the spatial distribution of nodes is assumed to be uniform, then the traffic load is equally distributed. Each first tier node will forward hardly the same amount of traffic, and all first tier nodes will die at times very close to each other, after the network is first put into operation. Once all of the first tier nodes are dead, no other node will be able to send data to the gateway node, and the lifetime of the network will be over. Increasing the number of nodes in the network accentuates this effect, since there is more traffic to forward and the first tier of nodes has a smaller share of the total energy budget.

In general, the network death in IoT can be associated with several cutoff criteria such as the first node death, the percentage of dead nodes, or the number of dead nodes rising above a level where the routing to the sink node is no more possible [34]. Nevertheless, as we are experimenting with the clustering based protocols, in which the energy is evenly distributed throughout the mobile IoT-network, we consider the first scenario for the definition of the network lifetime. Because, when the first node dies, the number of dead nodes increases in the later rounds, and within 5–10 rounds the whole network becomes nonoperational. According to preliminary results, non-position-based routing protocols outperform geo-based protocols in terms of network lifetime. The primary reason for this behavior is that location-based protocols consume energy in terms of localization services. Moreover, the number of control messages plays a vital role in the network lifetime.

2.3.4 Communication Model

Radio interference, antenna shape and orientation, distance and environmental factors may vary during the network lifetime and affect link quality between the sensor nodes [20]. Despite the fact that the locations of sensor nodes are fixed as well as every node is configured with the same transmission range, environmental variations result in asymmetric links between nodes [24]. Therefore, these routing approaches shall estimate link quality to find the optimal path. Considering that the communication is at varying-range scale, the study of the communication in very short range is essential. And hence, we consider the proposed path loss formula in [34] at Short-range communication, which has two parts: the absorption path loss and the spread path loss. Meanwhile, energy-aware frameworks depend heavily on two main principles in their communication design. First principle is the number of hops without delay constraint. Second, is the number of hops with the delay constraint.

Number of Hops without Delay Constraints

If there is no delay constraint on the system, the highest achievable transmission rate is given by equation 18 in [34].

$$C_{ref}(d) = \sum_{i}^{n} \Delta f \; log_2 \left[1 + \frac{S(f_i) A(f_i, d)^{-1}}{N_0(f_i, d)} \right] \tag{2.4}$$

The bandwidth is divided into i sub-bands the i-th sub-band is centered around frequency f_i, $i = 1, 2, \ldots$ and it has width Δf. If the sub-band width is small enough, the channel appears as frequency-nonselective and the noise p.s.d. can be considered locally flat. The resulting capacity in bits/s is then given by where d is the total path length, S is the transmitted signal p.s.d., A is the channel path loss, and N_0 is the noise p.s.d. Based on [35], we find the end-to-end capacity of N hops path by

$$C_{e2e} = C_I (1 - F_{AVG})^N \tag{2.5}$$

where C_I is the channel capacity contributed by the first hop, N is the hop count determined by the forwarding scheme and F_{AVG} is the average capacity loss factor per hop. The value of F_{AVG} is calculated as follows:

$$F_{AVG} = F \left(\frac{d_0 - d_{AVG}}{d_0} \right) \tag{2.6}$$

where F is the capacity loss factor and d_o is a constant that denotes the reference distance from source-to-sink [35].

Number of Hops with Delay Constraints

The predictions about the preferred number of hops made in the previous section were based on the assumption that the block lengths used by channel codes can be randomly large. In many applications there is a strict limit on the tolerable end-to-end delay. There are several factors of delay in short-range communication systems. In the following we list these factors:

- *Waiting* for the data source to emit enough bits to form a block of a desired length (for channel coding);
- *Processing* delay caused by encoding/decoding the information bits for transmission;
- *Transmission* and *reception* of the whole encoded message.

If the communication system involves multiple hops, the latter three elements are repeated several times, increasing overall delay. To compensate for this, shorter block lengths must be used at a cost of reduced error-correcting capabilities at each link [28].

2.4 Cognitive Energy-Efficient Approach (CEEA)

In this section we propose a novel energy aware data delivery approach for the energy-constrained IoT. Let's assume that a randomly selected sensor n by GCN, depending on the harvested energy, is to be used for data retrieval. The random number of relay nodes within the communication range of the sensor n can be characterized by a spatial Poisson process X. Let the sensor n be at point $z \in \mathbb{R}^2$ and define $l(z,X)$ as the shortest distance from the sensor location z to the nearest point of X such that $l(z,X) \leq r$, and only common control channels are considered. Since X is a spatial Poisson process, then $l(z,X) \leq r$, if and only if $RN(d(z,r)) > 0$, where $d(z,r)$ is a disc of radius r centered at z. Conclusively, the probability of having at least one relay neighbor within the transmission range of the sensor n is given as follows.

$$P(l(z,X) \leq r) = 1 - \exp(\beta_{harv} A_d(d(z,r))) \tag{2.7}$$

where A_d is the area of the disk $d(z,r)$, and where β_{harv} denotes the rate of harvested energy of a senor. Note that $exp(\beta_{harv} A_d(d(z,r)))$ denote the probability that no relay node is within the transmission range of the sensor n; i.e. the network lifetime of the neighborhood of sensor n is expired. When the lifetime of the neighborhood nodes is expired, the IoT-network is assumed dead. Thus, assuming $f(n_j)$ is the cost function of transmitting from RN_j to GCN, $g(n)$ is the energy of neighboring RNs, $h(n)$ is the minimum distance from a neighbor RN_j to GCN, $i(n)$ initial energy of the neighboring RN. Our approach relies on a "*cognition process*" that has three main criteria in data routing: 1) Evaluation criteria; $f(n_j) =$ Cost(Neighbor RN to GCN) and $h(n_j) = min(f(n_j))$, this is guaranteed by lines 11 to 18 in

Algorithm 2.1a, 2) Selection criteria; $g(h(n_j)) > i\,(h(n_j)) * 50\%$, this section is found between lines 19 and 21, and 3) Termination Criteria; all one-hop RNs are dead or $P(l(z,X) \le r) = 0$.

Algorithm 2.1a: Pseudo-code of the CEEA algorithm.

1. **Function: Cognition process in CEEA**
2. **Input**
3. Source RN
4. **Output**
5. RN index chosen by CEEA to deliver data towards GCN.
6. **Begin**
7. **Initialize**
8. Hop Count =0; //for RNs beginning of round.
9. Identify source RN as a start node for current round.
10. List all neighbor RN indices from source RN
11. **If** source RN has one-hop, send directly to GCN.
12. **Else If** there is least one RN connected with this RN
13. **For** each source RN index 'j' do
14. $f(n_j) =$ Distance (Neighbor RN to GCN)
15. $g(n_j) =$ Energy (Neighbor RN's energy)
16. $h(n_j) = min(f(n_j))$
17. $i(n_j)$= InitialEnergy (RN's initial Energy)
18. **End**
19. **If** $g(h(n_j)) > i\,(h(n_j)) * 50\%$
20. Chosen RN Index $= g(h(n_j))$
21. **End**
22. **End**
23. **End**
24. **Else**
25. There is no source RN
26. **End**
27. **If** RN's energy when connected with GCN < 0,
28. Then disconnect from path
28. **End**
29. Update neighbor energy information of source RN
30. **Termination Criteria**
31. $P(l(z,X) \le r) = 0$ in Eq. 2.7 is equal to 0.
32. **End**
33. **Return** $g(h(n_j))$

In the above Algorithm, elements of cognition in the utilized *cognition process* form the two main constituents of our proposed approach. The elements that help in implementing cognition in the cognitive nodes are: *reasoning* and *learning* elements.

2.4.1 Learning

Learning is used in our CEEA approach in order to determine the most appropriate paths towards the GCN that satisfy the IoT-network requirements. This cognition element uses a direction-based heuristic to determine the data delivery path through RNs that lie in the direction of the GCN. Hence, each time the *cognition process* has to choose the next hop, the direction-based heuristic eliminates RNs that increase the distance between the current RN and GCN. Knowledge of the positions of the CRN and its one-hop RNs is used by the heuristic to determine the set of such RNs, which we call forward-hop-RNs. Thus the forward-hop-RNs of a CRN identified by the direction-heuristic is constituted by those RNs that reduce the distance between the CRN and the GCN. This information is stored in the CRN for use in the next transmission rounds. Thus the direction-based heuristic, along with feedback from the network about the chosen paths helps the *cognition process* to learn data delivery paths to the sink, as the network topology changes.

Example 1

Assume S_1 and S_2 have data to be sent to destination nodes D_1 and D_2. R_n are all the available relays towards the destination. Out of these relays, it is determined that R_5 as shown in Figure 2.1 has the lowest link outage probability to D_1 and D_2. Therefore, S_1 initiates routing data to R_5. Meanwhile, S_2 also forward a high traffic of data to R_5 (depicted by solid paths in Figure 2.1). When multiple source nodes start routing their data to R_5 as well, the route to R_5 may get congested. A cognitive network with *learning* capabilities will be able to identify the congestion at R_5 (by observing the decrease in throughput). Sharing this observation with neighboring nodes, the cognitive IoT-network would be able to respond to the congestion proactively, by routing the data through a different path involving nodes R_4, R_8 and R_9 as shown in Figure 2.2.

2.4.2 Reasoning

In the CEEA approach, we assume a modified version of the Analytic Hierarchy Process (AHP) [36] for implementing the reasoning element of cognition in the IoT. AHP supports multiple-criteria decision making while choosing the data path. For example, if we have a delay-sensitive data, the node which provides the lowest delay, will be chosen even though it might degrade other metrics such as the network energy or throughput. If two next-hops guarantee the same delay then the next attribute to compare will be energy, and then, throughput, assuming that energy is the next desired attribute in the IoT-network.

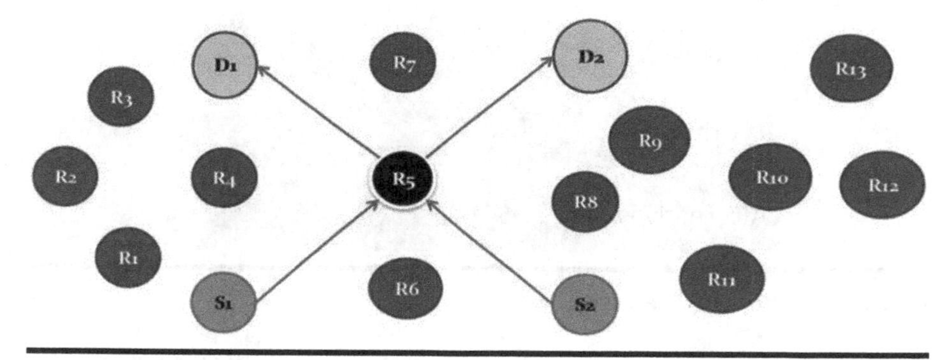

Figure 2.1 Classical routing in a sensor network.

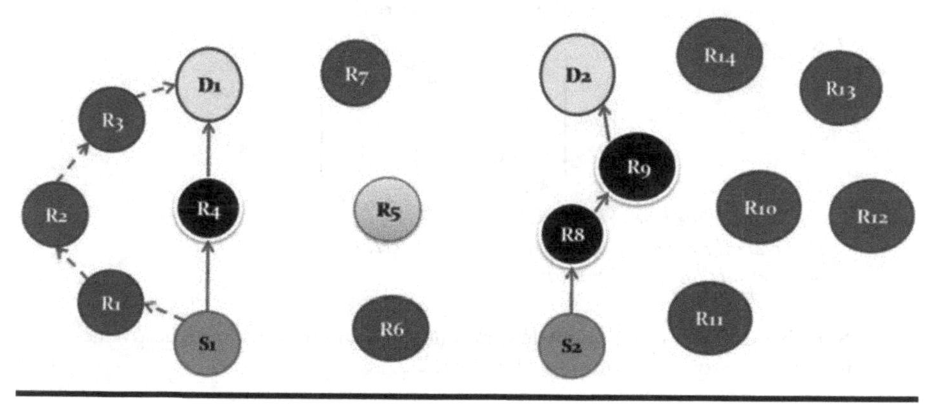

Figure 2.2 Cognitive routing in IoT-networks.

AHP provides a method for pair-wise comparison of each of the attributes and helps to choose the node that can provide the best network performance on the long run. The following subsequent example has more details on the utilized AHP. While AHP calculations help in deciding the next-hop, it also help in planning for future actions. The *cognition process* enables the CEEA approach to maintain the calculated values of the IoT-network attributes, which can be used in future transmission rounds. Hence, these values are not necessarily calculated at every transmission round.

Example 2

Assume a three level hierarchy in the AHP: *Goal, Criteria* and *Alternatives* as shown in Figure 2.3. A fundamental scale for pairwise comparisons is then used to set priorities for the IoT-network attributes/criteria at the CNs. Given the very limited energy constraint in IoT, we would assign the highest priority to energy, followed by *reliability* and then *throughput*. We tabulate the relative priorities of

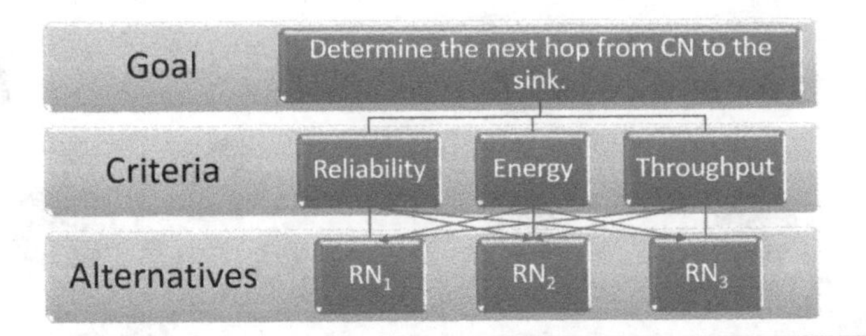

Figure 2.3 The AHP Hierarchy.

these attributes using pair-wise comparison in [36] and generate Table 2.1. From Table 2.1 we generate Table 2.2. Then, we apply the following steps:

1. *Represent the values of Table 2.2 in the matrix form*

$$A = \begin{bmatrix} 1 & 4 & 6 \\ 1/4 & 1 & 3 \\ 1/6 & 1/3 & 1 \end{bmatrix}$$

2. *Compute the Eigen vector of the matrix A,*
3. *Isolate the absolute, real values of the Eigen vector,*
4. *Compute the relative priority values.*

Note that our goal is to find the best next-hop, which provides the highest value for a specific attribute, as illustrated in Table 2.3.

Table 2.1 Pair-wise comparison of the IoT-network attributes

Energy (Kj)	4	Reliability	1
Energy (Kj)	6	Throughput (Mbps)	1
Reliability	3	Throughput (Mbps)	1

Table 2.2 AHP for QoI Attributes v/s Goal

Goal - Best attribute	*Energy (Kj)*	*Reliability*	*Throughput (Mbps)*	*Relative Priorities of the attributes*
Energy (Kj)	1	4	6	0.691
Reliability	1/4	1	3	0.2176
Throughput (Mbps)	1/6	1/3	1	0.0914

Table 2.3 AHP evaluating the overall priorities for all possible RNs

Best candidate for next hop RN_x	*Priority with respect to*			
	Energy (Kj)	*Reliability*	*Throughput (Mbps)*	*Goal*
RN_1	0.252	0.015	0.101	0.375
RN_2	0.2	0.018	0.11	0.329
RN_3	0.164	0.019	0.116	0.296

Algorithm 2.1b: AHP analysis for path selection in *cognition process*

1. **Function AHP (priorities of the attributes P)**
2. **Input**
3. P: End-user defined priorities on the attributes for requested data
4. **Output**
5. RN_x: Forward-hop $RN_x \in \{RN_1 \ldots.. RN_n\}$ with best P
6. **Begin**
7. **Initialize**: priority matrix for traffic type; Success=0;
8. **While** $P(l(z,X) \leq r) > 0$ in Eq. 2.7
9. *AHP_analysis*(Next-hop RNs v/s attributes)
10. Next hop $RN = RN_x$
11. Transmit data to next-hop RN
12. **If** (next hop = GCN)
13. Success=1;
14. **Else**
15. *Choose next-hop RN*
16. goto step 8
17. **End**
18. **If** *(Success==0)*
19. GCN Retransmits request
20. **End**
21. **End**

If energy consumption is measured as a function of the number of events taking place before the data packet arrives to the sink, the hop count can be used to approximate the energy cost. Accordingly, the modified AHP steps in prioritizing the IoT-network attributes and identifying the best next-hop are described in Algorithm 2.1b.

2.5 Performance Evaluation

In this section, we evaluate the performance of the proposed CEEA. We use MAEB, GEAR, ReInForM and LinGo algorithms as baseline evaluation algorithms.

Based on the aforementioned system models, we summarize these four baselines' categories as follows.

Geographic and Energy-Aware Routing (GEAR)

In this approach, sensor nodes must have a hardware component for positioning such as a GPS unit or a localization system. GEAR routing protocol is used to improve the efficiency in terms of energy consumption via forwarding queries to targeted regions. The forwarding scheme operates at two phases; *setup* phase and *operation* phase. Setup phase designed to assist sensor nodes in measuring their distances from the anchors. In the operation phase, a source sensor node incorporates its location information in a packet header. A receiving node checks its location, the destination location and source location for either forwarding or dropping the received packet.

Reliable Information Forwarding Using Multiple Paths (ReInForM)

ReInForM employs a probabilistic flooding procedure to deliver information-aware packets at a predetermined priority level. This leads to more reliable routing protocol at a proportional data delivery cost. The routing mechanism is based on local knowledge of network conditions, such as channel error, and hop-count to sink.

Movement-Aided Energy-Balance (MAEB)

MEAB has been chosen as a baseline due to movement and energy consideration. It has a neighbor discovery procedure which is conducted by the network cluster heads. They send their data packet to the Gateway following a forwarding rule, in which the distance and velocity to the Gateway and the remaining energy is recorded to select the neighbor cluster heads on the route towards the Gateway.

Link quality & Geographical beaconless OR protocol (LinGo)

LinGo introduces a different progress calculation approach compared to the aforementioned ones [37]. It takes into account both the progress of a given forwarding node towards the destination with respect to the last-hop, as well as the radio range. In this way, LinGO reduces the number of required hops on a data towards the destination node.

2.5.1 Performance Metrics & Parameters

To compare the performance of these five schemes, the following four performance metrics are used.

1. *Average Delay*: is measured in msec and is defined as the average amount of time required to deliver a data unit to the destination.
2. *Idel time*: this metric reflects the ratio of idle time every node spend while just waiting to forward a message. It is measured in μsec.
3. *Throughput*: is set here as a quality measure. It is the average percentage of transmitted data packets that succeed in reaching the destination reflecting the effect of node heterogenuety and delay in IoT setups over the utilized data delivery approach.

4. *Average Price*: this metric is used to observe the influence of the utilized data delivery approach on the overall price to deliver a data unit from source to destination on average. The price charged by each node $\boldsymbol{n}_i$ as $\boldsymbol{p}_i$.

$$p_i = \gamma_i * \left[\frac{E_{Tx}\left(D_k, n_j\right) + E_{Rx}\left(D_k\right)}{e_i} + \acute{\boldsymbol{\pi}}_{\boldsymbol{\imath}} + \acute{\boldsymbol{u}}_{\boldsymbol{\imath}} \right] \tag{2.8}$$

where $\acute{\boldsymbol{u}}_{\boldsymbol{\imath}}$ is the available buffer space at node i, and $\acute{\boldsymbol{\pi}}_{\boldsymbol{\imath}}$ is the power amount to be consumed per packet processing at node i. $E_{Tx}\left(D_k, n_j\right)$ *and* $E_{Rx}\left(D_k\right)$ are the mounts of energy used to transmit a data packet D_k from node i to j and receive a data packet D_k at node i, respectively. And $\boldsymbol{e}_i$ is the instantaneous available energy per node i.

Meanwhile, the three data delivery performance is assessed using the following three parameters:

1. The size of the network in terms of total node count. This reflects the application's complexity and the scalability of the exploited routing scheme. Knowing that larger node count in a data path raises the risk of node failure and, hence, dropped packets.
2. Average energy consumption rate per data unit $(\acute{\boldsymbol{\pi}}_{\boldsymbol{\imath}})$ as an indicator of the network power saving. This metric is measured in Kj.
3. PNF (%): It is the probability of a physical damage and/or a battery depletion for the deployed sensor node due to a disaster harsh-operational conditions. This parameter is chosen to reflect the impact in case of disaster scenarios or fragmented networks in IoT.
4. Cost (γ_i) to observe the influence of the charged price rate over the utilized data delivery approach. It is a pricing factor for each node in the IoT measured in $/byte. This is a factor that could be set as a flat rate per number of bytes transmitted, or computed based on the state of the current resources at node $\boldsymbol{n}_i$, where setting it to a relatively high value would diminish the chances of $\boldsymbol{n}_i$ to be selected for relaying the data packet D_k.

2.5.2 Experimental Setup

In order to limit our search space, we assume a virtual grid, where SNs are placed on the grid vertices. We assume up to 1500 total SNs communicate with one GCN via 36 RNs. We used NS3 as simulation tool for this purpose. The simulation is processed in three platforms which are Windows, Linux and OSX for validation purposes. We executed our simulation 100 times for each experiment and plotted the average results. More details about our simulation are summarized in Table 2.4.

Table 2.4 Simulation parameters and values

Parameter	*Value*
Targeted area	1000m x 1000m
Number of nodes	SNs: 350, RNs: 36, GCN: 1
Communication Range	SN: *142m*, RN: *300m*, GCN: *500m*
Initial Energy	SN: *31104J*, RN: *110160J*, GCN: *Unlimited*
Energy Consumption	SN and RN (Receiving): *31.2 uJ/bit* SN and RN (Transmitting): *53.8 uJ/bit*

2.5.3 Simulation Results

In Figure 2.4, the experienced delay in delivering data packet is plotted against the size of the network for the different simulated algorithms. We observe that RelnForM has the highest delay, while CEEA has the lowest delay as the number of nodes increases. Therefore, we can say that CEEA is more delay-tolerant in comparison to all of the sampled algorithms. We also observe that there is a monotonic increase in delay for RelnForM algorithm, while MAEB has a slightly higher delay than CEEA with a constant difference at every node. For CEEA and MAEB, we observe a steep increase between 100 and 200 nodes while RelnForM has its steepest slop between 150 and 200. For LinGo and GEAR we observe a fairly continues increase in delay as the number of nodes increase since they are more dependent on the network nodes' geolocations.

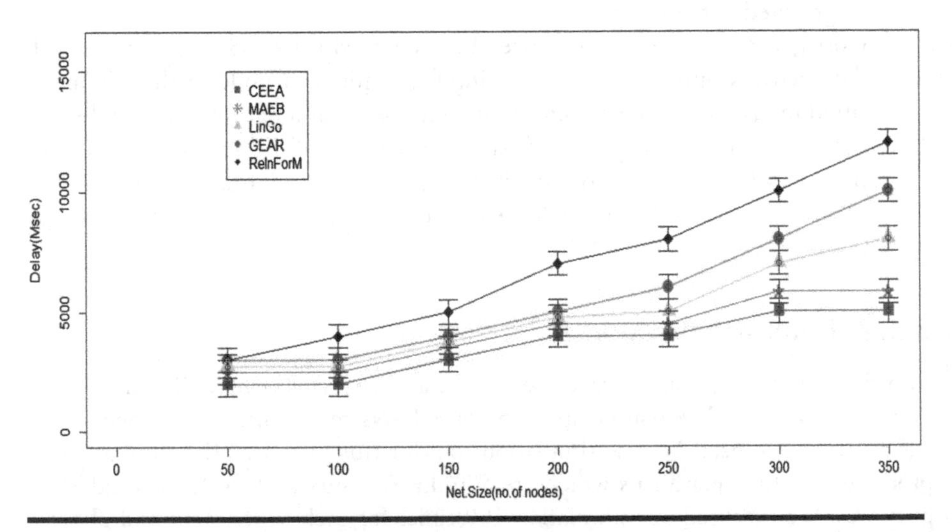

Figure 2.4 Delay vs. the number of nodes in an IoT network.

Figure 2.5 shows the experienced network throughput versus the number of nodes for the sampled algorithms. We can observe that there is a general increase in throughput of the sampled algorithms as the size of the network increases. MAEB, LinGo and GEAR have the same throughput until the size of the network is about 150 nodes, after which MAEB gives a higher throughput. Also, it's worth remarking here that LinGO adds redundant packets in order to increase the packet delivery probability while experiencing link error periods. This leads to significant increment in the overall throughput in comparison to GEAR and RelnForM methods. From the graph, we can also observe that in all instances, CEEA has a higher throughput than the others do, and RelnForM has the lowest throughput. And hence, we can conclude that CEEA has a better throughput as the network size increases compared to the sampled algorithms. This can be returned to the efficient retransmission approach in CEEA algorithm in comparison to other approaches in the literature. This makes it also the most scalable approach for the next generation IoT networks where the connected network nodes are dramatically increasing a day after a day.

Plotted curves in Figure 2.6 show the average consumed energy against throughput for the different examined algorithms. We notice that there is almost a liner increase in energy consumption while applying the RelnForM approach as the network throughput increases, while CEEA, GEAR, LinGo and MAEB forms a concave-like curves. We also observe that for every amount of energy consumed, RelnForM has the lowest throughput. While on the other hand, CEEA has the highest throughput for the same amount of energy. For this reason, we can conclude that CEEA is the most efficient algorithm in terms of energy consumption compared to the sampled ones. Moreover, we notice that when the energy budget is greater than or equal to 60 *Kjoule*, the network throughput is saturated due to other design factors such as the network size and cost factor (γ_i).

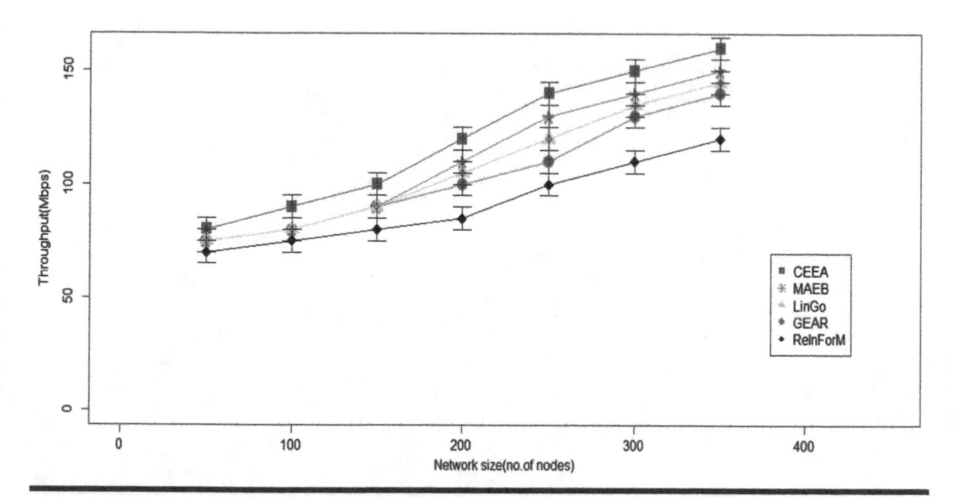

Figure 2.5 Throughput vs. the network size.

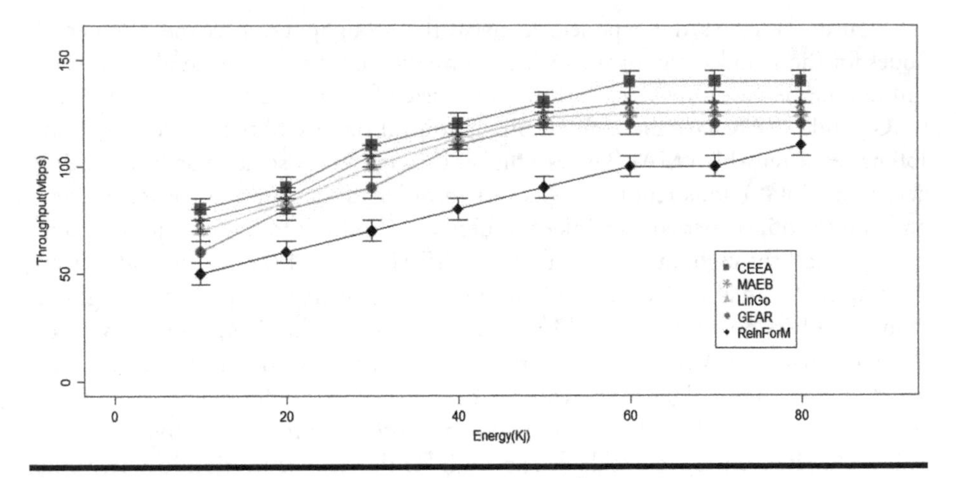

Figure 2.6 Throughput vs. the available energy budget.

Figure 2.7 shows the average charged cost (γ_i) per network node i against the experienced data delivery delay for all the simulated approaches. From this figure we can observe that with the increase in gamma, there is a general decrease in delay time for all the sampled schemes. Which is an expected network behavior as the flat rate charge increases per node. We also notice that all the schemes reach a certain threshold where the delay becomes constant at 2000 milliseconds. CEEA is the first to get to the threshold when γ_i equals to 40, while ReInForM is the last when γ_i equals to 60. GEAR and MAEB reaches the threshold at γ_i equals to 50. Consequently, we can say that CEEA is the most cost-effective scheme since it has the lowest delay time with the lowest γ_i.

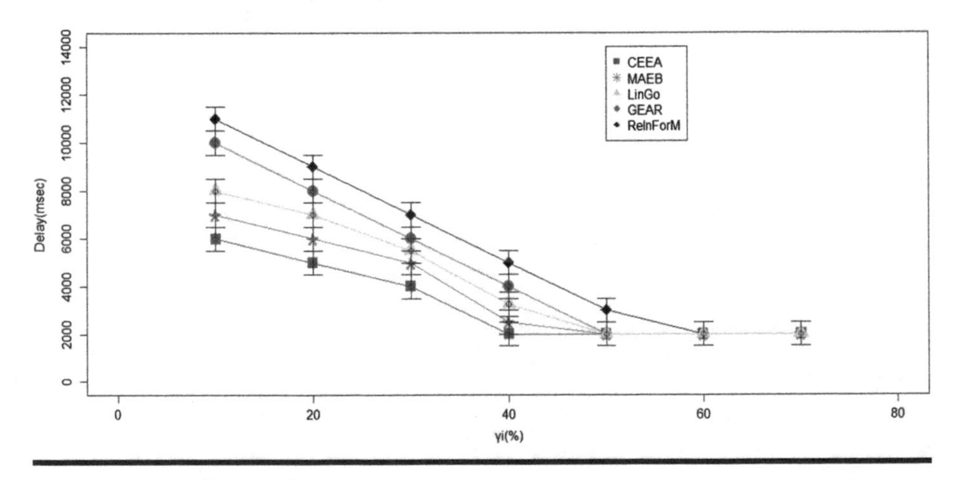

Figure 2.7 Delay vs. the average γ_i rate percentage.

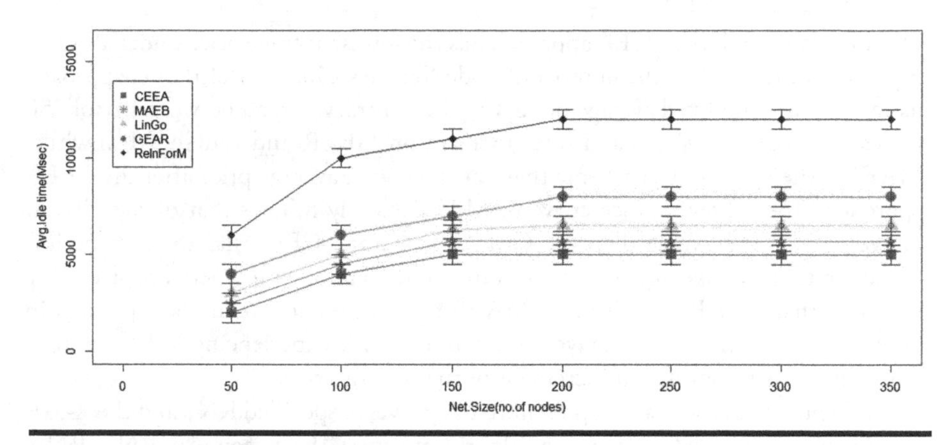

Figure 2.8 Avg. idle time vs. the network size.

In Figure 2.8, average idle time is compared under varying total count of network nodes. As the network size, or the number of SNs increases, there is a general increase in the average idle time. However, we observe that CEEA has the lowest idle time compared to other baselines. From Figure 2.8 we can also deduce that after a network size of 150 nodes, the average idle time of CEEA remains constant, which means it is not affected by the number of nodes. MAEB has a slightly higher idle time than CEEA whose difference to CEEA remains constant as the network increases in size. GEAR and ReInForM have an increasing idle time until 200 nodes, and then stay in a steady state. Therefore, we can conclude that CEEA is most efficient compared to the sampled baseline algorithms.

Figure 2.9 depicts the network size against the average price of all the schemes. From the figure, we can observe that ReInForM has the highest average price.

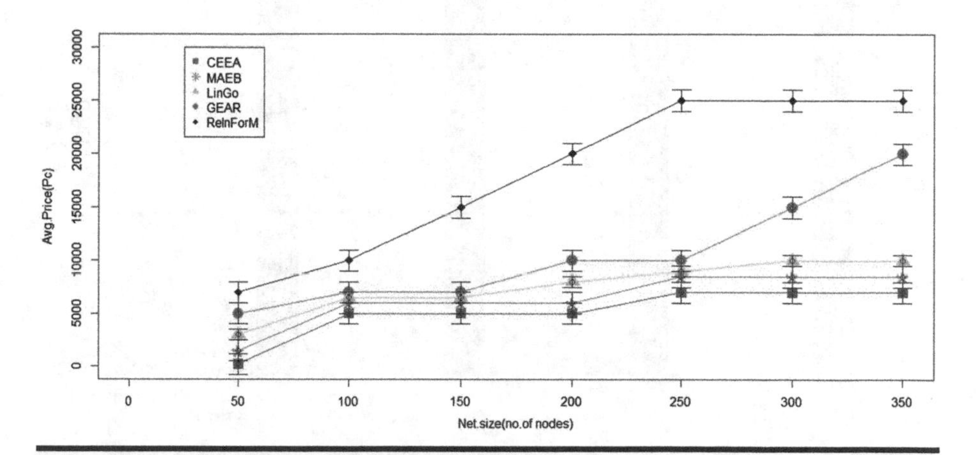

Figure 2.9 Avg. price vs. the network size.

On the other hand, the CEEA approach has the lowest average price under all varying node counts. When the number of nodes reaches 250, the ReInForM approach has a constant and fixed average price. On the contrary, after a network size of 250 nodes, we observe a sharp and liner increase on GEAR and LinGo. Meanwhile, MAEB is the second most scheme that has the lowest average price after the CEEA approach. The achieved price curve of MAEB closely follows that of the CEEA. However, it is still worse than the CEEA. Therefore, CEEA has the best performance in terms of average price as well under all experimented network sizes. The reason is that GEAR, LinGO and MAEB approaches add redundant packets in order to increase the packet delivery probability while experiencing link error periods. This leads to significant increment in the overall price.

In Figure 2.10, the Y-axis represents energy level of specified RN's and the X-axis represents the specified RNs and the algorithm types. The reason why RN_{14}, RN_{15}, RN_{20} and RN_{21} are selected is because these RNs have bidirectional connection with the GCN. To transmit packet to the GCN, one of these RNs must be used. We compare these RNs energy levels against the ReInForM, GEAR, LinGo and MAEB algorithms, since they are the most energy-efficient ones. Obviously, GEAR has the worst performance in this figure due to a fairness problem in this algorithm while relaying towards the sink node. Although the energy level of $RN_{14,}$ $RN_{15,}$ RN_{20} and RN_{21} are better for LinGo and MAEB, these RNs' energy levels are significantly outperformed by the CEEA approach. Thus, CEEA increases the network lifetime and it is better in energy saving. Furthermore, when we compare these algorithms in terms of the number of transmission rounds, it can be clearly observed from the simulation

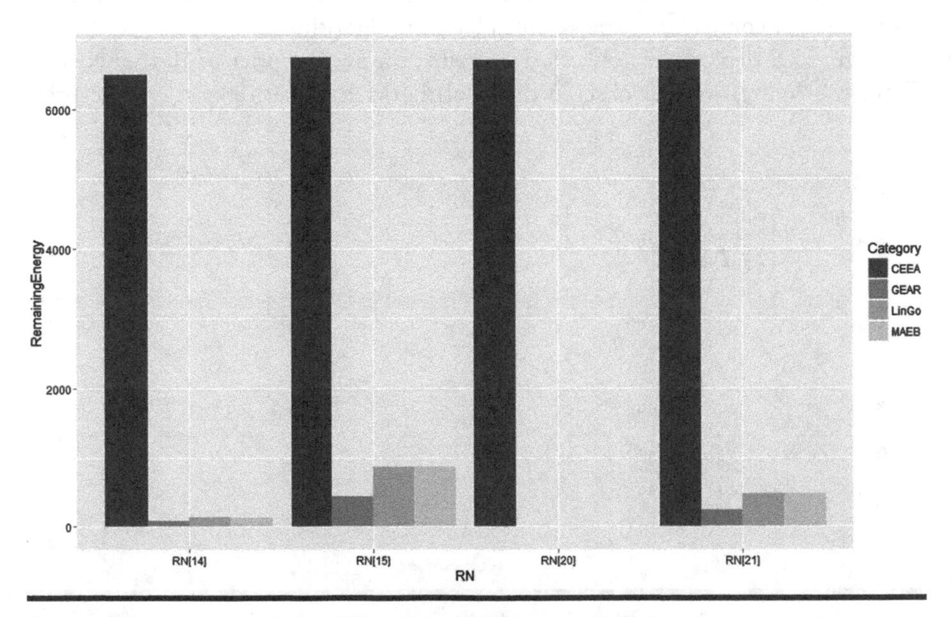

Figure 2.10 Comparison of One-hop RNs' Energy Level.

results in Figure 2.11, that CEEA outperforms GEAR, LinGo and MAEB. Notably, the more savings in terms of remaining energy shown in Figure 2.10 by applying the CEEA approach have led to prolonged network lifetime in Figure 2.11.

Moreover, we examined the four routing approaches; CEEA, LinGo, GEAR and MAEB in terms of the average delay impacts (Figure 2.12) while considering disaster scenarios and/or fragmented network, where failure of a critical node partitions the network into disjoint segments. Based on Figure 2.12, we notice a sever

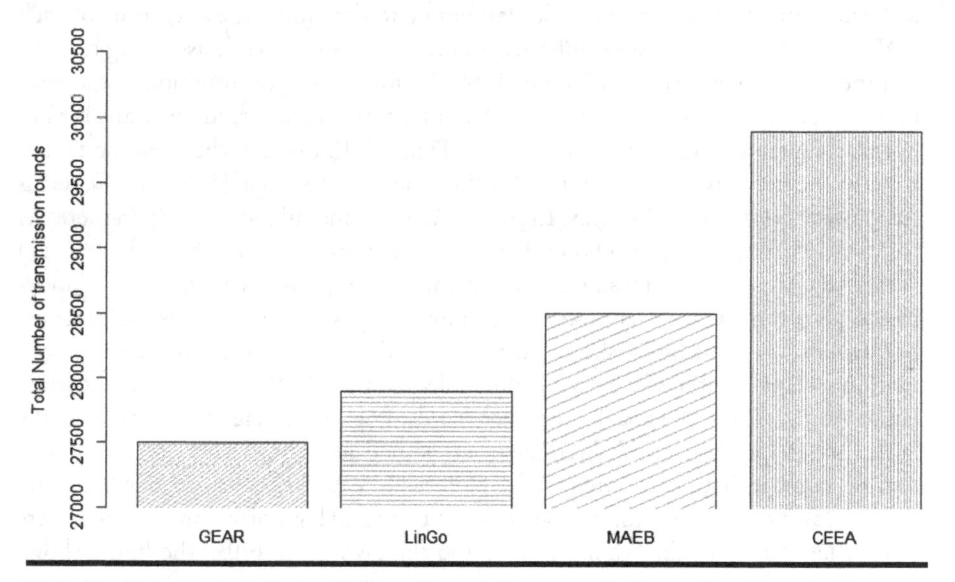

Figure 2.11 Comparison of the 3 data delivery techniques based on total number of transmissions.

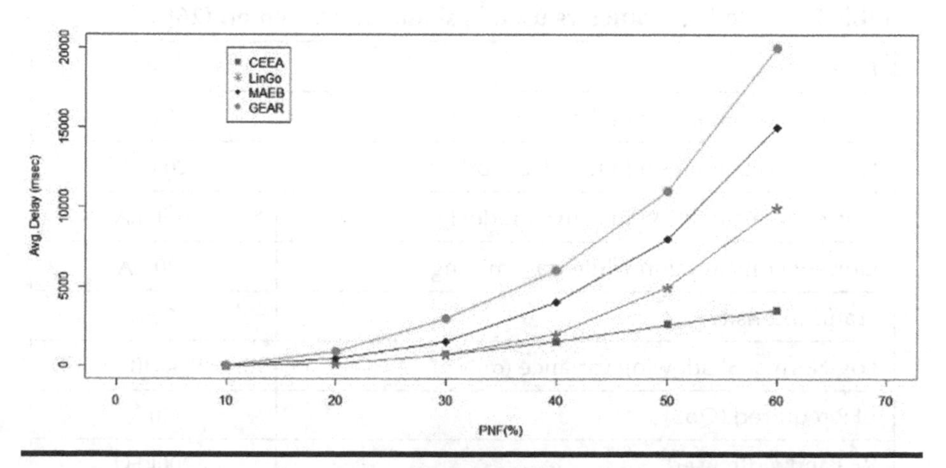

Figure 2.12 Average delay versus the probability of node failure in the network.

effect on the average delay while the probability of node failure (PNF) is increasing. We notice that all approaches are experiencing an exponential increase in the experienced delay as the network becomes disconnected. However, using the proposed CEEA approach the increment is going linear, which can be a very desirable feature in IoT while experiencing harsh operational conditions and sever mobility effects.

The good performance achieved by CEEA approach in this chapter can be returned mainly to the utilized cognitive elements that help a lot in disaster scenarios. In fact, the proposed CRNs make use of the received feedback about the utilized channel condition and modulation rate to determine the sleep time of each node. This concept has been emphasized more in Figure 2.7 while assuming realistic channel conditions as summarized in Table 2.5 for energy consumption of the relay node in four modes: sleep mode, receive mode, active mode (ready to transmit but not transmitting), and transmission mode. Figure 2.13 displays the mean node lifetime in the network using Adaptive Modulation and Adaptive Sleep (AMS) versus Adaptive Modulation (AM) only. In case of AM, the modulation level (parameter M in M-QAM modulation) is chosen for each packet according to channel condition (i.e. SNR). This case assumes no cognition and the sleep time is predetermined independently from the user requests and/or any changes to application requirements. Meanwhile, the same figure shows the average node life time using an adaptive modulation scheme combined with a scheduled sleep via the AMS mechanism that adapt based on the CRNs feedback. In other words, cognition here is employed at the MAC and PHY layers and sleep times are scheduled according to channel conditions and bit rates. In both cases high traffic patterns (mean packet arrival rate 90%) are assumed and simulated using Poisson distributions and log-normal shadowing where shadowing variance takes values from 0 (no shadowing) to 6dB. The figure shows that the cognitive approach significantly outperforms the non-cognitive one in terms of the average node lifetime. This is because M-QAM modulation, when carefully

Table 2.5 Node parameters used in simulation based on [26]

Parameter	*Value*
Current consumption in Sleep mode: I_{sleep}	1 μA
Current consumption in Receive mode: I_{rx}	20 mA
Current consumption in active mode: I_{ac}	100 mA
Current consumption while transmitting	120mA
Traffic intensity	90%
Log-Normal Shadowing variance (σ)	0, 2dB, 4 dB or 6dB
BER required (QoS)	10^{-4}
RF Bandwidth used	200kHz

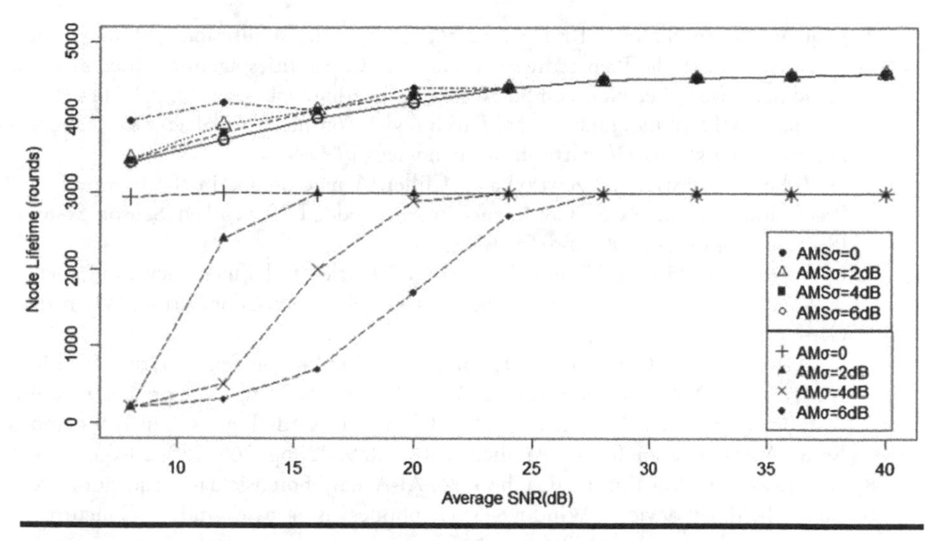

Figure 2.13 Node lifetime using Adaptive Modulation (AM) vs. Adaptive Modulation and Adaptive Sleep (AMS).

chosen, will require less transmission time, and thus, the cognitive system exploits this information to modify the sleep time accordingly. The improvement made by the cognitive approach is higher for both high traffic intensity and severe channel conditions (i.e. low SNR and high shadowing variance).

2.6 Conclusıons

In this chapter, we investigated routing techniques for the IoT paradigm in terms of energy consumption, cost and delay, while experiencing harsh operational conditions and sever energy limitations. We proposed a novel approach for sensor-networks in IoT, called CEEA. We found that CEEA can save considerable amount of energy. Moreover, we showed how the data delivery price can be affected by the network size for varying energy-based routing approaches. Furthermore, we showed how the CEEA algorithm provides the best price in comparison to other routing approaches in IoT. CEEA approach is recommended for disaster-inspired applications and it outperforms key other baseline approaches in terms of transmission and energy consumption.

References

1. F. Al-Turjman, H. Hassanein, M. Ibnkahla, "Efficient deployment of wireless sensor networks targeting environment monitoring applications", Computer Communications, vol. 36, no. 2, pp. 135–148, 2013.

2. Hijji, M., Amin, S., Iqbal, R., Harrop, W., (2015): "The Significance of Using Expert System to Assess the Preparedness of Training Capabilities against Different Flash Flood Scenarios", Lec Notes on Software Engineering, vol. 3, No. 3, pp. 214–219.
3. Y. Chang, MIT (2008, June 19). MIT tech review [online]. Available: www.technologyreview.com/s/410293/smarter-faster-nano-sensor/.
4. G. Tolle, J. Polastre, R. Szewczyk, D. Culler, A macroscope in the redwoods, in: Proceedings of the ACM Conference on Embedded Networked Sensor Systems (SenSys), San Diego, 2005, pp. 51–63.
5. L. Arved, P. Christian, "Water, Emergency Vehicles and Emergency Equipment," in Nanotechnology for Disaster Relief and Development Cooperation, Wiesbaden, Germany, 201, Ch. 2,5, pp. 27–47.
6. C. DOrazio, K. R. Choo, and L. T. Yang "Data Exfiltration from Internet of Things Devices: iOS Devices as Case Studies", IEEE Internet of Things, vol. pp, no. 99, 2016.
7. Y. Yang, H. Cai, Z. Wei, H. Lu, K. R. Choo, "Towards Lightweight Anonymous Entity Authentication for IoT Applications", vol. 9722, pp. 265–280, 2016.
8. N. Cahyani, B. Martini, K. R. Choo, M. Al-Azhar "Forensic data acquisition from cloud-of-things devices: Windows smartphones as a case study", Concurrency Computat.: Pract. Exper., 2016.
9. F. Al-Turjman, "Information-centric sensor networks for cognitive IoT: An overview", Annals of Telecommunications, vol. 72, no. 1, pp. 3–18, 2017.
10. S. Ali, S. Madani, "Distributed Efficient Multi Hop Clustering Protocol for Mobile Sensor Networks", The International Arab Journal of Information Technology, vol. 8, no. 3, pp. 302–309, July, 2011.
11. T. M. Cover and J. A. Thomas, Elements of Information Theory, MA: Wiley Press, 1991.
12. S. A. Nikolidakis, D. Kandris, D. D. Vergados and C. Douligeris, "Energy Efficient Routing in Wireless Sensor Networks Through Balanced Clustering", Algorithms, 2013, pp 6, 29–42. DOI: 10.3390/a6010029.
13. A. Aburumman, and K. R. Choo, "A Domain-Based Multi-cluster SIP Solution for Mobile Ad Hoc Network", SecureComm, vol. 153, pp. 267–281, 2014.
14. F. Al-Turjman and H. Zahmatkesh, "An Overview of Security and Privacy in Smart Cities' IoT Communications", Wiley Transactions on Emerging Telecommunications Technologies, 2019. DOI. 10.1002/ett.3677.
15. B. Deb, S. Bhatnagar, and B. Nath, "ReInForM: reliable information forwarding using multiple paths in sensor networks," in Proc. 2003 IEEE Conf. Local Computer Netw., pp. 406–415.
16. L. Villalba et al., "Routing Protocols in Wireless Sensor Networks", Sensors, 2009. DOI: 10.<pg>3390/s91108399.
17. M. Z. Hasan and F. Al-Turjman, "Analysis of Cross-layer Design of Quality-of-Service Forward Geographic Wireless Sensor Network Routing Strategies in Green Internet of Things", IEEE Access Journal, vol. 6, no. 1, pp. 20371–20389, 2018.
18. Maniak, T., Jayne, C., Iqbal, R., Doctor, F., "Automated Intelligent System for Sound Signaling Device Quality Assurance", Information Sciences, vol. 294, pp. 600–611, 2015.
19. Doctor, F., Syue, C-H., Shieh, J-H., Iqbal, R., "Type-2 Fuzzy Sets Applied to Multivariable Self-Organizing Fuzzy Logic Controllers for Regulating Anesthesia", Journal of Applied Soft Computing, vol. 38, pp. 872–889, 2016.
20. F. Al-Turjman, "Cognition in Information-Centric Sensor Networks for IoT Applications: An Overview", Annals of Telecommunications, pp. 1–16, 2016.

21. Kumar, N., Iqbal, R., Mistra, S., Rodrigues, J., "Bayesian Coalition Game for Contention Aware Reliable Data Forwarding in Vehicular Mobile Cloud", Journal of Future Generation of Computer Systems, vol. 48, pp. 60–72, 2014.
22. H. Fang, L. Xu and K. R. Choo, "Stackelberg game based relay selection for physical layer security and energy efficiency enhancement in cognitive radio networks", Applied Mathematics and Computation, vol. 296, pp. 153–167, 2017.
23. H. Fang, L. Xu, Jie Li and K. R. Choo, "An Adaptive Trust-Stackelberg Game Model for Security and Energy Efficiency in Dynamic Cognitive Radio Networks", Computer Communications, 2016. DOI: 10.1016/j.comcom.2016.11.012.
24. Baddeley, Adrian, Imre Brny, and Rolf Schneider. "Spatial point processes and their applications." Stochastic Geometry: Lectures given at the CIME Summer School held in Martina Franca, Italy, September 1318, 2004(2007): 1–75.
25. Tsioliaridou, Ageliki, et al. "CORONA: A Coordinate and Routing system for Nanonetworks." Proceedings of the Second Annual International Conference on Nanoscale Computing and Communication. ACM, 2015.
26. M-H. Zayani, and V. Gauthier, "Usage of IEEE 802.15.4 MAC–PHY Model", Online: http://www-public.it-sudparis.eu/~gauthier/Tools/802_15_4_MAC_PHY_Usage.pdf
27. Mahmud S., Iqbal, R., Doctor F., "Cloud enabled data analytics and visualization framework for health-shocks prediction", Journal of Future Generation of Computer Systems, Vol 65, pp. 169–181, 2016.
28. G. Singh, and F. Al-Turjman, "A Data Delivery Framework for Cognitive Information-Centric Sensor Networks in Smart Outdoor Monitoring", Computer Communications, vol. 74, no. 1, pp. 38–51, 2016.
29. Liaskos, Christos, and Angeliki Tsioliaridou. "A Promise of Realizable, Ultra-Scalable Communications at nano-Scale: A multi-Modal nano-Machine Architecture." IEEE Transactions on Computers, 64.5 (2015): 1282–1295.
30. S. Khan, A. S. K. Pathan and N. A. Alrajeh, "Wireless sensor networks", in Current Status and Future Trends, ed. 1, MA: CRC Press, 2012, ch. 1.
31. B. Baranidharan, B. Santhi, "An Evolutionary Approach To Improve The Life Time Of The Wireless Sensor Networks", Journal of Theoretical and Applied Information Technology, vol. 33, no. 2, pp.177–183, Nov, 2011.
32. F. Lewis, "Wireless Sensor Networks", in Smart Environments: Technology, Protocols, and Applications, New York, 2005, ch.2.
33. A. Al-Hourani, S. Kandeepan, "Cognitive Relay Nodes for Airborne LTE Emergency Networks", International Conference on Signal Processing and Communication Systems, Dec. 2013.
34. J. M. Jornet and I. F. Akyildiz, "Channel modeling and capacity analysis for electromagnetic wireless nanonetworks in the terahertz band," IEEE Trans. Wireless Commun., vol. 10, no. 10, pp. 3211–3221, Oct. 2011.
35. F. Al-Turjman, "Smart-cities Medium Access for Smart Mobility Applications in IoT", Wiley Transactions on Emerging Telecommunications Technologies, 2019. DOI. 10.1002/ett.3723.
36. G. Singh, and F. Al-Turjman, "Learning Data Delivery Paths in QoI-Aware Information-Centric Sensor Networks", IEEE Internet of Things, vol. 3, no. 4, pp. 572–580, 2016.
37. D. Rosário, Z. Zhao, A. Santos, T. Braun, E. Cerqueira, "A beaconless Opportunistic Routing based on a cross-layer approach for efficient video dissemination in mobile multimedia IoT applications", Computer Communications, vol. 45, pp. 21–31, 2014.

Chapter 3

Medium Access in Cloud-Based IoT

Fadi Al-Turjman
Antalya Bilim University

Contents

3.1 Introduction

Smart cities are becoming not only a reality but also more popular and spreading day by day. In fact, the deployment of smart and cognitive cities is expected to be very dominant in several regions of the world in the near future due to their advantages and merits in making our life easier, faster, simpler, and safer [1–4]. However, still thousands of people die in car accidents/collisions every year. For example, over 1.2 million died, for example, in traffic accidents around the world in 2016 [1]. There are a number of different reasons for road accidents/collisions in different countries. Nevertheless, the major cause of these road collisions is mainly driving the car under unpredicted weather conditions. Authors in Ref. [1] note that 50% of the fatal collisions happens while driving under a speed of 55 km/h. We therefore need a system where the speed limits are set according to the existing weather, traffic, and road conditions.

Smart cities are composed of a massive amount of smart, intelligent devices that have sensing, computing, actuating, and communication capabilities. These devices are designed in such a way that reduces human intervention through cognition and automation in communication. They are commonly termed as Internet of things (IoT) devices, which are expected to dominate the infrastructure of smart cities. Among the massive design requirements of IoT devices, low complexity with highly reliable communication channels comes as a key priority besides energy efficiency, latency, and security in emerging 5th generation (5G) services [5]. Of course, the impact of such an enabling technology is expected to be revolutionary. The new infrastructure for communication is expected to transform the world of connected sensors and reshape several existing industries. Such a revolution would, of course, require extensive research and development for the coexistence and device interoperability with 5G systems, and deployed sensor networks. This is a must for cooperative and smart sensing techniques, improved quality of service (QoS), and energy-efficient architectures in new intelligent transportation systems (ITS) infrastructure. This can be achieved by developing new sensor/5G protocols and standards, which can communicate with the Cloud reliably.

Focusing on optimization of the ways, the data is exchanged between the sensory devices and applications in IoT, and cyber-physical systems have led to the 5G/IoT integration as well. For example, studies on approaches to construct higher level abstractions of data at local gateways are proposed to reduce the traffic load imposed on the communication networks that provide the real-world data [5]. Test beds and real-world data sets are popularly employed to analyze the methods proposed.

With this recent revolution in wireless telecommunications, several advanced solutions relying on wireless communication standers have been proposed to provide ITS in the IoT paradigm. For instance, authors in Ref. [2] projected an automatic speed control (ASC) system that adjusts the speed of the vehicle according to the speed limit on the road. The feasibility of a smart box called "telematics," which has the ability to capture, analyze, and communicate, is being studied in cooperation with IBM's Engineering and Technology Services. Using multiple microprocessors and tiny sensors attached to the vehicle body, it is able to observe the vehicle's velocity, for example, and compare it to the upper speed limit of the road. In case the speed of the car is higher than the announced limit allowed by authorities, the box will verbally notify the driver. Moreover, a digital image processing system has been proposed by Baró et al. [3]while utilizing onboard cameras to read and recognize signs at the side of the road and send the warning signal to the driver and/or directly control the car. Different versions of this system have been investigated intensively all over the world.

The results in Ref. [4] have shown that this solution is able to cut down the accident rate by 35%. In the near future, the speed control system will be very dependent on the standard of IEEE 802.16 to locate each vehicle and satisfy the demands of the required real-time services such as voice and video [5].

In IEEE 802.16, there are different medium access control (MAC) scheduling services, such as unsolicited grant service (UGS), real-time polling service (rtPS), and non-real-time polling service (nrtPS) to provide better QoS. There are two commonly used schedulers for real-time traffic: the UGS and the rtPS [6]. However, these schedulers do not fulfill the requirements for the real-time services in smart cities. Hence, the suitability of a batch Markovian arrival process (BMAP) is analyzed [7] for modeling of IP-based data traffic. Accordingly, the BMAP has been found to be better in comparison with Markovian-based models. Hence, we propose a real-time-BMAP (RT-BMAP) model for real-time services. The objective of this approach is to provide the required QoS level with the minimum delay. Accordingly, the major contributions of this chapter are as follows:

1. We present the enhanced-rtPS (E-rtPS) integrated with RT-BMAP to solve the interference problem in the smart cities paradigm optimistically.
2. Our proposed approach has less computational overhead and better performance in terms of resource utilization.
3. The examination of real-time results shows that the proposed framework outperforms the existing IEEE 802.16 services in terms of delay, throughput, and reception percentages.

For more readability, a list of used abbreviations along with their definitions is provided in Table 3.1. The rest of this chapter is organized as follows: Section 3.2 overviews related works in the literature. Section 3.3 presents a detailed description of the proposed framework. Section 3.4 discusses extensive simulation results. Finally, Section 3.5 concludes the work in this chapter.

Table 3.1 Used Abbreviations and Acronyms

Abbreviations	*Definitions*
IoT BMAP RT-BMAP	Internet of things Batch Markovian arrival process Real-time batch Markovian arrival process
MAC	Medium access control
UGS	Unsolicited grant service
nrtPS	Non-real-time polling service
rtPS	Real-time polling service
QoS	Quality of service
ASC	Automatic speed control

(Continued)

Table 3.1 (*Continued*) Used Abbreviations and Acronyms

Abbreviations	*Definitions*
ITS	Intelligent transportation systems
LTE-A	Long Term Evolution-Advanced
E-rtPS	Enhanced-real-time polling service
SNR	Signal-to-noise ratio
WSN	Wireless sensor network
BS	Base station
RF	Radio frequency
OFDM	Orthogonal frequency division multiplexing
MCC	Management and Control Center
PSTN	Public switched telephone network
IRSN	IP real-time subsystem-network
GS	Grant size
MC	Mobile controllers
LC	Local controllers
VoIP	Voice over IP
MST	Minimum spanning tree
RS	Real-time server
RC	Real-time client
RSU	Road-side unit
BER	Bit error rate
FER	Frame error rate
CSCF	Call Session Control Function
HSS	Home Subscriber Station
D2D	Device-to-device communication

3.2 Related Work

By 2021, the mobile networks will triple the amount of used smart devices in 2016. In addition, mobile video services will increase nine times what we experienced in 2017. Given that our resources are limited and QoS is critical, the system base-station (BS) shall allocate the existing system bandwidth proactively. In smart cities, smart transportation systems, cellular network, wireless sensor network (WSN), device-to-device (D2D), and many of these smart paradigms involve the wireless mobile multimedia communications and/or transmissions.

Specifically, the D2D has received increased attention from wireless cellular networks because it can noticeably off-load the network traffic and reduce the transmission energy since the communicating devices are in close proximity to each other [2]. This close proximity makes it an eminent networking architecture for relay-based transmission employed by energy-constrained devices, such as smart-phones, tablets, and personal digital assistants (PDAs). Therefore, it is considered to be vital to improve the energy efficiency of D2D networks while maintaining their reliability in terms of the experienced throughput and latency. Most of the existing works have considered the direct link situation for the degradation of long-distance communication [2], poor propagation [3], and interference [4].

As relay-based transmission can improve the longer distance D2D communication, most of the researchers have given their attention for the multi-hop D2D communication. Authors in Ref. [5] deal with spatial density and power transmission to get better transmission capacity. However, it is declared without specifying a routing technique. Authors in Ref. [8] propose a multi-hop routing technique to maximize the hop count of the networking systems. But then, it is proven undependable to minimize the distance between the users. Since the existing relay-based techniques do not consider the energy efficiency for the D2D communication, there is a need to formulate a combinatorial optimization that improves the energy efficiency of such communication systems. Accordingly, technological advancements are balancing the act between bandwidth usage and time delay in any data delivery approach. In addition, low complexity with high-reliability communication comes as a first key priority besides energy efficiency, latency, and security [5].

There are two main categories of approaches that are focusing on addressing these challenges; the distributed versus centralized approaches. The distributed approach is a distributed routing in the network, which is very popular, and it has many sub-approaches. Distributed coding is also proposed and adopted to be used in future 5G and beyond systems as an indispensable channel coding scheme [9,10]. However, different channel codes are constructed and generated according to a pre-specified value of the signal-to-noise ratio (SNR). This issue becomes even more complicated with fading channel consideration due to multipath existence. In an attempt toward proposing reliable channel codes which considers fading conditions, several studies have been conducted and performed in the literature. In Ref. [11], a multilevel fading channels, which utilizes the nested coding, has been exhibited. It was demonstrated that this approach tends to tackle fading channels with a limited number of states. In Ref. [12], polar codes are connected to remote channels. A strategy for acquiring the Bhattacharyya parameters related to Rayleigh channels is exhibited. In Ref. [13], lower and upper limits on Bhattacharyya parameters were selected to develop competitive channel codes. In Ref. [14], the authors detailed the coding procedures for the radio frequency (RF) channel with realized channel state data at the two ends of the connection with known channel dissemination data. Another example on investigating the coding for block-fading channels can be found in Ref. [15].

The channel use is exhibited with the objective that codes could be grasped to be encoded over isolated squares in Ref. [16]. In Ref. [17], the creators analyze the general sort of the superfluous information trade and coding with multi-stage understanding. In light of this examination, they proposed a graphical structure methodology to create channel codes for picture hindrance. In Ref. [18], a clear system for the improvement of portable channel codes considering Rayleigh impacts was displayed. The sub-channels are shown as multipath obscuring channels, and their varying decent requests are pursued. In Ref. [19], codes are planned solely for square-based channels. The authors built channel codes customized for square fading channels while polarizing them with codes. The accomplished arrangements are shown to convey a significant increase contrasted with ordinary coding.

In Ref. [20], authors gave an elective structure of this methodology by treating the shadowing, fading, and coding of the channel as a solitary element. This empowers developing channel codes by mapping the codes with shadowing and fading effects. The acquired codes are adjusted to divert vacillation in versatile situations. Authors in Ref. [21] proposed a scheduling scheme of dynamic carrier aggregation (DCA) to provide higher energy efficiency in uplink communication. In Ref. [12], the authors have not considered the uplink real-time scheduling and resource allocation framework for the demand of the 3rd Generation Partnership Project (3GPP) Long Term Evolution-Advanced (LTE-A) networks. In Ref. [22], an orthogonal frequency division multiplexing (OFDM) approach under multipath fading channels is proposed. In this methodology, the codes are permuted with the goal that the code bits contrasting with the hardened bits are doled out to subcarriers causing ceaseless piece goofs. This change can be considered as a kind of interleaving that can distinguishably upgrade the channel coding. Nevertheless, changing the codes depending on the channel is not always required. That is on the grounds that it would result in a persistent change in the code of the correspondence channel type. This is viewed as a bulky, complex, and wasteful, particularly for Cloud-based applications. As a conclusion, the greater part of the previously mentioned methodologies accepts either the LTE-A or the IEEE 802.16 in their correspondence conventions. However, these measures have not been viewed as productive in planning. Subsequently, it is yet an open research issue [23].

In this work, we aim at proposing an adaptive framework for dynamic channel management in smart cities. We propose a novel design for the mobile vehicular Cloud infrastructure that takes into account varying weather, road, and traffic conditions. This framework utilizes the latest channel coding techniques while utilizing the existing cellular infrastructure to stream data, sound, and video with the least latency.

3.3 Framework Description

In this section, we recommend a complete depiction for the proposed vehicular Cloud framework. We define our framework components as follows:

- Management and Control Center (MCC): This component controls the velocity of the vehicle depending on the street, traffic, and surrounding environment conditions.
- Speed Limit Transmitter (SLT-x): This component is used to communicate the speed limit to the end-driver. The speed limits are transmitted as a wireless message. The MCC controls/adapts the speed of the vehicle based on road, traffic, and weather conditions, given that enough road-side units (RSUs) are located at predefined points on the road.
- Speed Limit Receiver (SLR-x): This component takes the transmitted street upper limit and displays it clearly in the vehicle. It will also be communicated to the driver's smartphone.
- Vehicle Speed Sensor: This component is required to measure the speed of the vehicle accurately. The speed measurement system that already exists in the car can be used as well.
- In-vehicle Microcontroller: Its main task is to compare the actual speed to the road speed limit, which is received by the SLR-x. Based on the vehicle speed, the controller may generate an audio warning or may communicate such incidence to the MCC through the 5G network via the driver's smartphone.
- 5G Modem: This modem is used to send a speed data to a central station, which is monitored and operated by any other governmental or private entity.
- Drivers' Records Server (DRS): It stores information about the drive. The DRS is updated when the Global System for Mobile Communications (GSM) modem sends a signal from the driver's car. The DRS can be accessible by third parties with the approval of the driver. Such third parties include parents, family members, and insurance companies.

3.3.1 Functioning of the Vehicular Cloud Framework

Figure 3.1 shows a simplified schematic for the planned vehicular Cloud system in normal conditions. The SLT-xs are connected to the MCC component through the public switched telephone network (PSTN) and the GSM/5G networks, which

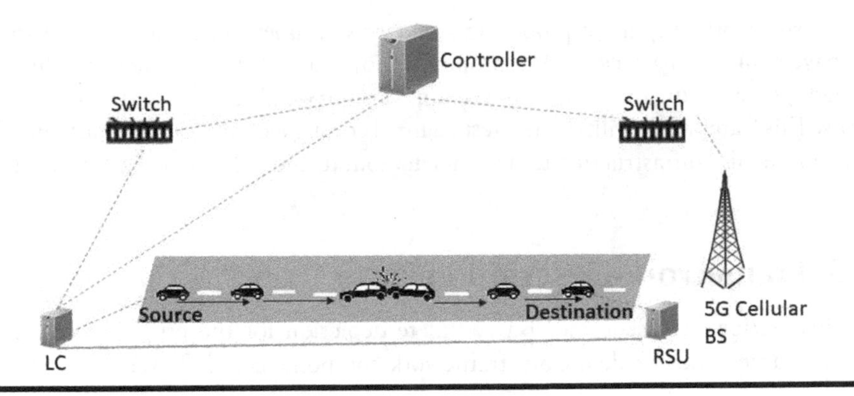

Figure 3.1 A schematic for the vehicular cloud in smart cities.

are usually connected to Cloud data centers. They are controlled by the MCC, which recommends changing the speed limit depending on the road, weather, and traffic conditions [24]. The MCC gets the required information from the patrol police, forecast stations, and driver's smartphones. The vehicle speed is continuously compared to the received speed limit, which is also displayed to the driver. Hence, the driver will always know the upper limit of the street he or she is on. This is a better method than the conventional speed limit signs located on the side of the road. If a driver is driving below the speed limit, there is nothing to do except that it might be a congestion case. If the vehicle speed limit exceeds the received one, a warning signal will be generated to alert the driver that he or she exceeded the speed limit. If the driver does not respond within a given period, the In-vehicle Microcontroller sends a note (violation) through the 5G modem to the DRS. The violation can also be sent to the driver's smartphone. The violation can also be recorded using various network-positioning techniques. Other forms of violations such as tail tracking and red light tracking can be motored in the car, which has other types of sensors. In such a case where a violation was recorded, the DRS can inform the driver instantaneously through his or her mobile/e-mail/mail, or by some other IEEE 802.16 standard means. We remark that this study focused on the Voice over IP (VoIP) service as an example of the exchanged real-time contents over the vehicular Cloud.

3.3.2 Registration Phase

We assume that RSUs are located in different places along a given road. Local controllers (LCs) are static BSs in the city. Mobile controllers (MCs) are sensing platforms attached to vehicles. We use a minimum spanning tree (MST) algorithm to select an LC at each route. Each MC is considered as a node in the MST, whereas all RSUs are considered as terminal nodes, as depicted in Figure 3.2. Accordingly, this registration phase is summarized in the following four main steps:

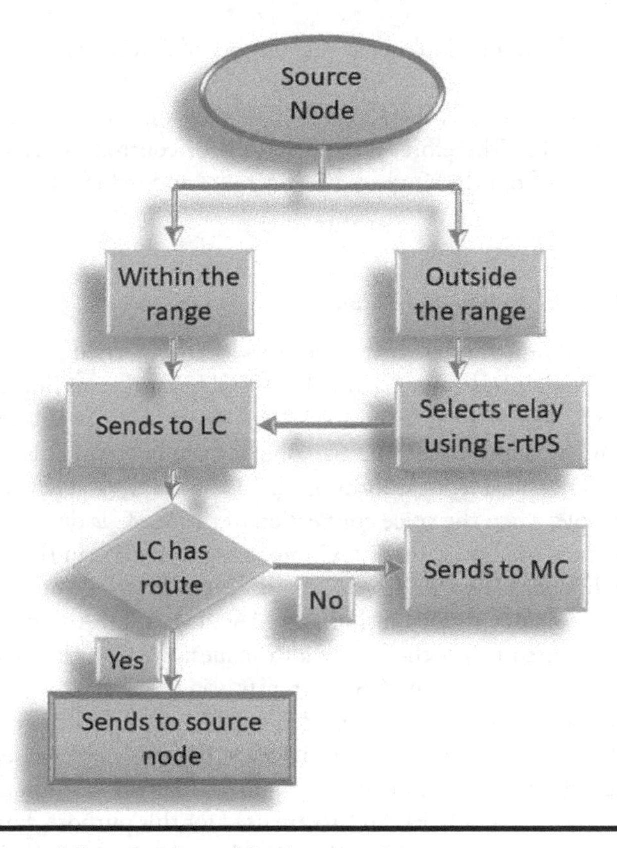

Figure 3.2 A workflow for the vehicular Cloud.

Step 1: The MC first transmits a control message such as a Hello message to all RSUs in range to determine the layout of the network as depicted in Figure 3.1.

Step 2: The RSU computes its associated delay and compares it with the neighboring RSUs in the same depth based on the IEEE 802.16 standard means. The RSU with the lowest delay declares itself as the LC.

Step 3: The selected LC broadcasts a message to the MC and all RSUs in range, with the updated information about all the nodes; these are vehicles and RSUs. All nodes contacted record the routes to an MC in their flow table.

Step 4: The previous three steps are repeated, and finally, the MC institutes a global layout and sends it to all LCs, RSUs, and vehicle networks.

We remark that RSUs can calculate the number of Hello messages sent previously as the vehicle moves and updates each other and the nearby RSUs. This helps in improving the stability of the vehicular Cloud system by averting sparse network conditions, thus selecting an LC with the highest connectivity. We similarly calculate the hop count by looking at the number of RSUs a message has passed by (see Figure 3.2).

At the end of this phase, the selected LCs create a localized global view at each depth from the MC. Therefore, different kinds of controllers are used to reduce the system burden from the single main controller and reduce the overall overhead and delay.

3.3.3 Channel Coding Phase

We assume an IP real-time subsystem-network (IRS-N)-based framework utilizing the Channel Quality Indicator Channel (CQICH) for demanding the bandwidth from the free wireless channel resources. We make use of the two reserved bits in the universal MAC header of IEEE 802.16. One of these bits, which is called the grant size (GS) bit, is responsible for informing the BS about the state of the voice transition. For example, when the voice connection of the MCC is on, the MCC ascribes the GS bit to *zero*. Moreover, the MCC imparts any changes in the voice transition using the traditional MAC header. So, the BS keeps informed of the changes happen to the voice state without causing any MAC overhead. Alternately, the uplink resources can be used to get the bandwidth request process while using the MCC and BS resources to monitor the GS bit transmission. The BS assignment considers the uplink resource to realize an enhanced RT-BMAP. Accordingly, many performance problems, such as delay, medium access, bit/frame error rate, and resource management, arise while supporting the multimedia services in a smart city. We assume the reserved bits of IEEE 802.16 are used for this purpose. Furthermore, the MCC is used to privilege the SC bit. The IRS-N employs the required bandwidth in order to hold the uplink usage and the frame of voice codec acquired by the network.

3.4 Results and Discussions

In this section, we present the results of our planned framework. It involves hardware and software implementation parts. The main purpose of the hardware part is to construct a simple test bed for such projects/ideas. This test bed has been proved to be very helpful with the cost approximation, and it provides critical information about design challenges such as the delay and system throughput. This simple test bed is composed of three subsystems: the in-vehicle subsystem connected via cellular networks, the IP-based network (Internet), and the MCC represented by the network client.

We assume the IEEE 802.16 model is built in the MAC layer to use UGS and E-rtPS as an online resource for voice services. Also, we deploy real-time client (RC) and real-time server (RS) as agents to support the voice call creation and dissolution. These agents use the Real-time Transport Protocol (RTP)/User Datagram Protocol (UDP) to control messages transmissions.

We remark here that we assumed ten cars while obtaining the results in this section for realistic verification purposes. The MCC is implemented via Ubuntu PC (laptop) for experimental purposes. The car toys were equipped with Arduino boards attached to the General Packet Radio Service (GPRS) modules. To evaluate the performance of the assumed resource allocation framework in the central cell, we consider using 5 MHz LTE-A with 19 cells working at 2.0 GHz as a wireless cellular network. For each user, the packet arrival process follows the Poisson distribution with a mean arrival rate of 1 kbps. In Table 3.2, our stimulation parameters are presented.

Figure 3.3 depicts the use of an IRS core function in Ubuntu, consisting of Proxy Call Session Control Function (P-CSCF), Interrogating Call Session Control Function (I-CSCF), Servicing Call Session Control Function (S-CSCF), and Home Subscriber Station (HSS) in order to monitor the quality of the online service. The system depicted in Figure 3.3 does not only possess the ability to allocate the network service resources but can also mask out roaming restrictions [25,26].

We use IRS core functions to transform voice connection to real-time service. The data can be transformed in serial or parallel mode to investigate the efficiency of E-rtPS with RT-BMAP compared to UGS and rtPS. The authors in Ref. [27] say

Table 3.2 Simulation Parameters of IMS Networks

Parameters	*Values*
Number of cells	19
Bandwidth	5 MHz
Shadowing standard deviation	8 dB
Number of RBs	25
Path loss	128.1 + 3.76log(R), R in km
R	500 m
Frequency	2 GHz
Uplink transmission power	24 bBm
Modulation scheme	QPSK, 16, 64 QAM
Proximity distance	10 m
Maximum transmission P_m	24 dBm
Threshold η	0.8
Channel model	200Tap

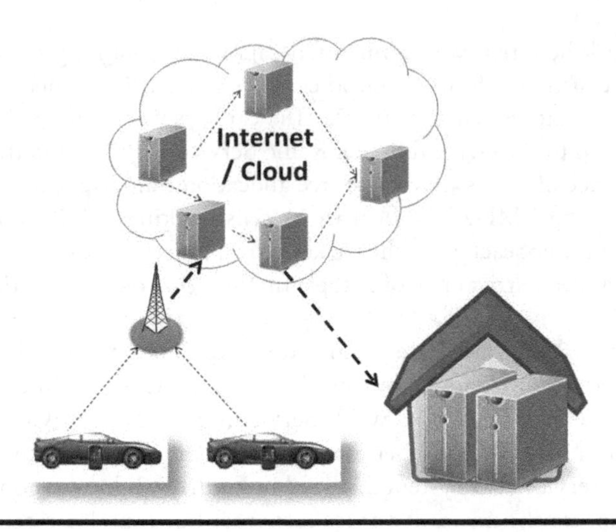

Figure 3.3 Experimental test bed for the vehicular Cloud system of IEEE 802.16d/e.

that through the knowledge of 3GPP, RTS provides a better service scalability to the devices connected; therefore, an IRS core system with features of LTE-Advanced has been deployed to provide better scalability and services. In this study, the core network is assimilated with HSS to analyze the stability of voice services across available networks. Moreover, the IRS core is embedded with its CSCF to generate a realistic test bed. Since the system is designed for real-time services, we gradually increase the voice connectivity in order to examine the system throughput and connectivity delay. Moreover, an emergency call is registered as an unidentified caller to analyze the session setup delay and throughput of real-time services. We perform an experiment to weigh the performance of the Session Initiation Protocol (SIP)-based VoIP compared to IRS core using wireless connectivity and also to inspect the voice connectivity delay and throughput of the network. We employ the use of IRS clients and packet analyzers in order to establish a connection between VoIP call and service connection.

As depicted in Figure 3.4, the throughput of the proposed algorithm is much higher compared to UGS or rtPS. Moreover, Figure 3.5 shows that the delay of the proposed algorithm is lower at all critical points in comparison with the other services. It is imperative to note that getting the predefined delay of a service is important in the system of IEEE 802.16d/e [5] for the packet with delay violations.

Figure 3.6 depicts the bit error rate performance versus the energy per bit of the wireless channel codes using our proposed framework. From this figure, we can conclude that the achieved performance over a multiway channel is almost the same as that obtained in Ref. [16]. The gain is also attained as a result of canceling the dispersion and fading effects. This can dramatically enhance the mobile communication channel performance in terms of reliability.

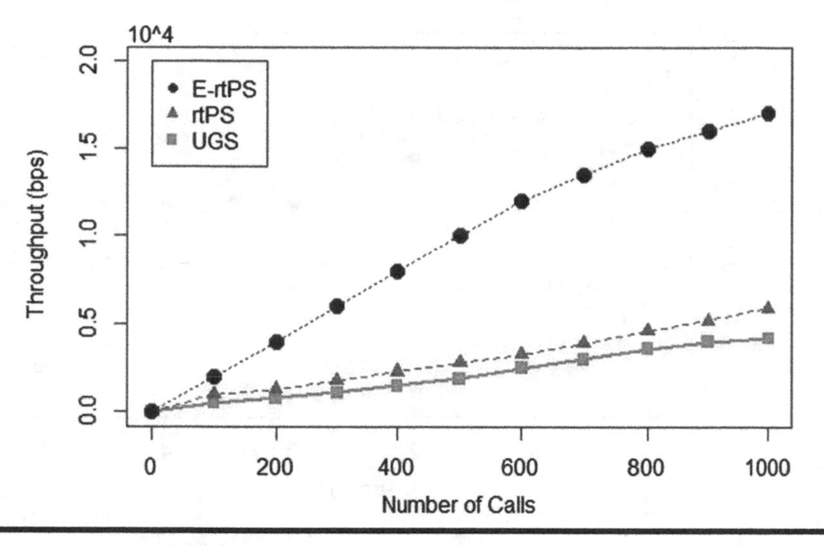

Figure 3.4 Throughput versus number of calls.

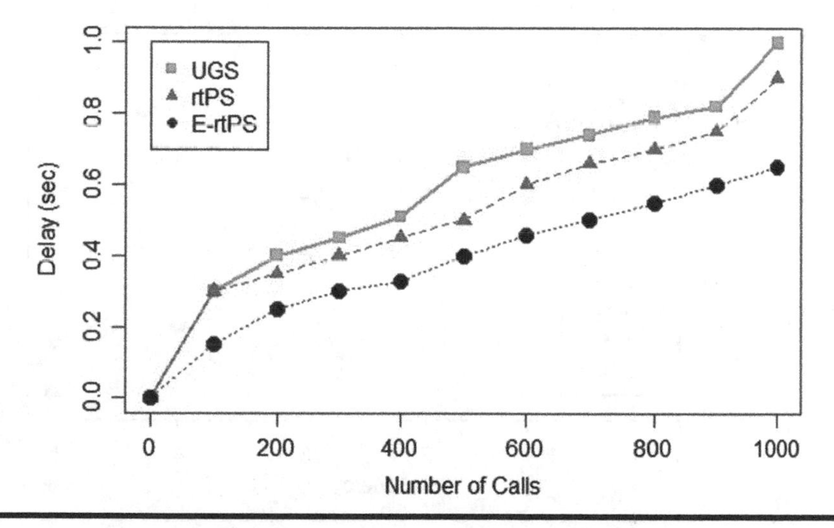

Figure 3.5 Average packet delay versus number of voice connections.

Meanwhile, Figure 3.7 depicts the frame error rate performance versus the energy per bit of the wireless channel codes at varying block sizes. Obviously, as the block length increases, the performance gets improved. It is worth mentioning that the obtained reliability in terms of bit and/or frame error rate of the proposed framework makes the designed codes more appropriate over multipath fading channels. However, this might be associated with a slight channel loss at the maximum exchange rate. Therefore, quantifying the amount of data exchange rate incurred by the proposed framework is recommended.

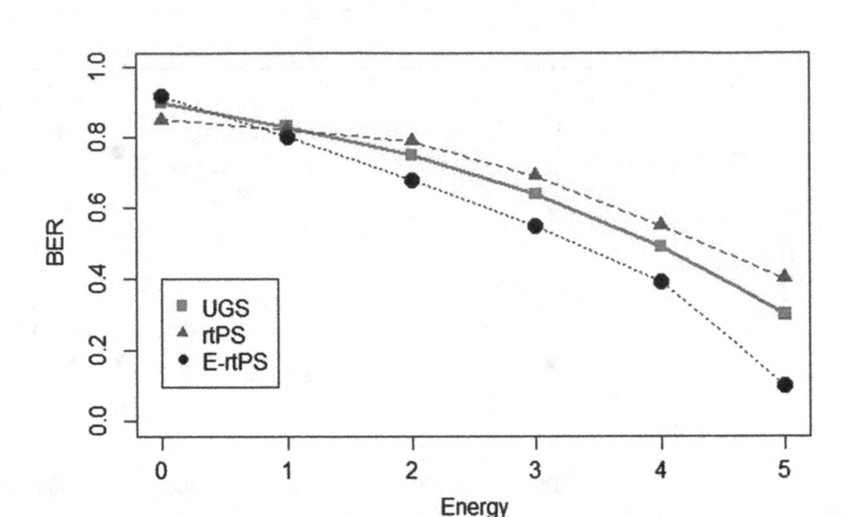

Figure 3.6 Bit error rate versus energy per bit.

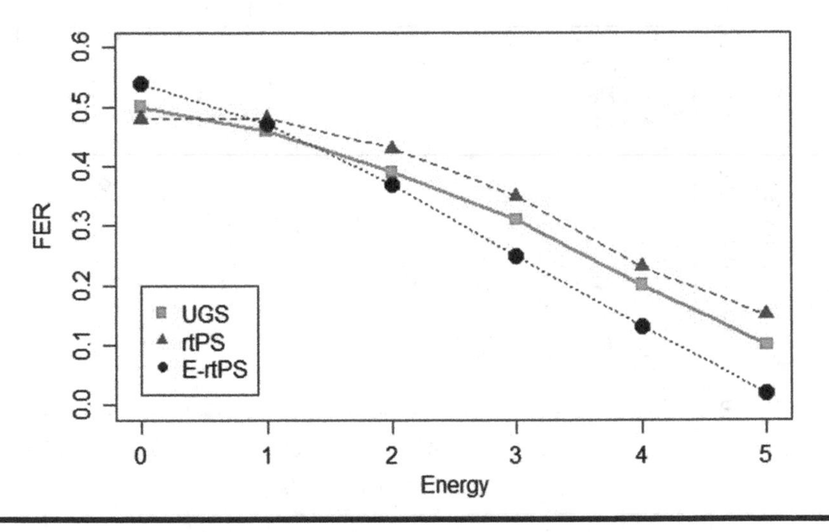

Figure 3.7 Frame error rate versus energy/power per bit.

3.5 Concluding Remarks

This chapter presents an adaptive framework for dynamic speed management in smart cities. The system makes use of the latest development in wireless communication and exploits the existing telecommunication infrastructures, which have been used in data streaming, sound, and video to maximize the system adaptability and reduce the price. A key component to the planned framework is the dynamic medium access approach for real-time communications. Accordingly, this chapter proposes the E-rtPS for the vehicular Cloud of IEEE 802.16. The proposed approach integrates RT-BMAP to analyze the throughput rate and average packet delay. It enables using the same coding design used for multipath selective channels while maintaining the same reliability performance in terms of delay and throughput.

In future work, data that is collected from the RSUs will be published in the Cloud after a preprocessing stage. The Cloud will be able to employ machine learning and prediction techniques towards more efficient decisions. These decisions will be communicated back to relevant drivers on the road.

References

1. R. Iqbal, T. A. Butt, M. O. Shafique, M. W. AbuTalib, and T. Umer, Context-aware data-driven intelligent framework for fog infrastructures in internet of vehicles, *IEEE Access*, vol. 6, 58182–58194, 2018.
2. S. Ali, A. Ahmad, R. Iqbal, S. Saleem, and T. Umer, Joint RRH-association, sub-channel assignment and power allocation in multi-tier 5G C-Rans, *IEEE Access*, vol. 6, 34393–34402, 2018.
3. X. Baró, S. Escalera, J. Vitria, O. Pujol, and P. Radeva, Traffic sign recognition using evolutionary adaboost detection and forest-ECOC classification, *IEEE Transactions on Intelligent Transportation Systems*, vol. 10, 113–126, 2009.
4. J. van de Beek, M. Sandell, and P. Börjesson, ML estimation of timing and frequency offset in OFDM systems, *IEEE Transactions on Signal Processing*, vol. 45, 1800–1805, 1997.
5. F. Al-Turjman and A. Abdulsalam, Smart-grid and solar energy harvesting in the IoT era: An overview, *Wiley's Concurrency and Computation: Practice and Experience*, 2018. DOI:10.1002/cpe.4896.
6. S. Alabady, F. Al-Turjman, and S. Din, A novel security model for cooperative virtual networks in the IoT era, *Springer International Journal of Parallel Programming*, 2018. DOI:10.1007/s10766-018-0580-z.
7. A. Klemm, C. Lindermann, and M. Lohmann, Modelling IP traffic using the batch Markovian arrival process, *Performance Evaluation*, vol. 54, no.2, 149–173, 2003.
8. F. Al-Turjman, H. Hassanein, W. Alsalih, and M. Ibnkahla, Optimized relay placement for wireless sensor networks federation in environmental applications, *Wiley: Wireless Communication & Mobile Computing Journal*, vol. 11, no. 12, 1677–1688, 2011.
9. D. Hui, S. Sandberg, Y. Blankenship, M. Andersson, and L. Grosjean, Channel coding in 5G new radio: A tutorial overview and performance comparison with 4G LTE, *IEEE Vehicular Technology Magazine*, vol. 13, no. 4, 60–69, 2018.
10. F. Al-Turjman, Fog-based caching in software-defined information-centric networks, *Elsevier Computers & Electrical Engineering Journal*, vol. 69, no. 1, 54–67, 2018.
11. F. Al-Turjman and S. Alturjman, Context-sensitive access in industrial Internet of things (IIoT) healthcare applications, *IEEE Transactions on Industrial Informatics*, vol. 14, no. 6, 2736–2744, 2018.
12. D. Deebak, E. Ever, and F. Al-Turjman, Analyzing enhanced real time uplink scheduling algorithm in 3GPP LTE-advanced networks using multimedia systems, *Transactions on Emerging Telecommunications*, 2018. DOI:10.1002/ett.3443.

13. A. Bravo-Santos, Polar codes for the Rayleigh fading channel, *IEEE Communications Letters*, vol. 17, no. 12, 2352–2355, 2013.
14. J. J. Boutros and E. Biglieri, Polarization of quasi-static fading channels, in *Proceedings of IEEE International Symposium on Information Theory*, Istanbul, Turkey, July 2013, pp. 769–773.
15. F. Campioni, S. Choudhury, and F. Al-Turjman, Scheduling RFID networks in the IoT and smart health era, *Journal of Ambient Intelligence and Humanized Computing*, 2019. DOI:10.1007/s12652-019-01221-5.
16. S. Choudhury and F. Al-Turjman, Dominating set algorithms for wireless sensor networks survivability, *IEEE Access Journal*, vol. 6, no. 1, 17527–17532, 2018.
17. P. Trifonov, Design of polar codes for Rayleigh fading channel, in *2015 IEEE International Symposium on Wireless Communication Systems*, Brussels, Belgium, August 2015, pp. 331–335.
18. F. Al-Turjman and S. Alturjman, Confidential smart-sensing framework in the IoT era, *The Springer Journal of Supercomputing*, vol. 74, no. 10, 5187–5198, 2018.
19. F. Al-Turjman, M. Z. Hasan, and H. Al-Rizzo, Task scheduling in cloud-based survivability applications using swarm optimization in IoT, *Transactions on Emerging Telecommunications*, 2018. DOI:10.1002/ett.3539.
20. I. Mehmood, A. Ullah, K. Muhammad, D. Deng, W. Meng, F. Al-Turjman et al., Efficient image recognition and retrieval on IoT-assisted energy-constrained platforms from big data repositories, *IEEE Internet of Things*, 2019. DOI:10.1109/JIOT.2019.2896151.
21. E. Arikan, Channel polarization: A method for constructing capacity-achieving codes for symmetric binary-input memoryless channels, *IEEE Transactions on Information Theory*, vol. 55, no. 7, 3051–3073, 2009.
22. R. Xu, M. Chen, C. Tian, X. Lu and C. Diao, Statistical distributions of OFDM signals on multi-path fading channel, in 2011 International Conference on Wireless Communications and Signal Processing (WCSP), Nanjing, China, 2011, pp. 1–6.
23. H. C. Hsieh and J. L. Chen, Distributed multi-agent scheme support for service continuity in IMS-4G-cloud networks, *Computers and Electrical Engineering*, vol. 42, 49–59, 2015.
24. M. Patel, SOS Uniform Resource Identifier (URI) parameter for marking of Session Initiation Protocol (SIP) requests related to emergency services, draft-patel-ecrit-sos-parameter-07.txt (October 26, 2009).
25. 3GPP TS 23.167, IP Multimedia Subsystem (IMS) emergency sessions, Release 11, 2013.
26. F. Al-Turjman and S. Alturjman, 5G/IoT-enabled UAVs for multimedia delivery in industry-oriented applications, *Springer's Multimedia Tools and Applications Journal*, 2018. DOI:10.1007/s11042-018-6288-7.
27. S. Alabady and F. Al-Turjman, A novel approach for error detection and correction for efficient energy in wireless networks, *Springer Multimedia Tools and Applications*, 2018. DOI:10.1007/s11042-018-6282-0.

Chapter 4

Data Caching in Cloud-Based IoT[1]

Fadi Al-Turjman
Antalya Bilim University

Contents

[1] **F. Al-Turjman**, "Fog-based Caching in Software-Defined Information-Centric Networks", *Elsevier Computers & Electrical Engineering Journal,* vol. 69, no. 1, pp. 54–67, 2018.

4.1 Introduction

A Software-Defined Network (SDN) is a virtual network capable of acquiring knowledge about its users/inhabitants and its surroundings, and uses such knowledge to help its inhabitants achieve their goals and desires in a context-sensitive manner [1][2]. This definitely improves inhabitants' quality of life, and helps in optimizing and controlling the dramatically increasing consumption rates of resources in smart environments. Inhabitants (users) of a large smart environment (such as a city) could be people, systems, devices, services, or agents, occupied with smart enabling technologies such as RFIDs, sensors, nano-technology, etc. With the evolution of the information-centric IoT the global data networks are interconnected and accessed over cloud systems. The increasing demand for highly scalable and efficient distribution of content/information has motivated the development of future Internet architectures based on named data objects (NDOs), for example, web pages, videos, documents, or other pieces of information. The approach of these architectures is commonly called information-centric networking (ICN). In contrast, current networks are host-centric where communication is based on named hosts, for example, web servers, PCs, laptops, mobile handsets, and other devices. Information-Centric Networks serves as a Data-based model which focuses on client's demands disregarding of the data's address or the origin of distribution. ICN is the next generation model for the *Internet* that can cope with the user's requests/inquiries regardless of their data-hosts' locations and/or nature. The current *Internet* model is suffering from the exchange of huge amounts of data while still relying on the very basic network resources and IP-based protocols. Meanwhile, ICNs promise to overcome major communication issues related to the massive amounts of distributed data in the Internet. ICNs adopt a Data-centric architecture which focuses more on the networked data itself rather than the meta-data. This kind of network architectures are known usually by the Content Oriented Networks (CONs) term [3]. Luckily, these CONs architectures match a lot with the emerging communication trend that aims at exchanging Big-data over tiny and energy-limited wireless sensor networks (WSNs) in order to realize numerous attractive projects such as the Smart-planet and the Internet of Things [4][5]. Thus, a new platform is needed to meet these requirements. A new platform, called Fog Computing [6], or, simply, Fog because the fog is a cloud close to the ground, has been proposed to address the aforementioned requirements. Fog is a Mobile Edge Computing (MEC) that puts services and resources of the cloud closer to users to be facilitated in the edge networks.

Unlike Cloud Computing, Fog Computing enables a new breed of light applications and services, that can be run at particular edge networks, such as WSNs. In order to enable WSNs to support this trend in communication and function in a large-scale application platform, such as the Fog Computing, we proposed the cognitive framework from our previous work [7]. In [7], we use smart in-network devices with the capabilities of making decisions based on the information obtained from WSNs to put forward a new information-centric system. The knowledge and reasoning used to dynamically determine the appropriate route where knowledge is defined using value and attribute, and reasoning is represented using the analytic hierarchy process (AHP) technique. The authors in [8] and [9] point out that the upcoming WSN properties such as reliability and delay shall use AHP in their Quality of Information (QoI) assessment. This cognitive Information-Centric Sensor Network (ICSN) framework is able to significantly outperform the *non-cognitive* ICSN paradigms. However, this cognitive ICSN framework did not consider yet the in-network caching feature. Caching in multitude of nodes in ICNs has pivotal role in enhancing the network performance in terms of reliability and response time. In this paper, we propose the use of Value of sensed Information (VoI) cache replacement strategy. It identifies the most suitable data to be replaced in order to maintain prolonged data availability periods while enhancing the network performance. However, authors in [10] and [11] claim that the conventional cache replacement strategy has been intended for IP-based networks and data-centers, which have different data positioning characteristics against the future networks, such as the ICSNs. Moreover, different caching strategies have different effects on the overall performance of the network, and hence, a given caching strategy can influence publishers' load, hit-ratio, and time-to-hit metrics. Numerous attempts in the literature have reviewed each of these metrics independently. However, a single ICSN has the ability to handle multiple users with different designs. Accordingly, a generic dynamic utility function with the ability to consider all the metrics mentioned above while emphasizing on the application itself should be used.

To this end, we provide a novel utility function that sets a value to each cached data item in an ICSN framework. This utility function can determine which data item to drop from the cache while experiencing limited hardware resources for caching. Furthermore, we provide a cache replacement strategy that depends on the VoI in choosing the most appropriate data to be replaced in the cache. We compare our VoI approach against three dominant cache replacement approaches: Node Role-based Caching (RC), Data-based Caching (CC), and Geo-based Caching (GC) with regard to various performance metrics under a variety of parameters including cache size, data popularity, in-network cache ratio, and network connectivity degree.

The rest of this paper is organized as follows. Section 4.2 provides an overview of the existing caching approaches in ICSNs. Section 4.3 talks about our ICSN-specific system model which we use to build the proposed VoI caching policy. We provide a detailed explanation for the VoI approach in Section 4.4. Section 4.5 presents the detailed simulation results obtained from comparing VoI against other caching approaches. Section 4.6 summarizes our concluding remarks.

4.2 Related Work

In Fog paradigm, data has to *be close* to the consumers/users. This is the purpose of caching approaches in this paradigm. Caching is associated usually with naming and data delivery approaches/architectures. For instance, the Data-Oriented Network Architecture (DONA) is coupled with naming tuples and labels. Other architectures differ in the basis of retaining data and which entity in the network can keep a copy of the data. Recently, data caching based on how long it was it was in the network is recommended. However, it is quite difficult to claim efficiency when the overhead messages cannot traverse the network. We look at the different Information Centric Sensor Networks caching approaches in this article. Also, we classify the existing caching in ICSNs as follows: A) Geo-based caching, B) Data-based caching, and C) Role-based caching.

4.2.1 Geo-Based Caching (GC)

In Geo-based caching, data is cached mainly based on the geographical location of the caching node. For example, Chai et al. in [12] recommended caching in less spaces in ICSN against caching everywhere. Their policy claims that data should only be cached in nodes with the highest cache-hit rate. Meanwhile, the Cache Aware Target idenTification (CATT) is a topology aware caching policy proposed by Eum et al. [13] where a downloading path is selected given that it has the highest connectivity degree. Nevertheless, this can make this kind of node behave like a geographical bottleneck in the network. Moreover, the authors in [14] have looked into the performance of topology based replica on internet router-level topology and concluded that the router-level fan-out is almost as good as the greedy placement of replica. The node degree used by the work done in [13–15] cannot be considered as a sufficient solution for replica replacement because most of the nodes contain similar, and relatively low degree or fan-out. The author in [14] proposed the use of self-organizing cache management systems, where nodes make globally similar decisions. This system has proven to have reduced delays against the conventional ways and smaller per-node cache. Li et al [21] proposed a selective neighbor caching system, in which a subset of neighboring proxies are selected such that the minimum mobility cost is experienced. This approach is grounded on caching data requests and their corresponding meta-data in a subset of proxies one hop away from the data publisher. Authors in [22] suggests a probabilistic approach for ICNs. They claim that the probability of a file being cached should be increased as it travels from source to destination by considering the following parameters: i) The distance between source and current node, ii) Distance between destination and current node, iii) Time-To-Live for the routed data content, and iv) the Time-Since-Birth. Authors also suggests redundancy in caching on a single path between source and distention. However, this degrades the ICN performance dramatically while experiencing limited caching spaces. Moreover, in [22], authors assume that all the network nodes has the capability of caching, which is not the case in practice with Fog systems. The proposed approach is weak as well due to considering static data request's frequency from a subnet where that data can exist. Nevertheless, we believe caching should be based on dynamic frequencies and location-independent.

4.2.2 Data-Based Caching (DC)

Data-based caching is another candidate category for caching ICNs, in which the data replacement decision is taken based on the content of the exchanged data. For example, authors in [16] propose an autonomic cache management architecture that dynamically (re)assigns data items to in-network caches. Distributed managers make (re)placement decisions, based on the observed data request patterns such as their popularity, in order to minimize the overall network traffic. In [17] also authors suggest that every cache manager should decide in a coordinated manner with other cache managers whether or not to cache an item. This approach assumes that every cache manager has a holistic network wide view of all the cache configurations and relevant request patterns. And thus, it adapts depending on the volatility of the user requests. It is evident that the network wide knowledge and cooperation give significant performance benefits and reduce significantly the time to convergence, but at the cost of additional message exchanges and computational overhead [18].

In the meantime, other authors are in favor of minimizing the traffic generated by the Internet Service Provider (ISP) and also minimizing the access of in-network devices by caching frequently requested data in the ISP-specific routers. Effective caching is the main problem that is being addressed here, where routers need to organize their data replacement strategy based on their content. The authors have given two data-based caching algorithms. Nonetheless, because the authors have assumed a single gateway in an ISP network, this system may not be practical. Shoa et al. [24] proposed WAVE, which is another data-based caching policy where the size of the cache is adjusted based on data popularity. A node that is in a higher level (called upstream node) suggests the number of chunks of data to be stored in the node at a lower level. This number increases exponentially as the number of requests increases so as to reduce communication and cache management overheard. Additionally, WAVE dispenses data to the network edge. This is where requests come from, putting into consideration the popularity of the data content and the distance relation. Authors in [19] have proposed a data age-based distributed caching system, whose main goal is to reduce the in-network delays and data publisher loads. This system allows for the simple cooperative mechanism so as to control where the ages of the data are updated. It distributes the prevalent content to the edges of the ICSN, while at the same time getting rid of the undesirable replicas in-between the ICSN nodes. However, this approach encounters issues from sustaining highly dynamic contents, and hence, nodes that reside far from the server take long periods of time to refresh their data.

4.2.3 Role-Based Caching (RC)

In this category, authors consider the role of the in-network caching in order to realize the full capabilities of an ICSN. Additionally, the data that needs to be cached at the control or management level shall be considered. Therefore, authors in [20] have established the effects incurred when handling caching decision at the data level and proposed a new method to deal with caching at the control level. The proposed method can be used to bring a balance between the benefits and cost overhead. Nonetheless, it can only be used in small scales, which means it cannot handle the extensive amount of data found in the internet.

Authors in [23] proposed *LocalGreedy* algorithm for caching in ICNs. They consider a cache cluster consisting of number of leaf nodes which are either directly connected or indirectly via first node as a common parent somewhere along the path to the root node. However, this approach necessitates a global knowledge of the in-network nodes' capability and this contradict with the Fog vision. The authors in [24] discussed the trade-off between caching in a distributed IP-based network and the new emerging systems used in Fog computing, such as the Content–Centric Networks (CCN). They applied their study on real-time traffic generated by functional resources such as the web, file sharing and multimedia streaming. It has been demonstrated that caching videos on the in-network routers increases the number of cache hits. Nonetheless, the other type of data will be better off cached in very large capacity storage area at the core of the network. Hence this type of caching is not efficient in ICSNs.

In recent times however, the *internet* has progressed towards an information centric sensor networking paradigm, where the focus is on delivering named blocks of data to users at the network edge rather than establishing end-to-end connections to the web server. So the design of the cache replacement policy in ICSNs must be a dynamic one, based on the user's request trends, and the application on hand. In this paper, although we need to use a content centric approach, the same cache replacement approaches cannot be applied to an ICSN. This is because of the unique resource constraints of the sensor network, the uncertainty of the wireless medium, and the need to be aware of user-requirements in the ICSN architecture. The resource limitations of the sensor network nodes include limited power supply, storage space, and heterogeneity in terms of the sensors used and the node functions. And this can dramatically affect the route discovery process while locating the cached data. It can significantly degrades the available limited energy in case an unreliable path is considered for data caching [15][23]. In addition, the same content (sensed data) cannot be replicated into multiple caches without associating them with location, because the sensed information may be different in different parts of the network, and it may change over time too, which is unlike the case of ICSNs. This makes the cache replacement trickier in information centric sensor networks. In addition, the replacement policy should take into account the type of user requests coming to the network, the sensor node availability at different locations (as nodes eventually die out), and also the sensing duration for different sensors on board the sensor nodes.

Unlike other attempts, this article proposes dynamic caching decisions by considering knowledge and reasoning elements. Moreover, by considering cognitive observations, we prioritize the cached contents in a hierarchal storage manner. At the same time, we propose a utility function that assigns a data item value based on its specific attributes such as popularity and age. This makes the proposed VoI scheme more appropriate for Fog networks.

4.3 System Models

In addition to ICSN's popularity, age and delay models, we also give a detailed explanation on the network model of the SDN-based ICSN in this section.

4.3.1 Considered ICSN Model

In our model, there are three main entities: a content provider (i.e., the data publisher/sensor), a service provider (e.g. Internet Service Provider (ISP)), intermediate caching nodes (called relay nodes (RNs) or local cognitive nodes (LCNs)), and a client (i.e., a user/destination for requested data). The elements of cognition found in LCNs such as knowledge, reasoning and learning, are assisting in data requests forwarding. They interact with publishers/sensors, RNs, and the end-user/sink, where all of the collected data are delivered. Data is sent from the publishers/sensors to the sink via RNs as a result of a user request. The sink node is also upgraded with cognitive elements, and hence now on we call it Global Cognitive Node (GCN). The ISP runs a web publishing service by displaying web-based pages for the data publisher, and the user visits the web pages. In general, both the ISP and the publisher will naturally like to encourage the client to reach the nearby web pages on the intermediate caching nodes which contain the targeted data via measuring the web activity [24]. There are many existing enterprises that try to provide services for measuring the activity of web sites. A partial list of these, in [25], includes companies like I/PRO, Nielsen, NetCount, RelevantKnowledge, and others. These companies use mainly two methods: sampling the activities of a group of web clients, and installing an audit module in web sites. Thus, we assume dynamic auditing or assurance services from the publisher and/or ISP side as our system is supposed to help the user finding the targeted data according to a computed value of information (VoI). The assumed system/software autonomy in the proposed approach is clarified based on software agent theory [24] as follows:

Client-Based Architecture

The architecture of the client-based side is depicted in Figure 4.1a. The figure shows the following main modules:

- **Input Proxy:** As our system could be installed on different devices (e.g., sensors, smart phones, tablets, laptops, etc.) that differ in their capabilities and adopted communication protocols.

- **Data AEC:** It is the module responsible for raw data Aggregation, Encryption, and Compression (AEC).
- **Network Connector:** It is the module responsible for sending the AEC raw data blocks to the SDN. So it should synchronize the blocks with the communication proxy located in the publisher side.
- **Output Proxy:** This module is responsible to send the notification and warning messages to the end-user based on the used device capabilities.

Caching-Node Architecture

The architecture of the caching node is depicted in Figure 4.1b. The figure shows the following main modules:

- **Request Handler:** It is the module responsible of getting the requests from the client to access a specific data and/or service. The request handler prioritizes the concurrent requests if any and to the most appropriate caching node or publisher in case the data was not cached locally.
- **Data collector:** It is the module responsible of monitoring all the user's interactions, and collecting the corresponding raw data, then sends it to the AEC module to be cached.
- **Configuration Manager:** It is the module responsible for data configuration, in which the caching node specifies which interactions and information to be monitored and collected, also it can specify the preferred cache size, encryption and compression options.
- **Notification Handler:** This module is responsible of collecting the notifications coming from the publisher side to be sent to the end user.

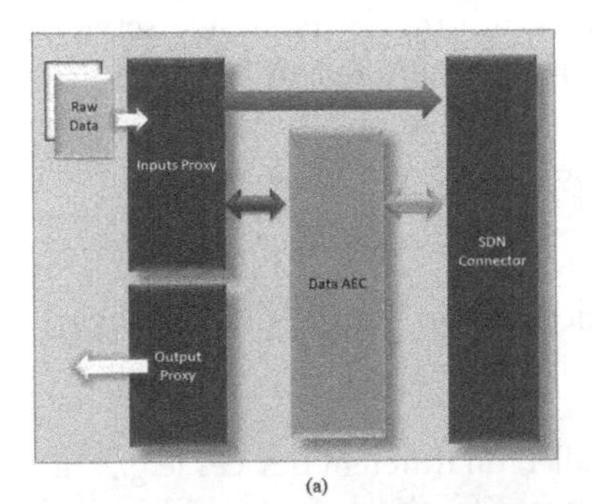

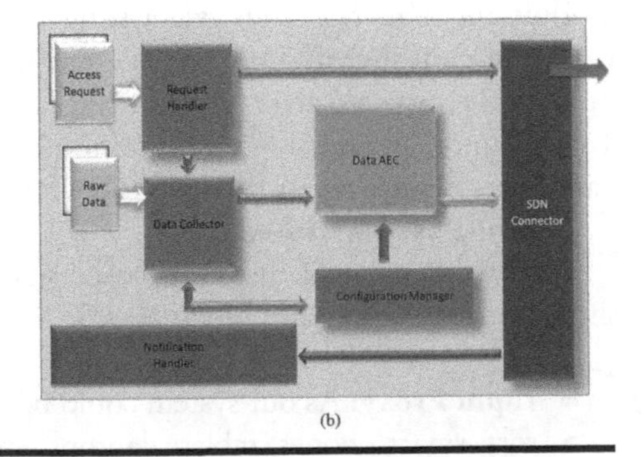

Figure 4.1 (a) Client node architecture, and (b) Caching-node architecture.

Publisher-Based Architecture

The architecture of the publisher-based side of our system is depicted in Figure 4.2. It shows the main modules described as follows:

- **SDN Communication Proxy:** the smart device might not be supporting the HTTP protocol (such as Contiki sensors, and other devices), hence this module must perform the mapping process between the mismatching protocols.
- **Data Processor:** It is the module responsible for collecting the data blocks and requests from the clients, then pass them to the history manager and observer modules. The data processor consists of a data collector and a data inspector. The data collector collects the data from different clients, while the inspector checks if the proper encryption and compression techniques are adopted, also it performs pre-processing steps on the collected data before passing it to the history manager module.
- **History Manager:** It is the module responsible for storing the inhabitants' data blocks and requests, so they can be accessed by the observer module for learning purposes.

Since the users of the information-centric SDNs including people, services, systems, or agents need to interact with each other to create a smart space that improves data availability and access, these interactions need identity verification in such a way that assures security levels. Existing solutions such as static identity approach discussed in [1] and [2] impose a risk of identity theft during the interactions among inhabitants. To maintain a strategic distance from such risks,

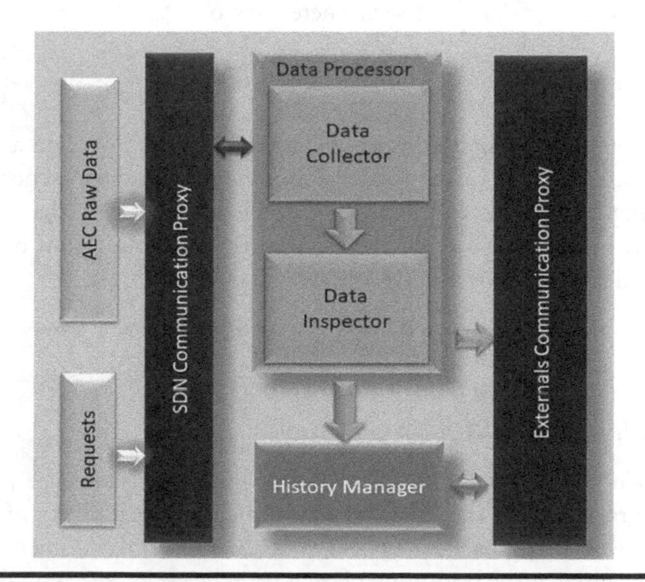

Figure 4.2 Publisher node architecture.

an innovative approach that uses an Objective-Driven programming has been exposed [26]. The secured objective-driven model can be created automatically at runtime by assuming Provisioning-Assurance-Auditing (PAA) Cloud Engine along with the XACML security annotation representation [26]. Where the later provides a secured separated abstraction layer for the cloud users on top of the programming model. In this research, we assume an encrypted and compressed user profile by using his cached activities' history and usage patterns of the environment's resources. And this allows creating an identity proxy to perform the verification required during the interaction.

4.3.2 Delay Model

For different sensors, there is a different amount of time that they should be open to the environment so as to better capture the sensed data. The author in [24] says that, the duration of exposure consequently affects the on-time of the sensor nodes, which in turn affects its life time. When the delay incurred when reading data is more, it is advisable to store the sensed data for long so as to increase the lifetime of the sensor node. This is called sensing delay. Additionally, there will be a propagation delay added if every time data is requested it has to be moved from the sensor node to LCNs. This is most likely to happen, if the sensor node is far from the sink. Hence, the delay components that we consider are sensing delay δ and the propagation delay τ. Furthermore, we limit the number of hops which are needed for data to be delivered at the sink so as to evade the unnecessary usage of energy. Accordingly,

$$\tau \propto n, \text{where} \quad n < 6 \tag{4.1}$$

$$\delta \propto \max(d_1, d_2, d_3, \ldots d_k) \tag{4.2}$$

Where k is equal to the total number of sensing elements found on a sensor node. And d_i is equal to a fixed sensing delay value of the sensor type i. Hence, the sensing delay is a function of the maximum delay. As shown in Eq. 4.3, the total delay (Δ) associated with delivering freshly sensed data to the sink is a sum of sensing and propagation delays as follows:

$$\Delta = \tau + \delta \tag{4.3}$$

4.3.3 Data-Age Model

The age model uses two approaches to decide which data should be plunged from the cache. First approach is one of the ways which uses periodicity of the periodic request. The second approach is applied when the cache is full. Since newly sensed data has to be given at the beginning of each periodic cycle, we can make use of this periodicity. Hence, when the cache is full at the end of each periodic request, old data from the cache can be flushed out. Hence, the time-to-live (TTL) gives the age of a sensed attribute-value pair. For our model, we don't cogitate on the use of historical data, and hence, cached data can be refreshed at the end of each cycle, provided that the data is transmitted to the sink/GCN at the end of each cycle. Therefore, the cache holding period becomes a function of the transmission cycle's periodicity according to the application needs to hold data for a longer period of time.

4.3.4 Popularity Model

A given set of sensed data can be of interest multiple times to a single ore many users, or a number of users can be interested in a given type of sensed data. Such sensed data is considered popular, and hence stored in LCNs' cache for a longer period of time. In order to keep data available when a sensor node start to die out, LCNs should retain data for longer periods of time. When a primary LCN begins to die, it is advisable to store the data in the neighboring LCN, that way they can be stored for longer. This process is managed by a traffic planning algorithm.

4.3.5 Communication Model

In this section we talk more on the assumed channel model in our wireless communication. T_{po} Represents the transmission power used by the $ICSN$ nodes, while T_r is the transmission range between BS and SN. The expression of the channel model is given below:

$$C_M = A\rho T_{P_0} T_r^{-\alpha} \quad (4.4)$$

where C_M is the transmission power of the BS, A is a constant gain factor for power from the antenna and amplifier gain, ρ is a small scale constant for the fading factor, and α is the path loss exponent. Accordingly, this model considers not only the available energy per node, but also the surrounding environment conditions.

4.4 Cache Replacement in SDNs

In this section, we describe our systematic approach followed in replacing the cache contents in an ICSN. First, all data that take a long time to be collected/found should be stored for longer periods of time so as to conserve more energy on the sensor nodes. Secondly, the storage of data must be a function of the periodicity. This periodicity will assist in keeping the old data version until the new one arrives, and working on requests from different types of traffic in a timely manner. Finally, the value of the data collected can be computed according to a utility function that adapts based on the targeted application. Hence, it is important to consider how old the data is when working on the requests for data on demand. Therefore, we can apply our cache replacement approach at the LCNs of the network since the criteria is known and fixed. Accordingly, our preposition VoI-based approach uses the abovementioned system models to realize the efficient way in caching while handling the following types of data:

1. Delay-based data: cached data has a delay sensitivity which is a parameter specified by the user to show how long they are willing to wait for the data. This delay-sensitive data can be found in areas which require emergencies such as disaster or health emergencies.
2. Demand-based content: This is how popular a set of data can be, which is obtained by how often the data is requested.
3. Age-based content: Some data are sensitive to time, for instance, if a user requires information about the city traffic for the next one hour, any information outside this time limit has no use.

Consequently, VoI cache management scheme used three factors to set VoI_{Si} for each sensor node S_i reading. This value depends on the content in each operational round. The function below is used at the beginning of every round as mentioned before to reset VoI_{Si}.

$$VoI_{Si} = \alpha * \Delta + \beta * TTL_{Si} + \gamma * Popularity_{Si} \tag{4.5}$$

where $\alpha, \beta,$ *and* γ are the factors that are itemized based on the user request and the type of traffic. The ability of VoI to get priorities based on ICSN gives it lead. Hence, for better priority caching, we must adjust some parameters to obtain an effective approach. We use the delay sensitivity parameters to minimize delay. The popularity factor is critical because it tells us which set of data has had the most frequent requests. The packet age factor is also important in that it lets us know which data has not been used for a long and replaces it with relevant data. The following algorithm provides steps to be followed by each node, if its cache is full so as to expel data with the minimum VoI_{Si}.

Moreover, VoI can be based on the combined value of the abovementioned QoI attributes (e.g., delay and popularity), and energy consumed during the process of delivering information to the GCN. VoI delivered to the end user is said to be maximized when data is delivered over links that provide the best effective QoI for each traffic type, while minimizing the energy consumed in the network while doing so.

$$VoI = \sum\nolimits_{n-hops} (Effective\ QoI) - \sum\nolimits_{n-hops} (Energy\ Cost) \tag{4.6}$$

Eq. 4.6 highlights that the lower the energy cost in delivering data to the sink, the higher the VoI associated with that data/information object. The QoI must be maximized and energy cost minimized to achieve the best VoI value. If energy consumption is measured as a function of the number of transactions taking place before data is delivered to the GCN, a simple metric - the hop count can be used to approximate the energy cost. If the information is transmitted from source to GCN over minimum number of hops, each link providing the best combined QoI for that traffic type, we can say that the information was delivered to the GCN with good VoI.

Algorithm 4.1: Drop least VoI_{Si}.

1. **Function VoI** (*content*)
2. **Input**
3. *content: A content item within the SDN.*
4. **Begin**
5. **for** each LCN node, **do**
6.   **for** each duty cycle, **do**
7.   **Set** *value* of each VoI_{Si} in the cache based on Eq. 4.5
8.   **if** *cache_full*
9.   Check history of user requests
10.   Drop the data content of the least VoI_{Si}
11.   **End if**
12.   **End for**
13. **End for**
14. **End**

In the above Algorithm, elements of cognition are implemented at the LCNs. These elements are: *Learning and Reasoning* elements.

Learning
Learning is used in our VoI approach in order to determine the most appropriate paths towards the GCN that satisfy the Fog network requirements. This cognition element uses a direction-based heuristic to determine the data delivery path through RNs that lie in the direction of the GCN. Hence, each time a LCN has to choose the next hop, the direction-based heuristic eliminates RNs that increase the distance between the current RN and GCN. This information is stored in the LCN for use in the next transmission rounds. Thus the direction-based heuristic, along with feedback from the network about the chosen paths helps the LCNs to learn data delivery paths to the sink, as the network topology changes.

Example 1

Assume S_1 and S_2 have data to be sent to destination nodes D_1 and D_2. R_n are all the available relays towards the destination. Out of these relays, it is determined that R_5 as shown in Figure 4.3a has the lowest link outage probability to D_1 and D_2. Therefore, S_1 initiates routing data to R_5. Meanwhile, S_2 also forward a high traffic of data to R_5 (depicted by solid paths in Figure 4.3). When multiple source

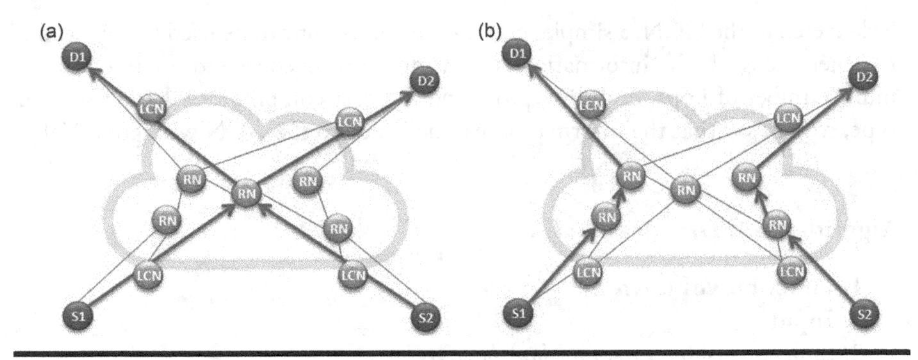

Figure 4.3 (a) Classical routing and (b) Cognitive routing in Fog: (a) Classical case in a sensor network, (b) Cognitive case in response to congestion.

nodes start routing their data to R_5 as well, the route to R_5 may get congested. A cognitive network with *learning* capabilities will be able to identify the congestion at R_5 (by observing the decrease in throughput). Sharing this observation with neighboring nodes, the cognitive Fog network (or ICSN) would be able to respond to the congestion proactively, by routing the data through a different path involving nodes R_4, R_8 and R_9 as shown in Figure 4.3b.

Reasoning

In the VoI approach, we assume a modified version of the Analytic Hierarchy Process (AHP) [7] for implementing the reasoning element of cognition in the Fog network. AHP supports multiple-criteria decision making while choosing the data path. For example, if we have a delay-sensitive data, the node which provides the lowest latency, will be chosen even though it might degrade other metrics such as the network energy or throughput. If two next-hops guarantee the same latency then the next attribute to compare will be energy, and then, throughput, assuming that energy is the next desired attribute in the Fog network. AHP provides a method for pair-wise comparison of each of the attributes and helps to choose the node that can provide the best network performance on the long run. The following example has more details on the utilized AHP.

The steps used in the AHP to establish priorities for the QoI attributes and identify the best next-hop path in delivering the application data to the GCN are illustrated in Algorithm 4.2.

Algorithm 4.2: AHP analysis to determine the data delivery path.

1. **Function AHP (QoI.priorities)**
2. **Input**
3. QoI.priorities: End-user defined priorities on QoI attributes for requested data
4. **Output**

5. RN_x: A selected best next-hop $RN_x \in \{ RN_1 \ldots.. RN_n\}$ to deliver data
6. **Begin**
7. **Initialize:** *Maxhopstosink*=k; *Expected QoI priorities on {L, R, T}; currenthop=1; flag=0*
8. **While** (*currenthop* < *Maxhopstosink)*
9. *AHP_analysis*(Next-hop LCNs v/s QoI attributes)
10. Next hop LCN = LCN_x //This is the LCN with QoI priority ratio consistent with expected values
11. Transmit and aggregate data at next-hop LCN
12. **If** (next hop = GCN)
13. **If** ((actual QoI < (.5) *expected QoI) for all attributes)
14. **If** (flag+1 <5)
15. Display: "Poor network performance";
16. **Else**
17. Display: "Network End of Life";
18. Exit;
19. **End**
20. **Else,**
21. Display: "QoI requirement met"
22. Exit;
23. **End**
24. **Else**
25. *currenthop= currenthop+1*
26. goto step 8
27. **End**
28. **If** *(currenthop* > *Maxhopstosink),*
29. GCN Retransmits request
30. Flag condition to learning algorithm in LCNs
31. **End**

Information about the relative priorities of the QoI attributes as desired by the user are received as input from GCN in steps 1–3. The output is a next hop RN that provides the best QoI as shown by steps 4–5. A maximum value is set for the hop count, within which data is expected to reach the GCN from its source. In step 9, AHP analysis identifies the best next-hop RN that satisfies these requirements. Actual values of latency, reliability and popularity are used during AHP analysis. The priority values are obtained by calculating the Eigen vector of the matrix and normalizing it. Thus, priorities for each QoI attribute that the RNs offer are obtained. Steps 13–19 help to identify conditions in which the network might be performing poorly, or is not able to deliver data with the expected QoI, in which case it considered as network's end-of-life as it is not providing useful information to the user. Steps 8–26 are iteratively run through till the GCN is reached. Steps 28–31 indicate specific restrictions on the data delivery path.

4.5 Performance Evaluation

In this section, we provide initial performance evaluation results for the VoI based cache replacement technique, which we have compared with the Role-based Caching (RC), Data-based Caching (DC), and Geo-based Caching (GC) techniques using NS3, a discrete event simulator. The caching schemes: RC, DC, GC, and VoI, are executed on 500 randomly generated wireless heterogeneous network topologies in order to get statistically stable results. The average results hold confidence intervals of no more than 5% of the average values at a 95% confidence level. We make use of the Cache Hit Ratio to compare the performance of the different cache replacement strategies. Cache hit ratio is defined as the ratio of the number of times requested data was found in the cache divided by the total number of times data was requested from the cache. The storage cache is implemented as a single storage level in one case (L1 cache) and as a hierarchy of two storage levels in another case (L1 and L2 cache). Simulation results are compared for VoI, GC, DC and RC replacement techniques. These simulations were run at a cache sizes ranging from 10 to 100, and the simulations end after serving 1000 packet requests. There are 100 different requests from which the packet requests are randomly generated.

4.5.1 Performance Metrics

To compare the performance of the proposed VoI approach, we track ICSN-specific metrics to achieve qualitative conclusions for the targeted in-network caching problem. We simulate the performance of an ICSN network with the detailed physical layer NS3 built-in parameters so that we achieve realistic simulation instances. The four considered performance metrics are as follows:

- Cache-hit ratio: is simply the fraction of time a request arrives at a node to which that cache is attached but does not contain the requested data item. It is the average hitting ratio over all the in-network caches.
- Time-To-Hit-data (TTH): is found by simply logging all the total costs of the request and response paths incurred by every sensor node.
- In-network latency (delay): this metric represents the end-to-end delay as described above. Note that we differentiate between latency to hit data and in-network latency since the two metrics may differ because of mobility or disruption conditions.
- Average Request per Publisher (ARP): this metric is measured in number of data request per hour (req/hr) and it represents the average load per publisher in an ICSN paradigm.

4.5.2 Simulation Parameters

Many of the ICSN paradigm parameters have to remain fixed while our simulation instances are generated. In particular, the parameters of our simulation are as follow:

- Percentage of nodes with caches (PoC): This parameter is our primary method for controlling the extent of caching in our ICSN. By varying this parameter, we can study the sensitivity of metrics like time-to-hit-data to the caching extent.
- Connectivity level (degree): It represents how much tightly connected is the ICSN network. We use the connectivity matrix, based on our described communication model in Section 4.3.
- Data Popularity: It indicates how frequent a specific data content is requested. This metric is measured in percentage with respect to other requested data contents.

4.5.3 Simulation and Results

The following figures depict the achieved results. Our first objective is to confirm that increasing the extent of caching in ICSNs, in terms of both size and number of levels will reduce time to meet data for all cache policies.

According to Figure 4.4, we can deduce that the VoI is not efficient in level one cache, however, RC and GC cache replacement techniques perform equally well in this scenario. In Figure 4.5, where we have two levels of caching (one on LCN and the other on RN) we find out that VoI surpass the other two replacement techniques. Since we have only one type of the considered requests, there is minimum performance gain when the cache size is increased beyond 30 *Mbyte*.

The next set of simulations' figures (Figures 4.6 and 4.7) are set to analyze the performance of the cache replacement strategies as the number of requests that a given network needs to serve increases from 500 to 5000 *req/hr*. The cache size is set to be 100 *Mbyte* and the number of request types are fixed at 4.

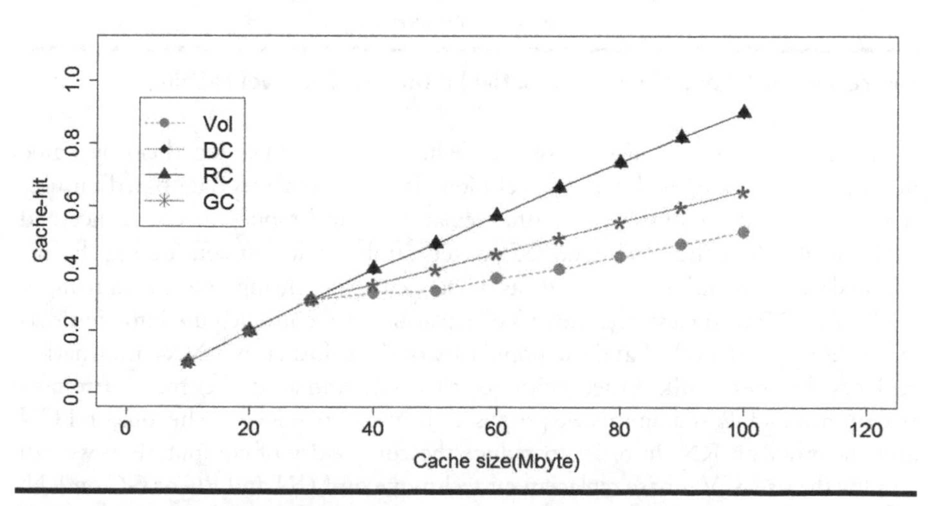

Figure 4.4 Cache size vs. the hit ratio with 1-level caching.

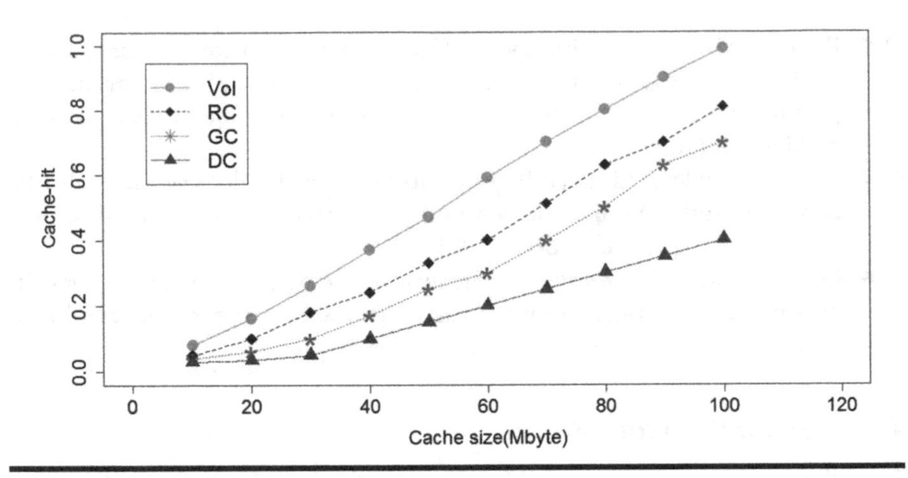

Figure 4.5 Cache size vs. the hit ratio with 2-level caching.

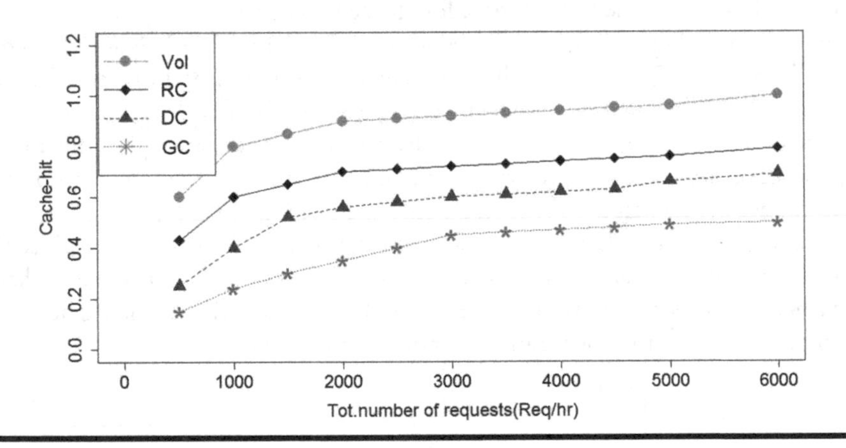

Figure 4.6 Total no. of requests vs. the hit ratio with 1-level caching.

From the Figures (4.4–4.7) we can deduce that the advantage held by Value of Information (VoI) replacement technique is that it replaces data based on user requirement and VoI of the data. Other replacement techniques are only concerned with the match of the data requested packet number, without considering the age of the data, its popularity or the delay associated with sensing and transmitting it to the sink. Nevertheless, the use of VoI replacement technique puts into consideration the age of data, VoI and the popularity of the information. New information replaces old ones, unlike other technique that only find a number match irrespective of their age. Based on this we proposed the use of two level cache, one on LCN and the other on RN, in order to reduce the complexity of computation, we can employ the use of VoI base replacement technique on LCN and RC or GC on RN. We can set the size of level 1 and 2 to 100 and the packet request can be 10000.

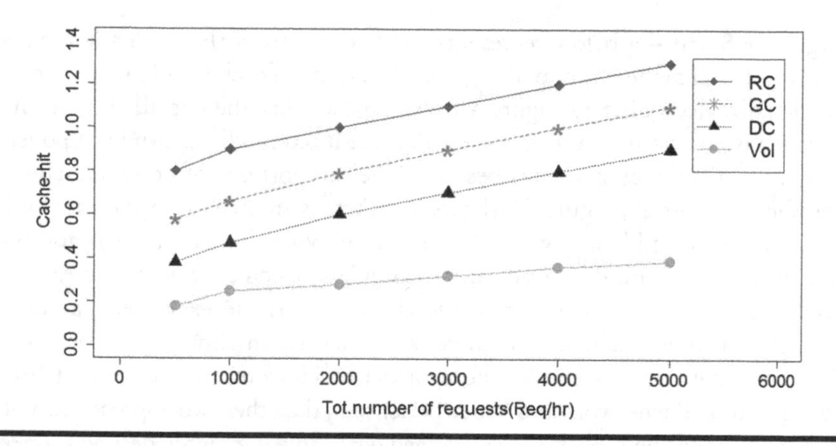

Figure 4.7 Total no. of requests vs. the hit ratio with 2-level caching.

Table 4.1 Two-level caching comparison

L1 Caching	*L1 HitRatio*	*L2 Caching*	*L2 HitRatio*	*Cumulative Hit Ratio*
VoI	0.900	**VoI**	0.009	0.817
VoI	0.811	**GC**	0.813	0.825
VoI	0.647	**RC**	0.779	0.610
GC	0.799	**GC**	0.500	0.751
GC	0.733	**VoI**	0.402	0.683
GC	0.900	**RC**	0.000	0.718
RC	0.897	**RC**	0.000	0.651
RC	0.900	**GC**	0.000	0.718
RC	0.899	**VoI**	0.000	0.618

We can set the size of level 1 and 2 to 100 *Mbyte* and number of packet requests is set to be 10000 (Table 4.1).

From Table 4.1, we can see that for the two levels of cache, the best possible combinations are: VoI based replacement strategy at L1 cache and VoI or GC based replacement strategy at L2 cache. Despite having a good hit ratio sometimes, RC is not reliable to serve the user needs considering the age of data and the delay required to retrieve the required data, because the decision making is only at LCNs. Implementing the VoI replacement technique at LCNs can greatly help save resources especially when a cache hit is found on the first level of the cache.

Figures 4.8 and 4.9 below represents the findings from the simulation experiment of the 2 level cache. From both figures, the extend of cache availability increases proportionally. According to Figure 4.8, we observe that the overall time to meet data, which is our main performance metric, is reduced in all performance policies. However, the Vol policy performs best at higher proportions of nodes attached to caches. On the contrary, Figure 4.9 shows that there is an increase in the data hit for all the approaches, and hence, we conclude that the Vol is better due to its ability to replace the most relevant data according to an ICSN-specific set of attributes.

In Figures 4.10–4.12, connectivity level (degree), is the examined parameter. From Figure 4.10, we deduce that there is an increase in time to hit data as the ICSN connectivity increases in all the approaches. However, we notice that VoI is less dependent on the network and hence better than the other two approaches. VoI is more dependent on the type of the exchanged data. This is a highly desired property

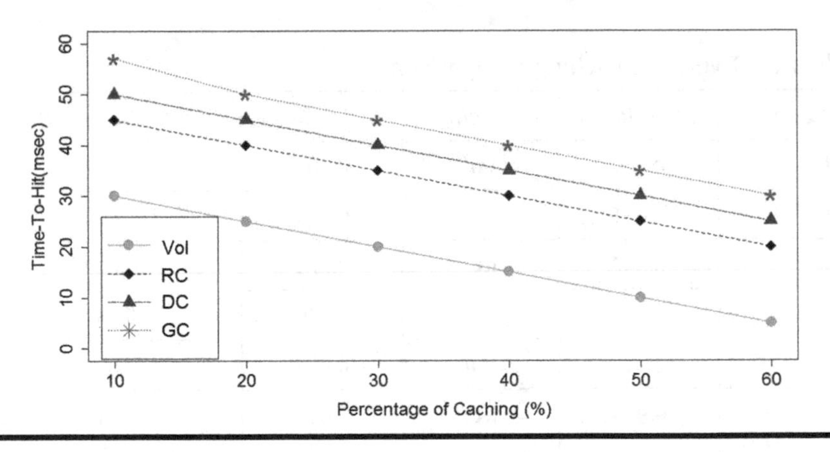

Figure 4.8 Time to hit ratio vs. percentage of nodes with caches (conn. degree = 30).

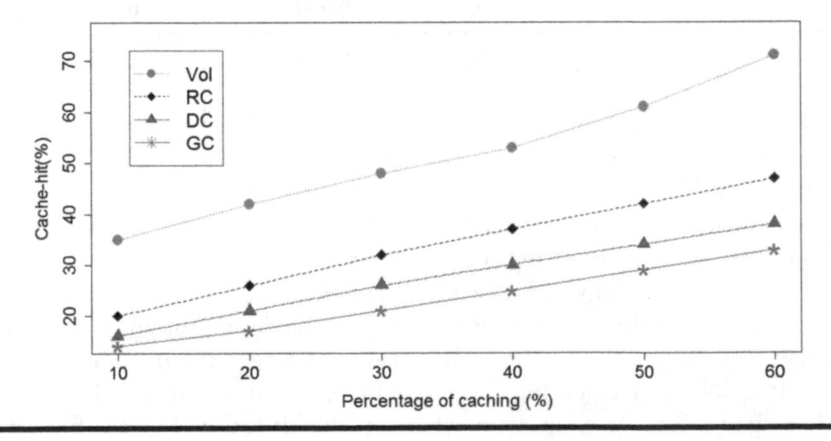

Figure 4.9 Hit ratio vs. percentage of nodes with caches (connectivity degree = 30).

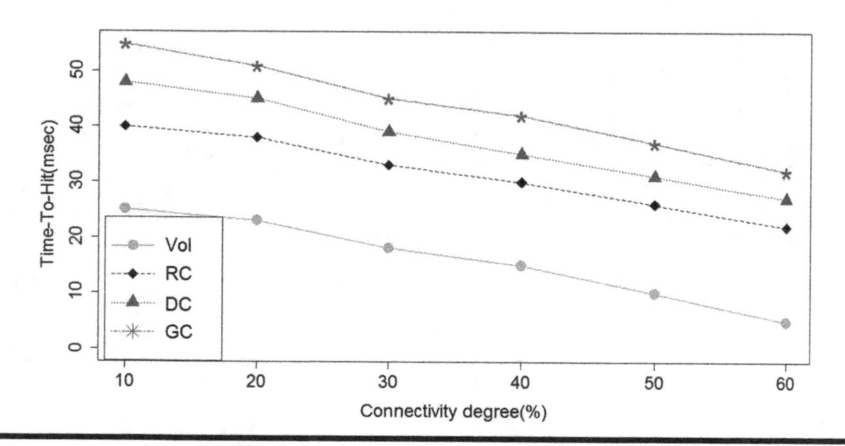

Figure 4.10 Time to hit ratio vs. the connectivity degree percentage.

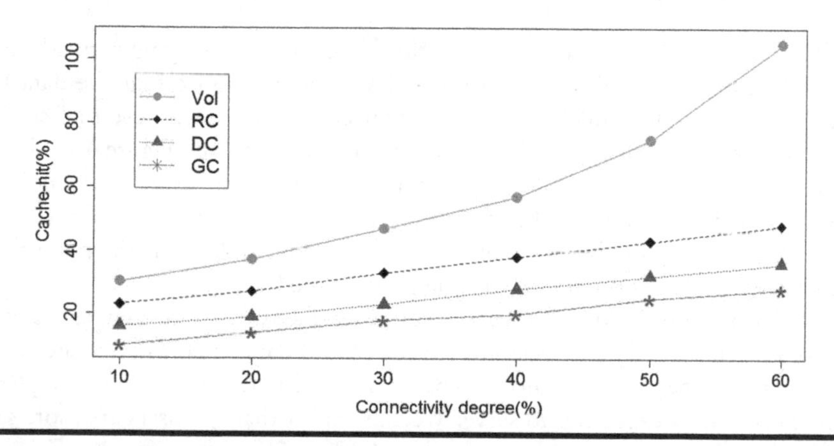

Figure 4.11 Hit ratio vs. the connectivity degree percentage.

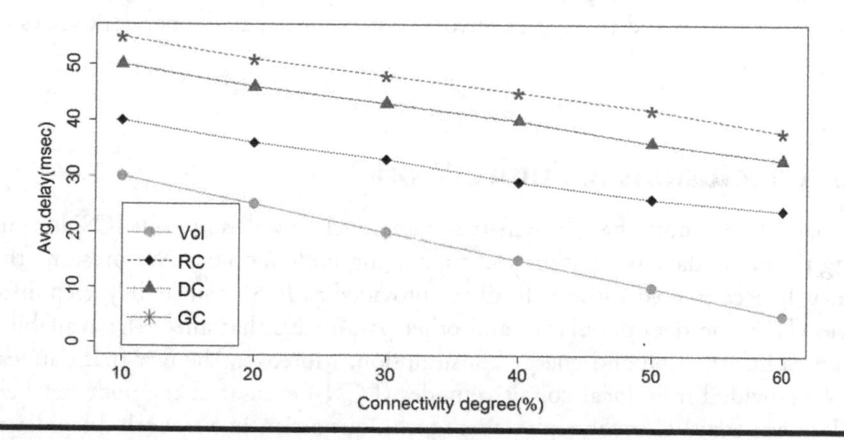

Figure 4.12 Avg. in network delay vs. the connectivity degree percentage.

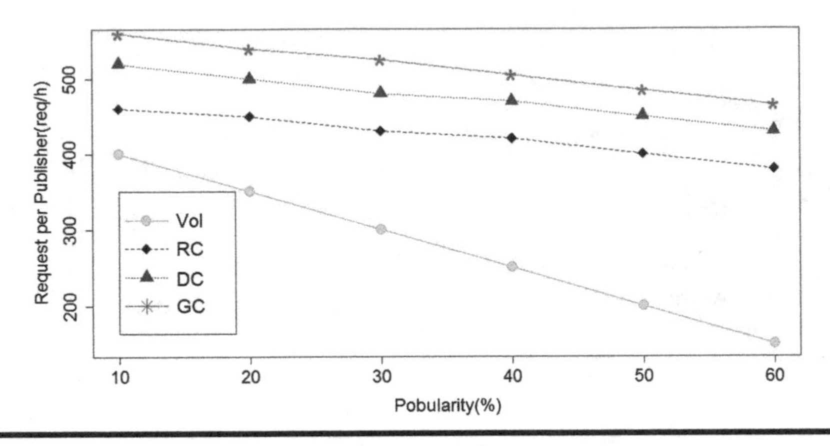

Figure 4.13 Publisher load vs. the data popularity.

in ICSN. Figure 4.11 shows the data hit performance against a varying network connectivity degree, and while applying the VoI scheme, we notice that the data hit increases exponentially while the network connectivity increases. Nevertheless, the data hit of the other two approaches increases linearly. Moreover, Figure 4.12 shows that VoI is the best in terms of delay. This can be attributed to the application of the delay factor while deciding which data to replace. Figure 4.13 shows the effect of data popularity in terms of publisher load. The VoI tops GC and RC as the popularity metric increases. This is a very desirable property in the ICSNs.

The VoI based technique would be suitable for the ICSN approach, as we propose to use a named data association for the sensed data, such as attribute-value pairs, and the cache size can be decided based on the different types of user requests that the network is expecting to serve. We can expect that the users are more satisfied with the response received from the LCN, as it retains information in its cache based on both data popularity and various parameters that affect gathering sensed information and the energy involved in doing so, as the network scales up to larger sizes.

4.6 Conclusions & Future Work

To conclude, we note that the VoI-based approach is well-suited in ICSNs while using the name-data association and supporting node mobility. We presume that users will be contented with the feedback provided by ICSNs since they keep information based on data popularity and other parameters that affect the availability of sensed information and energy consumption. Moreover, the fact that data can still be provided from local cognitive nodes (LCNs) even after the node has been died, proves that VoI cache replacement technique assists in a smooth dilapidation of the network. Additionally, we looked into the influence of changing network loads and the inter-LCN communication on how effective the cache replacement approach is in Fog networks. Subsequently, VoI has a delay-tolerant requirement at the edge of the network, and this provides a dynamic data replacement. By plummeting the load of the data publishers, VoI can maximize their gain as well. The consumer gain is maximized in reference to metrics such as delay, time-to-live, and data popularity.

One of the aspects of the information-centric capabilities, is its ability to cache information in cognitive nodes and use it collaboratively for information sharing across the network. While we acknowledged this advantage in this article, such caching would rise an issue with the mobility-enabled node presence, which we did not delve deeply into yet. Thus, exploring the role of caching in information access and data delivery, and the study of cache replacement techniques that suit the cognitive nodes in mobile ICSNs is still a direction to explore out of this work. Moreover, the idea of cognition can be investigated more while being applied to intermediate relays/switches of the current Internet infrastructure to realize the cognitive network concept in general.

References

[1] K. Lal, A. Kumar, ICN-WiMAX: An Application of Network coding based Centrality-measures caching over IEEE 802.16, Procedia Computer Science, vol. 125, 2018, pp. 241–247.

[2] F. Al-Turjman, "5G-enabled Devices and Smart-Spaces in Social-IoT: An Overview", *Elsevier Future Generation Computer Systems*, 2017. DOI: 10.1016/j.future.2017.11.035.

[3] I. Amo, J. Erkoyuncu, R. Roy, S. Wilding, "Augmented Reality in Maintenance: An information-centred design framework", Procedia Manufacturing, vol. 19, 2018, pp. 148–155.

[4] SmartSantander, Future Internet Research and Experimentation. [Online]. Available: http://www.smartsantander.eu/

[5] F. Al-Turjman, "Information-centric sensor networks for cognitive IoT: an overview", *Annals of Telecommunications*, vol. 72, no. 1, pp. 3–18, 2017.

[6] S. Chen, T. Zhang, W. Shi, "Fog computing", *IEEE Internet Computing*, vol. 21, no. 2, pp. 4–6, 2017.

[7] G.T. Singh and F.M. Al-Turjman, "A Data Delivery Framework for Cognitive Information-Centric Sensor Networks in Smart Outdoor Monitoring", *Elsevier Computer Communications*, vol. 74, no. 1, pp. 38–51, 2016.

[8] M. Z. Hasan, and F. Al-Turjman, "Evaluation of a Duty-Cycled Asynchronous X-MAC Protocol for Vehicular Sensor Networks", *EURASIP Journal on Wireless Communications and Networking*, 2017. DOI: 10.1186/s13638-017-0882-7.

[9] F. Al-Turjman, "Mobile Couriers' Selection for the Smart-grid in Smart cities' Pervasive Sensing", Elsevier Future Generation Computer Systems, vol. 82, no. 1, pp. 327–341, 2017.

[10] M. Chen; Y. Qian; Y. Hao; Y. Li; J. Song, "Data-Driven Computing and Caching in 5G Networks: Architecture and Delay Analysis", *IEEE Wireless Communications*, vol. 25, no. 1, pp. 70–75, 2018.

[11] X. Yue; Y. Liu; J. Wang; H. Song; H. Cao, "Software Defined Radio and Wireless Acoustic Networking for Amateur Drone Surveillance", *IEEE Communications Magazine*, vol. 56, no. 4, pp. 90–97, 2018.
[12] W. K. Chai, D. He, I. Psaras and G. Pavlou "Cache "Less for More" in Information-Centric Networks (Extended Version)", *Elsevier Computer Communications*, vol. 36, no. 7, pp. 758–770, 2013.
[13] S. Eum, K. Nakauchi, Y. Shoji, N. Nishinaga, M. Murata, "CATT: Cache aware target identification for ICN", *IEEE Communications Magazine,* vol. 50, no.12, pp. 60–67, 2012.
[14] M. Meddeb; A. Dhraief; A. Belghith; T. Monteil; K. Drira, "How to Cache in ICN-Based IoT Environments?", In Proc. Of the IEEE/ACS 14th International Conference on Computer Systems and Applications (AICCSA), Hammamet, Tunisia, 2017, pp. 1–6.
[15] F. Al-Turjman, "Energy–aware Data Delivery Framework for Safety-Oriented Mobile IoT", *IEEE Sensors Journal*, 2017. DOI: 10.1109/JSEN.2017.2761396.
[16] B. Nguyen, H. Phan, D. Ha, G. Nguyen, "An Information-centric Approach for Slice Monitoring from Edge Devices to Clouds", *Procedia Computer Science*, vol. 130, 2018, pp. 326–335.
[17] F. Al-Turjman, "Optimized Hexagon-based Deployment for Large-Scale Ubiquitous Sensor Networks", *Springer's Journal of Network and Systems Management*, vol. 26, no. 2, pp. 255–283, 2018.
[18] L. Gkatzikis, V. Sourlas, C. Fischione, I. Koutsopoulos, "Low complexity content replication through clustering in Content-Delivery Networks", *Computer Networks*, vol. 121, no. 5, 2017, pp. 137–151.
[19] Z. Ming; M. Xu; D. Wang "Age-based Cooperative Caching in Information-Centric Networks", *Int. Conf. on Computer Communication and Networks (ICCCN)*, 2014.
[20] W. Yaogong, K. Lee, B. Venkataraman, et al. "Advertising Cached Contents in the Control Plane: Necessity and Feasibility", *In Proc. INFOCOM Workshop on computer communications*, 2014.
[21] H. Li; H. Zhou; W. Quan; B. Feng; H. Zhang; S. Yu, "HCaching: High-Speed Caching for Information-Centric Networking", *In Proc. of the IEEE Global Communications Conference (GLOBECOM)*, Singapore, Singapore, Dec. 2017, pp. 1–6.
[22] A. Gupta, U. Shanker, SPMC-CRP:A Cache Replacement Policy for Location Dependent Data in Mobile Environment, Procedia Computer Science, vol. 125, 2018, pp. 632–639.
[23] T. Baker, B. Al-Dawsari, H. Tawfik, D. Reid, "GreeDi: An energy efficient routing algorithm for big data on cloud", *Ad Hoc Networks*, vol. 35, pp. 83–96, 2015.
[24] C. Zuo, J. Shao, G. Wei, M. Xie, M. Ji, CCA-secure ABE with outsourced decryption for fog computing", *Future Generation Computer Systems*, vol. 78, no. 2, 2018, pp. 730–738.
[25] R. Caropreso; R. Fernandes; D. Osorio; I. Silva, "An Open Source Framework for Smart Meters: Data Communication and Security Traffic Analysis", *IEEE Transactions on Industrial Electronics*, 2018. DOI: 10.1109/TIE.2018.2808927.
[26] M. Gomes, M. Pardal, "Cloud vs Fog: assessment of alternative deployments for a latency-sensitive IoT application", Procedia Computer Science, vol. 130, 2018, pp. 488–495.

Chapter 5

Smart Parking in Cloud-Based IoT

Fadi Al-Turjman
Antalya Bilim University

Arman Malekloo
Middle East Technical University

Contents

5.1 Introduction

The idea of smart parking was introduced to solve the problem of parking space and parking management in megacities. With the increasing number of vehicles on roads and the limited number of parking spaces, the congestion of vehicles is inevitable. This congestion would certainly lead to the polluted environment and driver aggression. Especially in peak hours where the flow density is at its maximum, locating a vacant parking spot is near to impossible. A recent report by INRIX [1] shows that on average, a typical American driver spends 17 h/year looking for a parking space; however, in the case of major cities such as New York, the figure is much higher. According to the report, New York drivers spend 107 h/year searching for parking spots, and taking into account the fuel spent during this period, this results in unnecessary emission of harmful gases, which can be easily avoided. Identifying these problems and trying to resolve them in a manner that is effective and at the same time sustainable is a challenging task. In the context of the smart city ecosystem, inputs from elements such as vehicles, roads, and users have to be networked and analyzed together in order to provide the best service in a fast and secure manner [2]. One of the reasons of marching towards a smart city ecosystem is to use the potential of existing technologies and infrastructures in providing the best utility to users and improving their future. With the help of Internet of Things (IoT) applications, mobility and transportation are considered to be the key influencing factors in sustaining our surrounding environments, especially those which utilize intelligent transportation systems (ITS) [3].

One of the key components in ITS is the smart parking system (SPS) which is based on analyzing and processing the real-time data gathered from vehicle detection sensors and the radio frequency identification (RFID) systems that are placed in parking lots to report the absence and/or presence of a vehicle. These sensors have their strengths and weaknesses in certain areas where they are deployed. Also there might be issues in data anomalies and discrepancies where the collected information does not always conform to the initially expected pattern. This could potentially lead to a less reliable system. Moreover, security and privacy issues of the data transmitted and/or received must be carefully treated. Several factors such as end-to-end communication and data encryptions must be considered in advance before implementing such systems. These subjects have to be seriously considered as the data collected from these sensors might be used in several critical scenarios such as the parking space prediction in an emergency situation and the path optimization in self-driven cars. Any vulnerability in these scenarios, no matter how significant/insignificant they are, can potentially lead to personal information leakage and increase the risk of security attacks. When all the aforementioned aspects are considered thoroughly, the SPS can significantly enhance the experience of both the parking lot operators (by maximizing their revenues) and the users (by easily searching, booking, and paying in advance for their parking lots). Accordingly, any smart application in the current smart city ecosystem has to be context-aware and has to dynamically adapt to surrounding changes.

In Ref. [4], the authors presented the main motivations in carrying smart devices and the correlation between the user surrounding context and the application. They focused on context-awareness in smart systems and space discovery paradigms. New generations of real-time monitoring in SPS have recently been discussed and gained attention as well. Vehicle ad hoc networks (VANETs), where vehicles that encompass On-Board Unit (OBU) communication system, can effectively communicate with other nearby vehicles as well as roadside infrastructures to provide a real-time parking navigation service [4]. The authors considered the development of a hybrid sensor and vehicular network that provides safety support in ITS. Another trend is the unmanned aerial vehicles (UAVs), also known as drones, which provide wireless connectivity in locations where the cellular range is limited or the existing infrastructure fails to operate. Equipping drones with multiple vehicle detection sensors can ultimately solve many problems that the current deployments of SPSs are facing. Furthermore, with the emerging long-range low-power wide area networks (LPWANs) and the fifth-generation (5G) networks, several IoT services can be provisioned. In Ref. [5], the authors describe the IoT paradigm as a dynamic global network infrastructure with self-configuring capabilities based on standard and interoperable communication protocols. With the increasing number of connected devices day by day, the need for faster and more efficient wireless communication trend in parking systems is expected. This is where 5G can fill the gap [6]. However, interoperability issues related to both software and hardware aspects in these solutions can dramatically degrade the system performance. And thus, ubiquitous IoT solutions need to be open and able to integrate with other existing platforms.

5.1.1 The Scope of This Chapter

The aim of this survey is to offer an insight into new parking paradigms in ITS. We look at different aspects of the SPS. First, we define and classify the existing smart parking attempts in order to provide the reader an idea about the concept of the SPS in general and its possible categories/alternatives. For a comprehensive study, we cover both hardware and software aspects in this system. In terms of hardware, we investigate all vehicular sensors that are currently in use via several use cases in terms of their strengths and weaknesses. Moreover, we list and overview the different physical communication technologies that are especially used in smart parking, while identifying the main elements influencing the parking system performability. Summary tables have also been added to offer a quick and rich overview about these sensors and communication technologies. On the other hand, relevant software systems and algorithms used to assure smooth and comfort quality of service to the end user have also been investigated. We outlined and analyzed potential algorithms in parking prediction, path optimization, and assisting techniques in the literature. We delved into the soft security issues and assessed the literature for countermeasures against the existing attacks. For more uniform and open access solutions, we discussed the system interoperability from the soft privacy point of view in the context of smart parking. We further present new emerging applications and software systems handling common parking problems with the help of VANET paradigms, in addition to recommending a novel conceptual hybrid SPS.

In general, this survey aims to gather and zoom in critical design aspects that might be of interest to readers from the academia, as well as the industry. It provides a multidisciplinary source for those who are interested in ecosystems and IoT-enabled smart cities. Also, it opens the door for further interesting research projects and implementations in the near future. For better readability, the abbreviations used in this survey are provided along with their definitions in Table 5.1.

Table 5.1 List of Used Abbreviations

Abbreviation	*Description*	*Abbreviation*	*Description*
ABGS	Agent-based guiding system	MQTT	Message Queuing Telemetry Transport
AI	Artificial intelligence	MSN	Mobile storage node
AMR	Anisotropic magnetoresistive	NAPS	Non-assisted parking search
APTS	Advanced Public Transport System	NB-IoT	Narrowband Internet of Things
AVI	Automated Vehicle Identification	OAPS	Opportunistically assisted parking search
CAPS	Centralized assisted parking search	OBU	On-Board Unit
CCTV	Closed-circuit television	O-DF	Open Data Format
COINS	Car Park Occupancy Information System	O-MI	Open Messaging Interface
DSRC	Dedicated short-range communication	PGIS	Parking guidance and information system
ECC	Elliptic curve cryptography	PLRS	Parking lot recharge scheduling
EV	Electric vehicle	PRS	Parking reservation system

(*Continued*)

Table 5.1 (*Continued*) List of Used Abbreviations

Abbreviation	*Description*	*Abbreviation*	*Description*
FCFS	First come first serve	QoL	Quality of life
FMCW	Frequency-modulated continuous wave	QoS	Quality of service
GAN	Generative adversarial network	RFID	Radio frequency identification
GIS	Geographic information system	RSU	Roadside unit
GPS	Global positioning system	SDN	Software-defined network
IIOT	Industrial Internet of Things	SPS	Smart parking system
ILD	Inductive loop detector	SVG	Scalable Vector Graphics
IoT	Internet of Things	TBIS	Transit-based information system
IOV	Internet of vehicles	UAV	Unmanned aerial vehicles
ITS	Intelligent transportation system	V2I	Vehicle to infrastructure
LDR	Light-dependent resistor	V2R	Vehicle to roadside
LoRaWAN	Long-range wide area network	V2V	Vehicle to vehicle
LPWAN	Low-power wide area network	VANET	Vehicle ad hoc networks
LTE	Long-Term Evolution	VMS	Variable message sign
MANET	Mobile ad hoc network	WIM	Weigh-In Motion
MAV	Micro aerial vehicle	WSN	Wireless sensor network
ML	Machine learning		

5.1.2 Comparison to Other Surveys

There have been similar attempts and surveys in the literature on smart parking solutions with their own merits and limitations. In this section, we overview these attempts and highlight how they contrast with our survey. For example, a very short survey on the parking lot reservations in cloud-based systems was discussed in Ref. [7]. The mechanism of each reservation technique was briefly discussed and explained with examples. A similar survey was provided in Ref. [8], however, without examples from the literature. The authors also touched slightly a few types of vehicular sensors. Nevertheless, it was not comprehensive and did not cover all types of possible sensors and limitations. The inadequacy of the previous surveys was covered and better explained with more examples and justifications in Ref. [9]. The authors presented some of the vehicular detection sensors with examples from the literature. However, they failed to demonstrate other aspects of the SPS such as the utilized communication protocols and/or software systems. With respect to communications, the authors in Ref. [10] explained the different implementation aspects of the wireless sensor communication protocols and later proposed an adaptive and self-organizing protocol. Despite this, they did not consider the emerging trends in LPWAN communication protocols and mostly relied on sensors and RFID systems while conducting their research. A multiagent, fuzzy, vision, and VANET-based smart parking method was overviewed in Ref. [11]. The authors talked about the used software systems in general, but they did not go much into the hardware details. A related, albeit brief, summary of smart parking software systems and applications with different advantages and disadvantages was presented in Ref. [12]. Like the other surveys, some necessary information and hardware details were omitted. Perhaps, the most comparable survey to our work was the one in Ref. [13]. This survey referred to reasonable aspects of the SPS such as information collection, dissemination, and deployment. However, it relies on outdated references and lacks the discussions of new enabling technologies related to drones and new emerging sensors. Our survey can be considered as the most comprehensive and up-to-date study in comparison with the aforementioned survey papers with additional insights and new trends which were briefly mentioned and/or totally ignored. In addition to covering the detailed hardware components, data interoperability, privacy, and security have been discussed and assessed. Moreover, hybrid solutions, with more comprehensive analysis in terms of communication networks and recent innovative parking applications, have been targeted in the smart city context. All the aforementioned surveys and contributions so far in the literature have been summarized in Table 5.2.

The rest of this chapter is organized as follows. Section 5.2 classifies existing SPSs into centralized versus distributed systems, and discusses their main features. In Section 5.3, we overview and compare the current state of vehicular detection technologies while focusing on critical parameters such as scalability, accuracy, and sensitivity to weather conditions. Utilized sensors are also classified into active and passive sensors. In Section 5.4, common design factors are listed under three

Table 5.2 Summary of Related Surveys

Reference	*Classification*	*Sensors*	*Communication Protocols*	*Software System*	*Security and Privacy*	*Interoperability and Data Exchange*	*New Applications*	*Open Issues*
[7]	–	✓†	–	✓†	–	–	–	–
[8]	✓*	✓*	–	–	–	–	–	–
[9]	✓*	✓*	–	–	–	–	–	–
[10]	✓†	✓†	✓*	✓†	–	–	–	–
[11]	✓*	✓†	✓†	✓*	–	–	✓*	–
[12]	–	✓†	✓*	✓*	–	–	–	–
[13]	✓*	✓*	✓*	✓	✓†	–	✓*	✓
Our survey	✓	✓	✓	✓	✓	✓	✓*	✓

✓ represents a comprehensive analysis.
✓* represents a general analysis.
✓† represents very little analysis.
– means not considered.

main categories: soft, hard, and interoperability factors. Varying use cases are presented and discussed in the context of both small-scale individual vehicular systems and large-scale implementations for smart city ecosystems in Section 5.5.1. Later in Section 5.5.2, we investigate new generations of the SPS utilizing VANET and drone paradigms, in addition to recommending a new conceptual hybrid model for the smart parking. In Section 5.6, we overview existing open research issues in SPSs ranging from sensors limitation to the communication network capabilities, as well as the need for energy harvesting and data interoperability. Finally, we conclude our survey in Section 5.7.

5.2 SPSs and Classifications

SPSs are categorized into various categories in which each of them has a different purpose and uses different technologies in detecting vehicles. SPSs benefit both the drivers and the operators. Drivers use the system to find the nearest parking spots, and the parking operators can utilize the system and the collected information to decide on better parking space patterns and a better pricing strategy. For example, since demand for parking is not fixed, assigning dynamic parking pricing depending on the time of day and even different customers can help the operators boost their revenue [14]. Smart parking enables services such as smart payment and reservation that could essentially enhance the experience of both drivers and operators. In addition, this smart system helps to prevent vehicle theft by improving the security measures on parking lots, and lastly by minimizing the delay in finding vacant spots, which results in a less number of vehicles on road, it can greatly take part in providing a clean and green environment by reducing the vehicle emissions [15].

There are commonly four layers of SPS architectures based on their functionalities [8,16]. First, the sensing layer is the backbone of the SPS, which is responsible for detecting the presence and/or absence of a vehicle in an area using different sensing technologies. These technologies are mostly comprised of receivers, transmitters, and anchors. Second, the network layer is the communication segment of the system, which exchanges messages between transmitters/receivers and the anchors. Third, the middleware layer is the processing layer of any SPS in which intelligent and sophisticated algorithms unique to each sensing technology are utilized to process the real-time data. It also acts as a data storage as well as a link between the end users requesting services from the lower layers. Fourth, the application layer is the top layer in the system, which interfaces the parking system with clients (end users) requesting different services from different mobile and/or stationary information panels as depicted in Figure 5.1. These multilayered parking systems are categorized into the following categories; Centralized assisted SPSs, Distributed assisted SPSs, and Non-assisted SPSs.

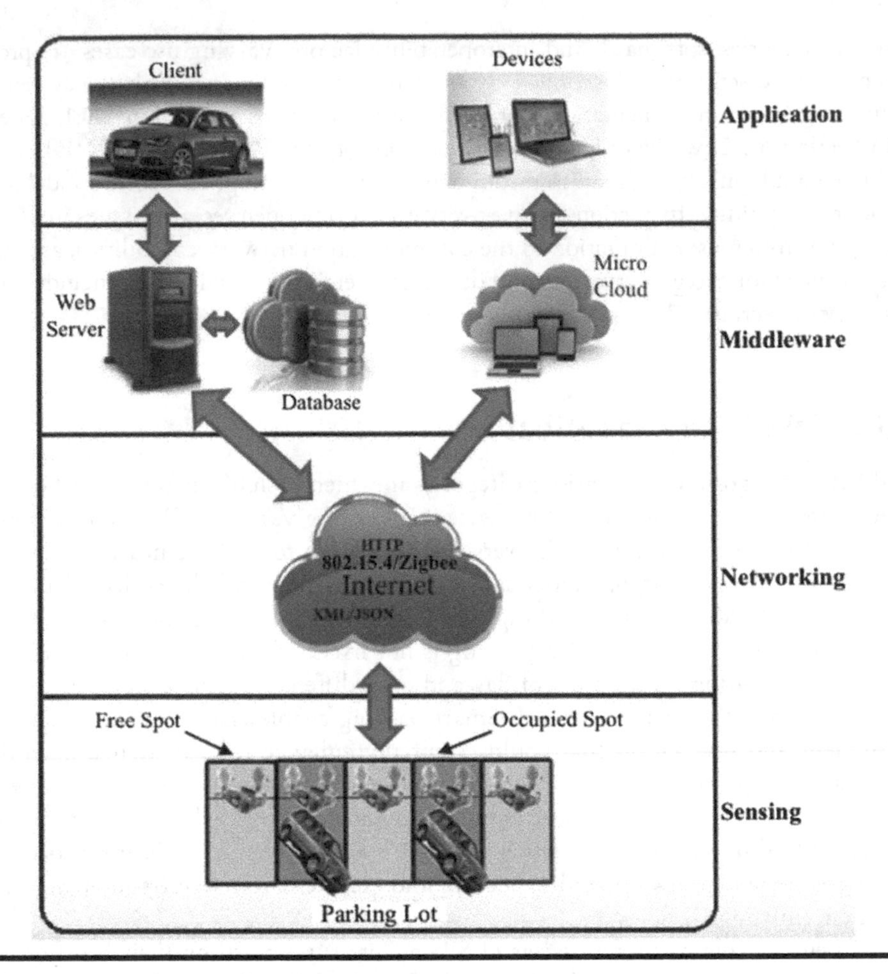

Figure 5.1 SPS architecture. (Adapted from Ref. [8].)

5.2.1 Centralized Assisted SPSs

In centralized assisted SPSs, a single central server collects the necessary parking information and processes it to provide services such as the parking lot reservation, allocation, and/or driver guidance. The following sample systems are generally implemented as a centralized system.

5.2.1.1 Parking Guidance and Information System

Parking guidance and information system (PGIS), also known as Advanced Public Transport System (APTS), works by collecting parking information dynamically mainly from loop detectors, ultrasonic, infrared, and microwave sensors to inform the drivers in a real-time manner about the vacancy of the parking lot via onboard guidance system or variable message sign (VMS) [15,17]. PGIS consists of four major subsystems, namely, the information collection, processing, transmission, and distribution [18]. It can be implemented in both citywide and/or individual parking lots, where in both cases, drivers can easily follow and navigate to reach the vacant parking space [19]. In Ref. [17], a combination of PGIS and dedicated short-range communication (DSRC) is presented, where DSRC-based PGIS provides a real-time, rapid, and efficient way of guidance; however, step-by-step implementation and concerns regarding an efficient PGIS algorithm and data safety as well as incorporation of different smart system may cause issues in implementing such systems. A combined PGIS with a mobile phone terminal with the help of global positioning system (GPS) to locate and predict vacant spots and to guide drivers to the destination was presented in Ref. [18]. In contrast, Shiue et al. [20] utilized both GPS and third-generation (3G) connectivity in combination. Reliability of GPS and 3G connectivity in a multilevel parking lot is an issue which may cause such systems to be impractical and ineffective. In Ref. [21], the authors proposed PGIS in combination with ultrasonic sensors and WSN, and in Ref. [22], the authors integrated RFID and ZigBee for smart parking solution. Both these sensors have their disadvantages in different areas, which are explained later in this chapter. In the communication of vehicles requesting parking and sensors in detecting the vacancy information, all the processing and decision-making are based on a central processor (server) [23].

5.2.1.2 Centralized Assisted Parking Search

In this sample, the first-come-first-serve (FCFS) approach is adopted where the first requester is guided towards a guaranteed vacant spot closest to the driver location. However, in this manner, other vehicles waiting in the queue are in constant movement until the server can satisfy them. This brings the issue of uncooperativeness between drivers, which can significantly degrade the performance of such systems. Furthermore, high initial and operating maintenance cost and scalability of centralized assisted parking search (CAPS) are serious concerns as well.

The FCFS scheduling approach is used in many applications of smart parking lot management. In Ref. [24], the authors proposed a parking lot recharge scheduling (PLRS) system for electric vehicles (EVs). The authors compared the performance of their approach against basic scheduling mechanisms such as FCFS and earliest deadline first (EDF). Their optimized version of FCFS and EDF outperforms the basic mechanism with regard to maximizing the revenue and number of vehicles in the parking lot. In another example, edPAS [25], abbreviation for event-driven parking allocation system, focuses on the effective parking lot allocation based on a certain event with respect to the event place while dynamically updating the communicator. This system utilized both FCFS and priority (PR) allocation scheme where PR consistently outperforms the FCFS.

5.2.1.3 Car Park Occupancy Information System

Car Park Occupancy Information System (COINS) utilizes video sensor techniques based on one single source to detect the presence or absence of vehicles. The status is then reported on information panels which are strategically placed around the parking lot [26]. COINS is categorized based on four different technologies: (1) counter-based, (2) wired sensor-based, (3) wireless sensor-based, and (4) computer vision-based; in the latter case, it provides a more accurate result of the exact status of the parking spot without deploying any other sensors in each individual spot [26,27]. In Ref. [26], COINS was developed and simulated in different environments with different parameters such as weather conditions and illumination fluctuations which add an extra layer of complexity to the system. Application of COINS in a multilevel parking lot may not be as effective as other systems using IoT technologies since scalability and coverage of such systems in these applications are the main concerns.

5.2.1.4 Agent-Based Guiding System

Agent-based guiding system (ABGS) simulates the behavior of each driver in a dynamic and complex environment in an explicit manner. The agent being an entity is capable of making decisions and has skills that are able to define the interaction between the drivers and the parking system based upon perceived facts from drivers and from different aspects such as autonomy, proactivity, reactivity, adaptability, and social ability. Meanwhile, SUSTAPARK was developed in Ref. [28] with the aim to enhance the searching experience while locating the parking space in an urban area with an agent-based approach. Their approach was to divide the problem of parking into manageable and simpler tasks that agents can follow. Another agent-based approach was introduced in Ref. [29] named PARKAGENT based on ArcGIS with similar functionalities to SUSTAPARK but including the effects of heterogeneity of the population of drivers. In Ref. [30], an agent-based system was used as a negotiator to bargain over the parking fee with capabilities to guide drivers to the optimal parking destination in the shortest path based on some perceived factors and interactions between other agents.

5.2.1.5 Automated Parking

Automated parking consists of computer-controlled mechanical systems which enable drivers to drive their vehicles into a designated bay, lock their vehicles, and allow the automated parking system to manage the rest of the job [9,15]. Stacking cars next to each other with very little space in between allows this system to work in an efficient way such that the maximum available space in the parking is utilized. The retrieval process of the vehicles is as easy as entering a predefined code or password. The process is fully automated which adds an extra layer of security and safety to the whole system to both drivers and vehicles, which makes it the greatest benefit for such a system [9]. Although the initial cost of automated parking is very high, for the services that you receive, the price is very competitive; in fact, 50% saving compared to conventional modes of parking in locations where the parking fee is high and limited is expected [27]. In these systems, employment of one or a combination of many services and sensors may be integrated to provide a fast, reliable, and secure mode of parking with little or no interactions between the drivers and the system. A general concern regarding such a system is that a universal building code does not yet exist, and in order to serve all customers, compatibility of the system with every vehicle model is not yet possible.

5.2.2 Distributed Assisted SPSs

In distributed assisted SPSs, many services are connected and are controlled by a single server. This is well explained in vehicular networks where one vehicle can exchange information to one or more vehicles effectively creating a distributed network of vehicles. Another example is in the systems where information processing and dissemination are generally based on roadside infrastructure. The following examples are considered as distributed and opportunistic SPS in the literature.

5.2.2.1 Transit-Based Information System

Transit-based information system (TBIS) is a park- and ride-based guidance system with similar functionalities to PGIS; that is, it not only informs drivers through VMS to guide them to a vacant parking spot, but it also provides real-time information of public transportation schedules and the routes, which enables drivers to preplan their journey. This system not only encourages drivers to use public transportation to help reduce unnecessary vehicle emissions, but it also helps to increase cities' revenue [9]. A field test in San Francisco [31] shows promising results in the effectiveness of TBIS; however, due to initial capital cost, such a system should be implemented mainly in large-scale applications to recover the system cost. Geographic information system (GIS) is also another way of providing information to users [32,33]. This system provides minimum travel time by optimizing the route and schedule of the operation of public transportation in real time and enables web-based GIS to be implemented for the convenience of users in planning their trip.

5.2.2.2 Opportunistically Assisted Parking Search

Vehicles with IEEE 802.11x communication standard in the ad hoc mode can share information of status and location of parking spots, which enables drivers to make a more informed decision while they are searching for parking spots. In this approach, drivers are guided towards the closest vacant parking space by analyzing time stamps and geographical addresses via GPS for example. Since opportunistically assisted parking search (OAPS) dissemination service does not impose global common knowledge of the status of parking spots, outdated time stamps and the infrequent update interval could cause delays and question the effectiveness of this approach [23]. Another issue as explained in Ref. [34] could be the misbehaviors of drivers where they enjoy the shared information from other drivers but show selfishness in sharing theirs which could increase the distance between the destination and parking spot as well as increasing parking search time.

5.2.2.3 Mobile Storage Node OAPS

Instead of normal vehicle nodes, the inflow of information is channeled through mobile storage nodes (MSNs), which enables sharing of information to and from other mobile nodes acting as a relay between vehicles. Similar issues in dissemination could also be observed as the status of parking changes overtime where the accuracy of data has a tendency to drop. In Ref. [34], the authors suggested that MSN could improve the performance of OAPS; however, they have no effect while used by selfish drivers.

5.2.3 Non-Assisted Parking Search

In non-assisted parking search (NAPS) approach, there is no inflow of information from any vehicles or servers, and the complete decision is solely depended on the driver's observation in the parking lot or from the former experience considering the traffic flow and the time of the arrival in the parking lot. Drivers wander around a parking lot and check for empty spots in sequence until an empty one is found, which is then allocated to the driver who reached the spot first [23,35]. Usually, it is performed with the minimum technology involvement. That's why we call here "non-assisted" parking.

Scalability and cost design issues associated with the aforementioned categories of smart parking are mainly justified/claimed by the amount and type of the utilized sensors and/or enabling technologies in detecting vehicles and parking lots. For example, for a large city-wide smart parking application (e.g., multistory shopping mall parking lot), where multiple sensors of one or different type(s) are required depending on the need and the scope of application, the overall cost may exceed the cost of using the same parking system in a smaller application. Therefore, as noted in Table 5.3, the cost of some SPSs might be dependent on the scale and the scope of requirements per application. Table 5.3 tabulates the aforementioned parking systems based on these varying parameters and requirements.

5.2.4 Use Cases in Practice

In this section, we outline three common use cases of the SPS: (1) the smart payment system (SPS), (2) the parking reservation system (PRS), and (3) the E-parking system for more comprehensive discussion. They are mainly presented to provide the required in practice background for the forthcoming sections in this chapter.

Table 5.3 Classified Parking Systems

Parking System Classification	*Sensors*	*Broadcast*	*Guidance*	*Scalability*	*Coverage*	*E-Parking*	*Reservation*	*Cost (O&M incl.)*
PGIS	All	VMS	✓	*****	City wide, local	✓	✓	Scale dependent
TBIS	All	VMS	✓	*****	City wide, local	✓	✓	Scale dependent
CAPS	All	VMS	✓* (FCFS)	***	Local	✓	✓	*****
OAPS	Vehicle	V2V, V2I, MSN	✓	**	Local	✓	–	***
NAPS	–	–	–	*	–	–	–	None
COINS	Video & image processing	Information panel	✓	**	Local	✓	✓*	***
ABGS	All	Agent	✓	****	City wide, local	✓	✓	Scale dependent
Automated parking	Limited	Information panel	–	–	Local	✓	✓*	****

* represents a rating system, one star (*) means *poor*, and five stars (*****) means *excellent* in terms of their used metric (e.g., accuracy, scalability, etc.), whereas, for the cost metric, more stars means more expensive.
✓* indicates only applicable in certain cases (applications) as discussed in each section.

5.2.4.1 Smart Payment System

Conventional parking meters were always slow and inconvenient to use, but with the advancement of technology and use of IoT, smart payment system (SPS) was introduced to ensure a reliable and fast method of payment [15]. This system employs contactless, contact-based and mobile devices-based modes of transactions. In contactless mode, smart cards and RFID technology such as Automated Vehicle Identification (AVI) tag; in contact mode, credit and debit cards; and lastly in mobile devices, mobile phone services can be employed to collect payments [8,15]. In Ref. [36] authors proposed image processing technology in conjunction with SPS by utilizing RFID technology. This enables drivers to recall their parking spot, contains the duration information of parking, and also enables the calculation of parking fee. Internet-connected parking meters could also be used as a tool for parking patterns and parking predictions especially for on-street parking where the data could be valuable for predicting parking spots with the help of machine learning (ML). However, the technical issue with SPS is the reliability and integrity of the system in case of attacks such as signal interception or routing protocol attacks which could compromise confidential information [37].

5.2.4.2 Parking Reservation System

PRS is a new concept in smart ITS that allows drivers to secure a parking spot particularly in peak hours prior to or during their journey [38]. The objective of PRS is to either maximize parking revenue or minimize parking fee. This system could easily be achieved by a simple formulation of min–max problem. Implementation of PRS requires several components, namely, parking reservation operations information centers, a communication system between the users and the PRS, real-time monitoring of the availability of parking, and an estimation of the anticipated demand [38]. Drivers can later use a variety of communication services such as SMS, mobile apps, or web-based applications to make the reservation of a parking space. SMS-based parking reservation was implemented in Ref. [39], where the integration of micro-RTU (Remote Terminal Unit) and the microcontroller and password requirement safety feature make it a smart solution in PRS. Such a system is also scalable and capable of handling multiple requests from drivers. CrowdPark system is another example of PRS proposed in Ref. [40], where the system works by crowdsourcing and reward point to encourage users to use the system to report parking vacancy. Malicious users and accuracy of the parking lot are the concerns in these types of systems; however, in the case of CrowdPark, 95% success rate on average in both situations is reported in San Francisco downtown area. ParkBid crowdsourced approach was proposed in Ref. [41], where this system, unlike other crowdsourced applications, is based on a bidding process which provides numerous incentives about the parking spot information that enables urgent drivers to find and reserve the closest parking spot.

5.2.4.3 E-Parking System

E-parking, as the name suggests, provides a system that users can electronically obtain information about the current vacancy of parking lots from other services and sensors, and make reservation and payments all in one go without leaving the vehicle and prior to entering the parking lot. The system can be accessed via mobile phones or the Internet. In order to identify the vehicle making reservations, a confirmation code is sent to the user's email or mobile phone through SMS, which can then be used to verify the identity of the vehicle [15]. Majority of the smart parking deployments that are introduced in this chapter are an example of E-parking where information about the parking vacancy can be achieved in advance. ParkingGain [42] is another example of E-parking where the authors presented a smart parking approach by integrating the OBUs installed on drivers' vehicles to locate and to reserve their desired parking spot. Moreover, their system offered value-added services to subsidize parts of the running cost of the system and to create business opportunities as well.

5.3 Sensors Overview

In this section, several vehicle detection technologies are discussed. They are the integrated part of information sensing in SPS, and in order to choose the best in each application, several factors should be taken into account as different sensors have their strengths and weaknesses. Vehicle detection sensors are mainly categorized into two different sets [43]: intrusive or pavement invasive and non-intrusive or non-pavement invasive. Table 5.4 summarizes the list of the sensors that are typically used in SPSs as well as several parameters that can influence the decision of the type of sensors to be used in smart parking applications.

The costs accompanying the sensors typically vary with the accuracy and the complexity of the data that they produce as well as the maintenance and/or the replacement they need. Moreover, depending on the installation requirement, the overall cost may rise. As in the case of piezoelectric sensor which will be discussed later, the higher accuracy added on top of the difficulty in installation increases the overall cost. Therefore, the best sensor is the one that can best fill the need of the parking owners while maintaining an overall lower cost without sacrificing accuracy and comfort. Sensors are classified into active and passive sensors, which are described in the following sections.

5.3.1 Active Sensors

Active sensors are those which need an external power source to operate and perform their task. In the following, we overview the most common parking sensors which are classified as active ones.

Table 5.4 Comparison of Different Parking Sensors

Sensor Type	*Intrusive*	*Count*	*Speed Detection*	*Multilane Detection*	*Weather Sensitive*	*Accuracy*	*Cost*
Passive IR	–	✓	–	–	✓	**	**
Active IR	✓	✓	✓	✓	–	**	**
Ultrasonic	–	✓	–	–	✓	***	*
Inductive loop	✓	✓	✓	–	✓*	****	**
Magnetometer	✓	✓	✓	–	–	****	*
Piezoelectric	✓	✓	✓	–	✓*	****	***
Pneumatic road tube	✓	✓	✓	–	–	****	****
WIM	✓	–	–	✓	–	**	****
Microwave	–	✓	✓	✓	–	***	***
CCTV	–	✓	✓	✓	✓	****	****
RFID	–	–	–	–	–	**	*
LDR	–	✓	–	–	✓	**	*
Acoustic	–	✓	✓	✓	✓*	*	**

* represents a rating system, one star (*) means poor, and five stars (*****) means excellent in terms of their used metric (e.g., accuracy, scalability, etc.), whereas, for the cost metric, more stars means more expensive.
✓* indicates only applicable in certain cases (applications) as discussed in each section.

5.3.1.1 Active Infrared Sensor

Active infrared sensors are configured to emit infrared radiation and sense the amount that is reflected from the object, that is, detecting empty spaces in a parking lot. While these sensors can accurately take advantage of multilane roads (active) and discover the exact location and speed of a vehicle on a multilevel parking lot, they are prone to weather conditions such as heavy rain, dense fog, and very sensitive to the sun [13].

5.3.1.2 Ultrasonic Sensor

Ultrasonic sensors work in the same way as infrared sensors do, but they use sound waves as opposed to light. They transmit sound energy with frequencies between 25 and 50 kHz, and upon reflecting from a vehicle, they can detect the status of parking [44] as well as receive several other useful information such as the speed of the vehicles and the number of vehicles in a given distance [45]. Similar to IR sensors, they are sensitive to temperature and environment; however, because of their simplicity in installation and low investment cost, they are widely used in smart parking applications to identify vacant parking spaces [44,46].

5.3.1.3 Closed-Circuit Television and Image Processing

Closed-circuit television (CCTV) technology combined with image processing software can effectively be used in many parking lots to determine the presence or absence of a vehicle. The stream of video captured from a camera is transmitted to a computer where it is digitized and can be analyzed frame by frame using image processing software to recognize changes in the frames over time. Since CCTVs are already in place in many parking lots for surveillance purposes, the implementation of these systems is suitable; moreover, a single camera can analyze more than one parking spot simultaneously, which makes it even more effective in the case of wide-area implementation, since one camera can share its information with another effectively reducing the number of cameras needed for that area [8,46,47]. Although CCTVs are reliable investments in SPS applications, they require to be positioned in an area where the field of view is not obstructed, and enough lighting is present. Moreover, weather conditions can affect the efficiency of these systems [38,44,47].

5.3.1.4 Vehicle License Recognition

In conjunction with CCTV and image processing unit, vehicle plate of the ingress and egress vehicles can be captured and analyzed to give an estimation of the number of the vehicles currently present or exiting the parking lots in real time.

Furthermore, continuous monitoring of vehicle movement in the parking lot until the vehicles reach the predesignated parking spaces that were allocated to them is a feature of these systems [38]. This also enables smart payment to be deployed which allows the drivers to exit the parking bay without any delay since the required information could be forwarded to the automated gate controller. However, bad weather conditions can disrupt the functionality of plate recognition systems. Additionally, privacy concerns regarding storing the information of vehicles on a database are another flaw of these systems.

5.3.2 Passive Sensors

Passive sensors are those which rely on detecting and responding to inputs from the surrounding physical environment without the need for dedicated power supply units. Common examples of this category are discussed in the following sections.

5.3.2.1 Passive Infrared Sensor

Passive infrared sensors work by detecting the temperature difference of an object and the surrounding [48]. In parking applications, they identify the vacancy of a parking space by measuring the temperature difference in the form of thermal energy emitted by the vehicle and the road. Passive infrared sensors, unlike the other types, do not require to be anchored or tunneled into the ground or wall, but rather they are mounted to the ceiling or the ground [49]. They are prone to weather conditions.

5.3.2.2 Light-Dependent Resistor Sensor

Light-dependent resistor (LDR) sensors detect changes in luminous intensity. By assigning a primary source of light such as the sun, a secondary source such as the moon, and other surrounding light sources in the case where the sun is setting, the vehicle that is parked in the parking space creates a shadow that causes the light sensor (deployed in the center of the ground of the parking spot) to detect luminous intensity changes, hence indicating the presence of the vehicle in the parking spot. Weather conditions such as rain, fog, and the change in the angle where the sunlight is assumed to arrive could affect the performance of these kinds of sensors and can lower the detection accuracy [50].

5.3.2.3 Inductive Loop Detector

Inductive loop detector (ILD) is normally installed on road pavements in circular or rectangular shapes [46]. They consist of several wire loops, where the electric current is passing through and generates an electromagnetic field. This field inductance excites a frequency between 10–50 kHz in the presence and/or passing of a vehicle over the sensor. It actually causes a reduction in the inductance that leads the electronic unit to oscillate with higher frequency which ultimately sends a pulse to the controller indicating the passing/presence of the vehicle [8,44,46]. Resurfacing of the road, multiple detectors for better accuracy, sensitive to traffic load and temperature variations, and disruption of traffic in case of maintenance are the major drawbacks of using ILD [44].

5.3.2.4 Piezoelectric Sensor

Piezoelectric sensors detect mechanical stress that is induced by pressure or vibration of objects passing over them by converting them into an electric charge. The value of the generated voltage is directly proportional to the weight of the vehicle exerting a force on the sensors. For accurate measurements, multiple sensors should be used; however, they are susceptible to high amount of pressures and temperature [9,44].

5.3.2.5 Pneumatic Road Tube Sensor

Pneumatic road tube sensors operate by air pressure burst which causes a switch to be closed producing an electric signal recognizing a passing vehicle. These sensors are very cost-effective and offer quick and simple installation, but in case of passing long vehicles such as a bus or large trucks over them, they lead to inaccurate axle counting resulting in less reliable parking vacancy information [44].

5.3.2.6 Magnetometer

Magnetometer functions in the same way as loop detectors. It senses the changes in the earth magnetic field that is caused by metallic objects, such as vehicles, passing over them. The cause for such distortion is that the magnetic field can easily flow through ferrous metals than air [51]. There are two types of magnetometers: single-axis and double-/triple-axis magnetometers; in the latter case, the accuracy of detecting vehicles is much higher due to the fact that it uses two/three axes to identify the presence of a vehicle. Both types are unaffected by weather conditions. Lane closure, pavement cut, and in some cases short-range detection and inability to detect stopped vehicles are considered as shortcomings of magnetic sensors [44].

5.3.2.7 Weigh-in-Motion Sensor

Weigh-in-motion (VIM) sensors can precisely determine the weight of a vehicle and the portion of the weight distributed to axles. The data gathered from these sensors are extremely useful and heavily used by highway planners, designers as well as law enforcement. There are four distinct technologies used in VIM: load cell, piezoelectric, bending plate, and capacitance mat [44,47]. Load cell uses a hydraulic fluid that triggers a pressure transducer to transmit the weight information. Despite its high initial investment, load calls are by far the most accurate VIM system. Piezoelectric VIM system detects voltage variations as pressure is being applied on the sensor. Such a system is consisted of at least one piezoelectric sensor and two ILDs. Piezoelectric sensors are among the least expensive sensors in the market; however, their accuracy of vehicle detection is lower than that of load cell and bending plate VIM system. Bending plate uses strain gauges to record the strain or change in length when vehicles are passing over it. The static load of the vehicles is then measured by dynamic load and calibration parameters such as speed of the vehicle and pavement characteristic. Bending plate VIM system can be used for traffic data collection. Its cost is lower than the load cell system, but its accuracy is not at the same level. Lastly, the capacitance mat system consists of two or more sheet plates where they carry equal but opposite charges. When a vehicle passing over them, the distance between the plates becomes shorter and the capacitance increases. The changes in the capacitance reflect the axle weight. The advantage of using this system is that it can operate for multilane road; however, it has a high initial investment cost.

5.3.2.8 Microwave Radar

Microwave radar generally transmits frequencies between 1 and 50 GHz with the help of an antenna which can detect vehicles from the reflected frequency. Two types of microwave radar are used in this sector: Doppler microwave detectors and frequency-modulated continuous wave (FMCW) [44,47]. In the former type, if the source and the listener are close to each other, the listener would perceive a lower frequency, whereas if they are moving apart from each other, the frequency would get higher. If the source is not moving, Doppler shift would not take place. In the latter case, a continuous range of frequencies is transmitted that is changing over time. The detector would then measure the distance between the detector and the vehicle to indicate the presence of a vehicle. Microwave radars are effective in harsh weather conditions, and they are also able to measure the speed of the vehicle and conduct multiple-lane traffic flow data collections. However, measurement of speed in Doppler detectors requires an additional sensor to function.

5.3.2.9 Radio Frequency Identification

RFID can be used for vehicle detection in many parking lots. RFID units consist of a transceiver, a transponder, and an antenna. RFID tag or transponder unit with its unique ID can be read via a transponder antenna [46]. RFID tag can be placed inside vehicles, and the reader antenna that is placed in the parking lot can read the tag and change the occupancy of the parking spot to be occupied. With this system, the delay can be minimized, and the flow of the traffic in the parking lots can be smooth. Although due to the range limitation of RFID, reading two tags placed on two side-by-side vehicles cannot be achieved [52]. RFID in hybrid systems has been shown to be effective and proved to be a more reliable option than a stand-alone RFID system [53].

5.3.2.10 Acoustic Sensor

Acoustic sensors can detect sound energy that is produced by vehicular traffic or the interaction of tires with the road. In the detection zone of the sensor, a single processor computer algorithm can detect and signal the presence of a vehicle from the noises generated. Likewise, in the drop of the sound level, the presence of the vehicle signal is terminated. Acoustic sensors can function on rainy days, and they can also operate on multiple lanes; however, cold weather and slow-moving vehicles can decrease the accuracy of such sensors [44,46].

5.4 Design Factors

In this section, we look at the design factors that influence the performance of the SPSs in practice. We divide these factors into three main categories. First, we overview soft design factors that deal with software aspects of the system. Several soft solutions that have been found in the literature are presented and correlated. Second, the experienced hardware issues and critical design aspects related to utilized sensors and communication networks in SPSs are intensively studied and discussed. Moreover, we investigate the system components interoperability and data exchange with a more focus on relevant smart city applications. This category is unique in nature and has been rarely overlooked in the literature, although it can make a significant effect in terms of the system performability and cost. We believe that the discussions in this section can be beneficial to all relevant smart solutions in a smart city ecosystem. Proposed examples and tabulated conclusions can significantly assist those who are interested in this area to quickly grab the fundamental information they seek.

5.4.1 Soft Design Factors

This category of the relevant design factors deals with software aspects of the parking system, as well as the processing of data collected via the aforementioned sensors. We also discuss under this category potential influences of the experienced privacy and security aspects from the collected data.

5.4.1.1 Software Systems in Smart Parking

Software systems play an important role in smart parking applications. They are used to manage the collected data from the sensors, analyze, and finally distribute them efficiently. Analyzing step is based on algorithms that vary in intricacy depending on the scale and the complexity of the application. Furthermore, using the collected data and applying certain algorithms, they can be used to predict parking vacancy and path optimization using ML methods that could enable the parking operators to manage their parking lots as efficient as possible, maximizing the parking revenue. In this section, the software aspect of SPSs and some interesting ideas such as path scheduling, path optimization, prediction, and parking assignment are discussed.

Managing the information gathered from multiple sensors on a multilevel parking lot requires a robust software system that can manage, monitor, and analyze the data in an efficient manner. Many of the smart parking applications, which will be discussed later, are using a centralized server to store and manage the data. After analyzing the information, many services such as parking reservation and guidance can be implemented. In Ref. [54], the authors presented a smart parking IoT and cloud-based system using real-time information. Services such as payments, reservation, and confirmation are all processed in their mobile application. In case of overshooting the parking time, their system offers extension of reservation, and failing to do so will impose certain fees on the driver. Context-aware SPS based on smart server, smart object, and smart mobile was proposed in Ref. [55]. Smart server integrated city context information and all the related information regarding the parking status and registered users; it then relays the information to the smart object to make changes in the availability of parking space where it finally informs the user on the user interface on a smart mobile application in which the user can search, book, and pay for the parking.

Guidance software systems are now becoming smart, and many factors are now taken into consideration that old PGIS sacrificed. As experimented in Ref. [56], the authors applied Stackelberg game theory to the PGIS so that dynamic changes in the behavior of drivers can be modeled. Their game strategy of balancing parking revenue of single vehicle and entire vehicles suggests that their model can effectively reduce the average time of parking. A utility-based guidance approach for parking in a shopping mall was discussed in Ref. [57]. Their approach made use of an improved A* shortest path algorithm that generates a two-stage optimal route based on user preference and the utility of parking based on six different factors. Message Queuing Telemetry Transport (MQTT) protocol parking guidance software was considered in Ref. [58], where the system can share real-time parking lot vacancy information to at least 1,000 requests from the users. Their web application in JavaScript presents the information on the layout of the shopping mall drawn by Scalable Vector Graphics (SVG). Internet of vehicle (IOV)-based guidance system was proposed in Ref. [59]. Onboard hardware in vehicles enables IOV to interact with everything around (vehicles, pedestrians, and sensors) which can then be used with the optimal path and parking algorithms embedded in the system to guide the drivers to the nearest parking spot. Continuous license plate recognition using video with gray-level changes and Dijkstra algorithms for locating and guiding to the nearest parking spot was experimented in Ref. [60]. The combination of license plate and gray-level changes detection ensures that their system is viable in case of passing pedestrians on the parking spot or when they cover the license plates. The use of GPS in smartphones with the combination of a genetic algorithm to locate and navigate to the closest parking lot was discussed in Ref. [61]. The authors presented their solution and were able to obtain accurate results in several case studies. Instead of using exploration algorithms for locating the nearest parking space, a learning mechanism was used in Ref. [62]. The authors used reinforced learning algorithms in conjunction with Monte Carlo approach to minimize the expected time to find a parking place in an urban area. They compared their algorithms with the method of tree evaluation and random method, and deduced that their algorithms are less complex and more efficient. In Ref. [63], the authors propose a cache replacement approach for fog applications in software-defined networks (SDNs). This approach depends on three functional factors in SDNs: the age of data based on the periodic request, the popularity of on-demand requests, and the duration for which the sensor node is required to operate in the active mode to capture the sensed readings. These factors are considered together to assign a value to the cached data in an SDN in order to retain the most valuable information in the cache for longer time. The higher the value, the longer the duration for which the data will be retained in the cache. This replacement strategy provides significant availability for the most valuable and difficult-to-sense data in the SDNs. Decisions based on when and where to park are largely based on drivers' observations. Many factors such as accessibility, fee, and availability of parking influence the decision of the drivers. On the other hand, decisions based on experience when drivers locate free parking space with or without prior information of the availability always tend to congestion of that particular space and increase the searching time and cause long queues. But if the availability of the parking space can be predicted and disseminated in time, the driver's experience in locating the most suitable location can be enhanced. Moreover, the prediction of parking availability offers parking operators to perform short- and long-term system checks, which ultimately enables them to take preventative decisions in case of system failure.

Nowadays, prediction is as easy as to collect data from the sensors and apply algorithms. Previously, prediction was based on historical data and surveys that were used to create a model of parking vacancy information. In Ref. [64], data collected from off-street parking garages was used to create a short-term model that could forecast the changing characteristic of the parking spaces by the wavelet neural network method. The authors compared their research method with the largest Lyapunov exponents in terms of accuracy, efficiency, and robustness. Although the model was successfully tested, other criteria such as driver's behavior or environmental characteristics were not taken into consideration. In another study [65], the authors proposed a real-time availability forecast (RAF) algorithm based on drivers' preferences and the availability of parking that is iteratively allocated using an aggregated approach. Their algorithm is updated upon each arrival and departure to predict the dynamic capacity and parking availability. Their test simulation in a parking facility in Barcelona showed promising results with small errors. The results were then compared with the numerical method in which they observed no significant difference between the two approaches. Prediction of parking space availability of Ubike system in Taipei City was experimented in Ref. [66]. The authors used regression-based models (mainly linear regression and support vector regression scheme) to forecast the number of bicycles in Ubike stations. Their model due to a constraint that the bicycles are only circulating around the station is different from the parking system used for cars. However, a similar approach could be used for car park facilities. A prediction mechanism based on three feature sets with three different algorithms for comparison—regression tree, support vector regression, and neural networks—was developed in Ref. [63]. With their datasets from San Francisco and Melbourne parking facilities and their model, they concluded that the least computationally intensive algorithm, regression tree, performed better.

Parking search and space optimization are another part of the software system in smart parking, which utilize the collected information to minimize the searching time and to maximize the number of parking spaces in a given parking lot. The authors in Ref. [67] showed that their adaptive multicriteria optimization model effectively reduced the searching time by 70% in an urban area. The criteria (acceptable walking distance, price, and driving time) were based on drivers' preferences that were presented by adequate utility function, and the objective was to maximize the expected utility. In a study conducted in the University of Akron [68], the authors used direct random search method to perform their optimization model based on mainly different classroom assignment options. However, other factors such as parking search behavior, arrival and departure distributions, and the location of different buildings and parking facilities were also taken into account. Their model managed to reduce the parking search time by about 20%. Cooperative parking search between vehicles searching for parking space using vehicle-to-vehicle (V2V) and vehicle-to-infrastructure (V2I) mode of communication was studied in Ref. [69]. The authors found that when vehicles search cooperatively, a search time reduction of up to 30% is observed. They also concluded that drivers would benefit more if they exchange information both before and after reaching the destination. An intelligent hybrid model for optimizing the parking space based on Tabu metaphor and rough set-based approach was used in Ref. [70]. Tabu search was used as a complement to heuristic algorithm, whereas the rough set was used as a tool to manage noisy and incomplete data due to traffic conditions.

5.4.1.2 Data Privacy and Security in Smart Parking

It is estimated that the number of connected IoT devices would reach to 20 billion by 2023 [71]. These devices are in constant communication, and a large number of data that is being processed, aggregated, and shared with users can be intercepted and used for nefarious intentions [72]. The most crucial part of any smart applications is ensuring that the network supports end-to-end encryption and authentication. In the case of interconnected IoT services and devices, any vulnerability, no matter the size is, can interrupt one side of the system and cascade it to the rest of system [73]. Some key considerations for security and privacy protection for the integrity of the system and ultimately the users are as follows: (1) data collection, which if limited to a certain extent could greatly help to mitigate any risks, for example, storing a large amount of data can elevate security breaches and collection of huge amount of personal data may be used in a way that it is out of the scope of consumer's expectations from the system; (2) optimization and data analysis for any smart-based application that requires information sharing; therefore, service providers and the technology partners should come into an agreement for secure data handling and techniques to ensure user's privacy such as de-identification; (3) reliability of servers, encryptions, and digital signature are also as important as risk management protocols and physical security of the system; (4) human errors, intentional or unintentional, can elevate security risks; therefore, policies and procedures for training are required to mitigate oversight issues; and (5) lastly, transparency of any smart system ensures the integrity of such system, which offers accountability and clear policies in regard to data security and privacy.

A good smart parking application requires end-to-end communication between the end user and the server. Since the majority of smart parking solutions are established on either web-based or mobile-based applications for determining the vacancy of a parking lot which may also provide the ability of reservation and payment, they require users in such systems to enter personal information such as their home/business address. Since these systems also keep track of the history of transactions including credit card information, they are therefore considered as the critical aspects in data privacy and security of the existing SPSs [72]. P-SPAN or privacy-preserving smart parking navigation system was developed in Ref. [74]. The navigation system for locating and guiding drivers to a vacant parking spot using Bloom filter and vehicular communications by introducing a privacy preservation mechanism has been shown to be an effective SPS with low computational and communication overhead. Another VANET based approach similar to the previous study with privacy-preserving in mind was discussed in Ref. [75]. The system provided a secure navigation protocol with one-time credentials. Some communication protocols lack data encryption or require high computational resources in order to function securely; however, in Ref. [72], the authors employed elliptic curve cryptography (ECC) with LPWAN protocol as an alternative to other cryptography techniques in devices where there exist hardware limitations. The aim of the work performed in Ref. [76] was to design a typical network security model for cooperative virtual networks in the IoT era. This chapter presents and discusses network security vulnerabilities, threats, attacks and risks in switches, firewalls, and routers, in addition to a policy to mitigate those risks. It provides the fundamentals of secure networking system including firewall, router, Authentication Authorization and Accounting (AAA) server, and Virtual Local Area Network (VLAN) technology. It presents a novel security model to defense the network from internal and external attacks and threats in the IoT Era. In Ref. [77], the authors proposed a context-sensitive seamless identity provisioning (CSIP) framework for the Industrial Internet of Things. CSIP proposes a secure mutual authentication approach using hash and global assertion values to prove that the proposed mechanism can achieve the major security goals of the WSN in a short time period.

Furthermore, in Ref. [76], the authors proposed a solution for secure data collection where they used a repository of sensing information acting as a sink for the sensing data and a mirror of reservation database which is synchronized with the repository. In this fashion, drivers are the only elements that can access the mirror database for payment for checking the vacancy of parking lots and reservation on their mobile devices.

5.4.2 Hard Design Factors

The communication networks surrounding the SPSs from the legacy to LPWAN communication protocols for small- and large-scale applications have been addressed in this section from the hardware perspective. In addition, we discussed the influences of the experienced sensor errors on the designed parking system.

5.4.2.1 Communication Networks

The employed IoT sensors in SPSs vary in terms of the used communication protocols. However, they all can be categorized into long-range LPWANs or short-range wireless networks. LPWANs are long-range and low-power method of communication, and the compatibility of the existing cellular technology with cellular IoT (designed explicitly for LPWAN) makes it so that no further infrastructure is required [13]. The work performed in Ref. [40], for example, provides an overview of modeling traffic and deployment strategies in femtocells and a review for the use of femtocells and their smart applications in the IoT environment. In addition, it presents open research issues associated with IoT-femtocell-based smart applications. Major LPWAN standards are LoRaWAN, Sigfox, Weightless (SIG), Ingenu, Long-Term Evolution (LTE)-M, and NB-IoT; on the other hand, short-range WSN standards are Bluetooth, Wi-Fi, and ZigBee. A comparison of range and bandwidth for both modern and legacy communication protocols is shown in Figure 5.2. Moreover, Table 5.5 summarizes the technical parameters for both short- and long-range methods of communications [77–84]. Libelium,[1] a wireless sensor network platform provider, has used both LoRaWAN and Sigfox in their Plug & Sense platform which uses magnetic sensors to detect vehicles in parking spots. Huawei[2] solution for smart parking which resulted in 80% energy reduction in their Czech Republic and ZTE[3] trials in China claiming 12% and 43%

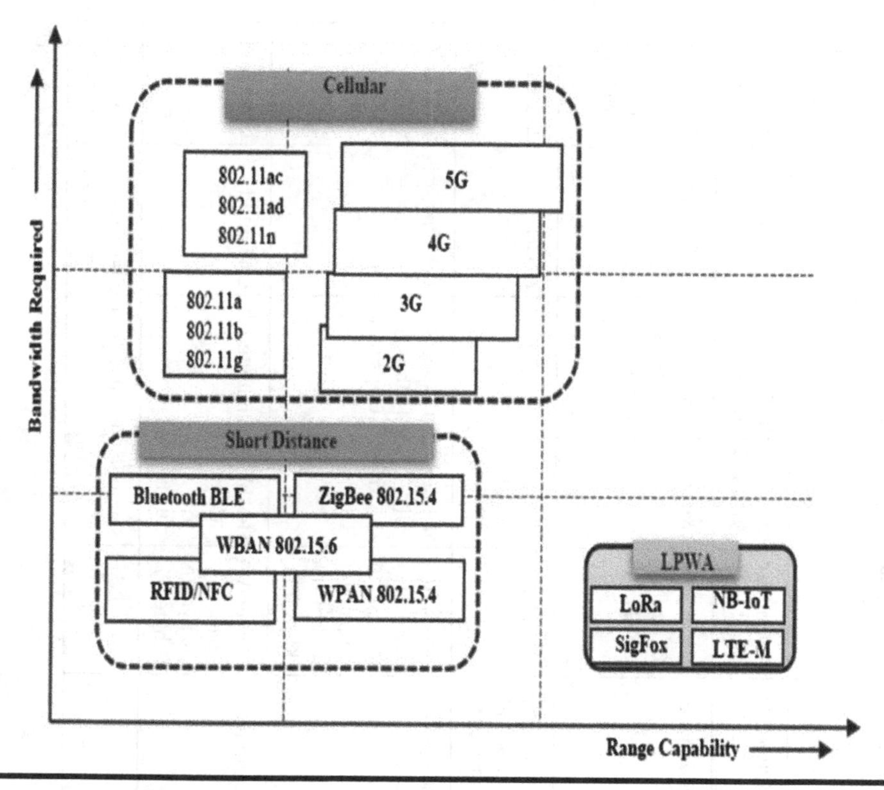

Figure 5.2 Comparison of range and bandwidth of LPWAN and other protocols [51].

reduction in congestion and the time spent searching for free spot, respectively. These are some of the examples of NB-IoT-based smart parking solutions. China Unicom Shanghai smart parking developed by Huawei uses 4.5G LTE-M communication protocol in their parking network; moreover, Nwave[4] and Telensa[5] solutions to smart parking use Weightless N protocol along with magnetic sensor for detecting vehicles. Deployment of NB-IoT- and third-party payment platform-based SPS was studied in Ref. [85]. The authors proposed a cloud and mobile application platform that utilized an NB-IoT module that can provide SMS and data transmission services over a long range with low power consumption.

As it can be observed, the applications of LPWAN protocol are limited; this is because the standard has not yet been adopted in many areas and regions. Unlike LPWAN, the legacy protocols, WSN, are the first choice in many smart parking solutions, but as the population of vehicles increases, the need for longer range, more reliable, faster, and more secure mode of communication is expected. This promises LPWAN to develop more to overcome the current existing challenges such as high complexity of interoperability between different LPWAN technologies, coexisting with other WSNs, and lack of models for large-scale applications [81,86].

Table 5.5 Comparison of LPWAN and Other Communication Technologies

Protocol	*SigFox*	*LoraWAN*	*NB-IoT*	*LTE-M*	*Weightless W/N/P*	*Ingenu*	*Wi-Fi*	*ZigBee*	*Bluetooth/ BLE*
Standard/ Mac layer	SigFox	LORa™ Alliance	3GPP rel. 8 and 13	3GPP rel. 13	Weightless Sig	IEEE 802.15.4 k	IEEE 802.11 b/g/n	IEEE 802.15.4	IEEE 802.15.1
Spectrum bandwidth	100 Hz	125/250/500 KHz	180 KHz	1.4–20 MHz	5 MHz (W) 200 Hz (N) 12.5 KHz (P)	1 MHz	80 MHz (2 antennas) 20 MHz (1 antenna)	9/868/915 MHz	79 channels 1 MHz. 40 channels 2 MHz (BLE)
Frequency band	ISM EU: 868 MHz US: 902 MHz	ISM 433/868/780/915 MHz	Licensed LTE bandwidth (7–900 MHz)	Licensed LTE bandwidth (7–900 MHz)	Sub GHz	ISM 2.4 GHz	2.4/5 GHz	2.4 GHz	2.4 GHz
Data rate	100 bps UP 600 bps DL	290 bps–50 Kbps	234.7 Kbps UP 204.8 Kbps DL	200 Kbps–1 Mbps	1 Kbps–10 Mbps (W) 30 Kbps–100 Kbps (N) 200 bps–100 Kbps (P)	78 Kbps UP 19.5 Kbps DL	11 Mbps (b) 54 Mbps (g) 1 Gbps (n/ac)	250 Kbps	1 Mbps (v. 1.2) 24 Mbps (v. 4)
Number of messages per day	140 12-byte	Defined by user	Defined by user	Unlimited	10 byte (W) Up to 20 byte (N) 10 byte (P)	Flexible (6 byte to 10 kbyte)	Unlimited	Unlimited	Unlimited
Topology	Star	Star of Star	Star	Star	Star	Star, tree	Point to hub	Star, Cluster Tree, Mesh	Star–bus network

(Continued)

Table 5.5 (*Continued*) Comparison of LPWAN and Other Communication Technologies

Protocol	*SigFox*	*LoraWAN*	*NB-IoT*	*LTE-M*	*Weightless W/N/P*	*Ingenu*	*Wi-Fi*	*ZigBee*	*Bluetooth/ BLE*
Battery life	8–10 years	8–10 years	7–8 years	1–2 years	<10 years	+20 years	7 days–up to 1 year (AA battery)	100–1,000 + days	Years on coin cell battery
Power efficiency	Very high	Very high	Very high	Medium	Very high	Very high	Medium	Very high	Very high
Range	10 km urban 50 km rural	2–5 km urban 15 km suburban 45 km rural	1.5 km urban 20–40 m rural	35 km 2G 200 km 3G 200 km 4G	5 km (W) 3 km (N) 2 km (P)	15 km urban	Up to 250 m	Up to 100 m	80–100 m
Scalability	Yes	Yes	Undetermined	Yes	Yes	Yes	Limited	Yes	Limited
Latency	1–30 s	1–2 s	1.4–10 s	10–15 ms	Low	>20 s	<50 ms (95th percentile)	15 ms	3 ms (BLE)
Cost	Medium	Low	High	High	Low	Medium	Low	Low	Low

As Table 5.5 suggests, almost all the LPWAN communication modules have longer than 7–8 years of battery life with high power efficiency as well as longer range in both urban and rural areas. In rural areas, the range of the communication is longer due to a smaller number of interferences and the absence of skyscrapers, which interferes with the quality of the data sent or received. Furthermore, the most common topology used in LPWAN is of the star type where all the nodes are directly connected to a central computer or server. Every node is also connected to every other indirectly in this topology. Short-range wireless networks are mostly connected in a mesh topology to extend their range, but the development cost and their energy usage for a large number of distributed devices make it ineffective in large case implementation [81] such as multilevel parking lots. This is where LPWAN technology comes into play to overcome the limitations of the previous generations of wireless networks. Latency or the delay of the information from sensor nodes to the central server is quite small for legacy protocols compared to Sigfox or LoRaWAN, but this does not necessarily mean that they are not effective in SPSs. However, in large case smart parking applications where low latency is a must, NB-IoT and LTE-M are among the best options.

5.4.2.2 Errors in Data Collection and System Reliability

Typical sensors in detecting vehicles in parking lots may inaccurately report the wrong number of vehicles or fail to detect at all [87]. As the authors discussed in their paper, one of these errors is double counting, which is apparent in the cases where drivers park their cars in between the designated parking spot causing the sensor to detect two vehicles instead of one. This behavior of drivers also causes the sensors to not report the vehicle in the spot at all. This is mainly observed in the locations where the range of the sensor is limited; therefore, miscounting can also be an issue. Another type of error is occlusion where large vehicles can block the line of sight of detectors, thus causing the smaller vehicle next to it to be excluded. This is mostly an issue in the parking lots where two or more vehicles are entering at the same time. The last error is phantom detection error where sensors cannot detect the correct position of the vehicles since the signals are bouncing off walls. Other than these physical constraints of detecting vehicles properly, there may be cases where the data collected from an event or observations do not conform to the initially expected patterns. Data collection is, in fact, useful when we are dealing with parking prediction and pattern analysis, which provides better management system to the operators of parking lots and ultimately an enhanced experience to the end users.

However, there is a caveat that not all sensors can behave according to a predetermined pattern. This brings the issue of data anomaly, or as the author in [88] defines it "outliers". Outliers, in general, can be caused by node malfunctioning that requires inspection or maintenance. In order to identify these misconducts, authors looked at a dataset provided by Worldsensing. They discovered that by looking at the dataset as a whole, the chance of missing outliers is high. They found that there exist similar data points in the dataset that share similar characteristics. And by clustering these data, they were able to easily identify the outliers while applying certain sophisticated algorithms.

Moreover, majority of presented smart parking solutions do not or unable to perform system reliability checks. The system reliability is related to, but not limited to, software, hardware, and other elements that make up the whole system that ensures an efficient and satisfactory performance of tasks in any given condition that it was designed or intended to. In Ref. [89], the authors proposed a sensor-based smart parking solution with a system reliability check. They performed certain system checks to detect any errors that may be caused by hardware failure or drivers' behavior. They created false positive and false negative criteria for different time intervals to ensure the reliability of the system. In another example [90], a framework was introduced that comprised a reputation mechanism in their mobile application-based smart parking management system where each time a vacant spot is reported empty by the user, they receive reputation score that reflects the truthfulness of the information. By doing so, they ensured that the collected data is reliable and can be used as a mode of determining parking vacancy. A similar application in Ref. [91], UW-ParkAssist, integrated the collected data from the mobile phone sensors and the official data from the parking officials or the police to reduce the data manipulation by the users. The proposed integrated system functions such that in case of false information by the drivers, officials can manually override the status of the parking spot in the hidden settings of the application database. With this verification method, they ensure that their system is able to provide improved data quality, gathered from the users' inputs.

5.4.3 Interoperability and Data Exchange

For seamless integration of different services and technologies, the ubiquitous IoT needs to be open and able to exchange data with other platforms. In today's IoT ecosystem, we are mostly dealing with closed and vertically oriented heterogeneous systems with no vibrant collaboration [92]. Imagine groups of SPS that can share their data with other smart city platforms in a single multi-service network (air quality, traffic, etc.) where those platforms are entirely based on different types of software. Then, for example, by combining the gathered information, drivers can predetermine the best route and time to reach the parking lot in the best and efficient way as possible. By achieving this, one service, therefore, can be used on top of the already existing platform in different regions allowing the same operation to be performed. In order to reach interoperability, the data structure and interface format (i.e., syntactic interoperability), in addition to meaning and ontologies or the relationships between exchanged the data (i.e., semantic interoperability), need to be first clarified. Not only data exchange between platforms is required, but rather cross-application domain access of the platforms is also essential. Such an ecosystem can allow the confirmation of parking reservation and parking finder services in an utter smart parking service [92]. Realization of interoperability in smart parking with open standards such as Open Groups Open Messaging Interface [6](O-MI) and Open Data Format [7](O-DF), which have now been widely used for interoperable smart parking solutions. The authors in Ref. [93] offered an overview of European Union's bIoTope Horizon 2020 project for economically viable IoT platforms with their proof of concept of smart parking lot management in the coming FIFA World Cup 2022. Another proof of concept under the same EU program for Electric Vehicle (EV) charging with smart parking was introduced in Ref. [94]. In the following, we discuss the smart parking interoperability from the perspective of security and privacy issues.

5.4.3.1 Interoperability with Privacy and Security Aspects

Having a unique software platform and a communication backbone in the interconnected IoT era seems to be a little farfetched. And trying to make things uniform is rather speculation. Efforts have to be spent towards embracing and managing the fast-paced IoT advancement such that there are minimal effects on the existing infrastructures and more utilization out of it. Making IoT an open world in the current heterogeneous and fragmented market is not impossible, but there are certainly some challenges that need to be addressed first. Such challenges can be (1) the lack of resources, (2) the exchanged data complexity, (3) the system proprietary, and (4) the most important security and privacy. There are already some constraints in the existing closed and vertical IoT solutions as discussed in the previous sections. Making them interoperable and able to exchange information on both lower and higher levels is admittedly a hurdle. The breach of information could pose great dangers to both the users and the companies responsible for the smart solution. The capabilities of traditional Internet securities embedded in the WSNs or LPWANs may not provide full protection against cyberattacks. Trying to unify one system while maintaining the low-cost approach of IoT solution, with the current capabilities especially in terms of the computational power, is another security concern. Interoperability of devices also brings the question of total anonymity, which is still one of the implications of developing IoT services. Many innovative IoT-related solutions depend on tracking and profiling users' movements and activities [95]. Any vulnerability could potentially lead to personal information leakage and increase the exposure of users to other attacks, which ultimately threatens the privacy of the system users.

5.5 Solutions in Practice

In this section, we investigate various implementations of the SPS. Moreover, another aspect of the SPS in smart city applications such as the parking lot management for EV charging stations and huge city events which are more complex and broader in size and density have also been discussed and analyzed.

5.5.1 Implementation of Smart Parking Systems

Various implementations of the SPS in the literature based on their vehicular sensors as well as their mode of communication and other relevant criteria are discussed and classified into a single vehicular system versus a large-scale ecosystem.

5.5.1.1 Single Vehicle Detection Sensors

A custom infrared sensor was used in Ref. [96] to detect vehicles entering the parking spot where the results were sent to the router using Arduino by means of a wired connection, Ethernet. Drivers are automatically connected to the parking network, and to check for parking spaces, an android mobile application based on Java Script Object Notation (JSON) was used to determine the vacancy. This was also used to guide drivers when they are leaving the parking spot as well as enabling smart payments. On the other hand, in Ref. [97], the authors used Raspberry Pi as the central server to send the information to the cloud-based mobile application. In these two examples, the scalability of these systems for multilevel parking lots is of concern where the system reliability and effectiveness are the major points to be considered. An example of a reliable implementation is presented in Ref. [98] where the system is comprised of both passive infrared and magnetic sensors to detect vehicles where the information was sampled in Java using TinyOS-2.x, and the results were reported to drivers as a web-based application. An SPS for commercial stretch in cities using passive infrared sensors and image detectors was discussed in Ref. [99]. They also introduced smart payment and reservation system in addition to guidance using GPS in their mobile application.

Ultrasonic is the most used sensor to detect vehicles in many areas. As the authors in Ref. [100] presented their smart parking solution with ultrasonic sensor to detect vehicles and ZigBee standard for communication module between the sensors and the RabbitCore microcontroller which controls the multiplexers sharing the information from many sensors to the controller. Their system uses shortest path algorithms to find the nearest parking spot and exit location near to the location of the driver as they enter or leave the parking lot. This system requires the drivers to follow the directions given by the system, and any deviation results in the failure of the system. Similarly, in Ref. [45], ultrasonic was used as the method for detecting vehicles in a multilevel parking lot; however, additional horizontal sensors mounted on the wall were used for detection of improper parking where an alarm would be triggered notifying the driver of his or her improper parking position. In this way, using three sensors per parking spot would not be feasible, and further studies should be conducted for cost-effectiveness. Ultrasonic was also used in mobile systems. ParkNet vehicles [101] comprised of GPS receivers and mounted ultrasonic sensors on passenger side doors of the vehicle which can detect parking spot in urban areas as the vehicle drives along a street. This can create a real-time map of parking information along that street. ILDs are mostly used for traffic surveillance [13]; however, in Ref. [102], the application for detecting vehicles was tested, and 80% success rate in detecting the magnetic signature was observed. IDLs are mostly used in conjunction with other sensors to increase the reliability and to provide traffic parameters such as speed, volume, and gap, which could be useful in the analysis of the performance of many parking lots.

Magnetometer sensor is also a widely used sensor in parking lots due to its accuracy and reliability in many conditions. In Ref. [103], the authors presented their SPS that is comprised of magnetic sensors placed on each parking spot to detect the vacancy. The SPS shares the results through T-sensor to base station via a ZigBee module, which later presents the parking vacancy information through VMS. Their system also offers a self-healing mechanism as well as a battery lifetime of over 5 years. In another example, in Ref. [104], the authors used triaxial magnetic sensors for detecting vehicles as they enter the parking lots. Background noise produced from multiple vehicles as they approach the sensors is the issue that should be worked on. Social network car park occupancy information was implemented in Ref. [105] where the system works by detecting vehicles via magnetic sensors placed at the entrance and exit of a parking space that can inform the drivers of the available car parks via the LED indicators placed in several locations that are connected via RS245&485 to the server and also through their Twitter account. A three-axis anisotropic magnetoresistive (AMR) sensor used for occupancy detection in a parking lot via a full-fledged detection mechanism in three phases—entering, stopping, and leaving the parking—was proposed in Ref. [106]. The authors achieved 98% detection accuracy compared to other AMR-based detection algorithms. Image processing and vehicle license recognition techniques are used in combination with several smart parking implementations. In Ref. [107], video cameras are used to detect vehicles as they enter the parking lots which are sequentially checked with the help of the Prewitt edge detection technique. The combined use of image processing and license recognition algorithm as previously discussed was studied in Ref. [108]. The system functions such that the characteristics of vehicles (e.g., color and license plates) are recorded on a database. Upon searching for parked vehicles by inputting the license plates, drivers can easily locate their vehicles. Since full recognition of a vehicle in some cases is not possible, classification probabilities based on the similarities of plates and colors are used for retrieval of vehicles. In Ref. [36], the authors proposed image processing in conjunction with RFID for retrieval purposes. They used RabbitCore microcontroller and ZigBee module, and provide A* shortest path to assign vacant spot to drivers where they can be informed of the location of the spot via VMS or the printed map on the ticket. An adaptive and self-organizing algorithm for WSNs with RFID for locating, reserving, and charging the parking space with security measures against theft in place was offered in Ref. [10]. However, this necessitates a dedicated infrastructure in practice which can be mostly expensive in implementation.

5.5.1.2 Smart Parking in Smart City Ecosystem

Existing smart parking attempts in the smart city paradigm are either utilizing a single detection sensor, or they fall in the large-scale implementations. In large-scale scenarios, multiple technologies must orchestrate together in exchanging data to achieve a smooth interoperability between the parking lots in IoT ecosystems. Smart parking in a smart city paradigm can be defined as a system that can facilitate interoperability among its subservices to provide a kind of quality of life (QoL) for urban citizens [109]. An example of a complex and large parking system can be the huge city event parking scenario. Large events such as sporting events require many services to work together in order to provide an effective smart parking solution. For instance, the initiative to create an open IoT smart parking ecosystem for FIFA World Cup 2022 (in Qatar). The authors in Ref. [110] created a proof of concept with consideration of open standards for facilitating data interoperability and data exchange, namely, O-MI and O-DF. Their concept of open IoT is applicable at a stadium level as well as the city. Another concept of a city-wide parking lot management is described in Ref. [55]. By utilizing a novel distributed algorithm and cloud computing, the authors provided an SPS that is context-aware and can gather information from the city or from the citizens for accurately determining the status and/or predicting the parking vacancy.

Towards cleaner environment and more energy savings, EVs have now become favored in worldwide. However, the lack of charging stations or their locations have led to the drivers' inconvenience. Although in the future more homes and offices will be equipped with charging services, the need for a smart system to manage the charging of EVs considering time, location, and other criteria should be premeditated. Scheduling such charging stations in the smart parking context that can deliver convenience to the users and at the same time be profitable to parking owners should be carefully analyzed. The authors in Ref. [111] delivered such a scheduled system that can adapt to the traffic pattern and maximize the total utility for the owner by adjusting the time-of-use (TOU) pricing strategy while providing the best quality of service. In order to integrate and expand the EV charging stations to the smart city ecosystem for full interoperability, like the example stated earlier, a careful consideration of using the open standards must be made. To address such problems, the authors in Ref. [94] presented a case study of EV charging stations via a mobile application to search and reserve these charging stations in an SPS. In the aforementioned examples, seamless integration of the ubiquitous IoT services and technologies must be achieved in order to realize an effective system in the ecosystem. A bilevel optimization of a distribution company for the parking lot owners was examined in Ref. [112]. The optimal scheduling of EV charging stations for the best of the company and the owners was evaluated by stochastic programming due to uncertainty in the parking lots. It shows the effectiveness of bilevel (distributed) approaches in comparison with centralized models. Another parking lot management system for EV was introduced in Ref. [24]. The 24-h trace-based model can track the mobility/

patterns of EVs for the maximum revenue and the presence of EVs in charging stations. Problems such as overstaying in charging stations, or unexpected behaviors of the battery in charging, can effectively disrupt and interfere with scheduling solutions. Therefore, more robust and flexible system should be researched.

5.5.2 New Applications in Smart Parking System

In this section, various emerging techniques for the SPS are highlighted and discussed for more comprehensive study. These techniques can be categorized into VANET- and UAV-based techniques.

5.5.2.1 Vehicle to Everything

With the advancement of technology and wireless communication improvements over the years, there has been a new trend in communication technology between vehicles and infrastructures known as VANET, which is a subgroup of mobile ad hoc network (MANET) that uses vehicles as mobile nodes [113]. They are classified into three different types: V2V, V2I, and lastly vehicle to roadside (V2R). The MAC layer standard of VANET is IEEE 802.11p standard with a connectivity range of 100–1,000 m and 27.0 Mbps data rate that promises a fast and reliable method of communication; however, they are more expensive than other protocols [114]. OBUs installed in vehicles by manufactures and roadside units (RSUs) installed near roads providing both road safety and a mode of communication between the vehicles and the roadside infrastructure are the requirements in this type of systems [4]. VANET is also used in SPSs where vehicles can detect parking occupancy and report the location to other vehicles or to RSUs in a two-way communication [13]. As the authors in Ref. [115] presented their smart parking solution, DIG-Park, with the combination of V2I and RSU along with distance geometry algorithms to find the nearest parking spots that could be used both indoors and outdoors in combination. They provide real-time parking navigation as well as antitheft and anticollision protection that make use of OBU and RSU. A similar work was carried out by [4]; however, their model represented a large parking lot, and the complexity of the model was simple compared to the previous work. SDN was also overviewed and examined in Ref. [116] to provide a more flexible, programmable network in the VANET paradigm that can overcome some of the difficulties and complexities in the existing vehicular environment. VANET possibilities are enormous. However, some problems and difficulties arise when using this system such as the problems in frequency and bandwidth spectrum [117]. Because VANET is operated in the band 75 MHz at 5.9 GHz where in some countries this band is used for military and radar systems. Another issue lies within the routing protocol where multi-hop communication between vehicles makes the current routing protocols incompatible with VANET such as the current protocol in MANET. Therefore, the current research is mostly focused on creating a new VANET-based routing protocol.

5.5.2.2 Unmanned Aerial Vehicles

Unmanned aerial vehicles (UAVs), also known as drones, are also becoming a trend in today's IoT applications. They are mainly used in wireless communication and serve as a backbone for network connectivity in places where communication range is limited, or they fail to work in case of a disaster. Equipping drones with the vehicle detection sensors as discussed previously not only can provide the benefit of having a one-package SPS with all the advantage of regular SPS, but it can also overcome the shortcomings of the sensors in certain areas. Larger detection range, ability to carry multiple sensors, high precision, flexibility in deployment, high mobility, and many others are the few examples showing ascendency of drones compared to static vehicle detection sensors. A UAV-assisted smart parking example was studied [118]. In this paper, the authors utilized generative adversarial network (GAN) for detection and prediction of parking vacancy in a vision-based approach. High precision of their UAVs shows the superiority of their detection algorithm and obstacle avoidance technique. Another similar solution was proposed in Ref. [119]. The authors in this study used the markers printed on the asphalt as their parameter to check for parking vacancy from their single drone application. Low light, harsh weather conditions, and collision avoidance of team of drones are some of the problems that need be examined while implementing a drone-based smart parking solution. The coextensive UAV-assisted VANET in the future of smart parking in the IoT era could be one of the approaches that offer a synergistic combination of the two described technologies.

5.5.3 Hybrid System

A hybrid model is a system in which more than one sensor is used for detecting vehicles, or in case of communications, both wired and wireless modes are used in conjunction. Integrating multiple data source into one provides a more reliable and faster way for detecting vehicles where in case of failure of one sensor node, the integrity of the system would not collapse. Streetline application [120] in hybrid smart parking is the perfect example of such a system. Authors incorporated magnetometer, light sensors, CCTV, and mobile phone application to detect and report the vacancy of parking spots for a large parking lot. They also offer dynamic pricing strategy for their payment systems. In another implementation of hybrid smart parking in Ref. [121], the authors presented the micro aerial vehicle (MAV) indoor application for SPS which is equipped with two ultrasonic sensors and two cameras for detection of parking spots. Near Field Communication (NFC) tag and Quick Response (QR) codes are also used for locating the MAV, for recording and updating the latest status of the parking spot, and as a way for drivers to record their parking information. In the case of hybrid wired and wireless communications, the authors in Ref. [16] presented a hybrid system that uses sensors and RFID tags for billing purposes, both of which are provided with wired and wireless capabilities, 802.15.4/ZigBee.

Table 5.6 Comparison of Hybrid and Single-Sensor SPSs

	Parking System Model	
Characteristics	*Hybrid [10,89,67,69,70]*	*Single Sensor [3,21,54–59,87–91]*
Large-scale application	+	Limited
Flexibility	+	Limited
Cost-effectiveness	+ (large scale)	Sensor dependent
General reliability	+	Low
Service providers	+	+
Overall recommendation	+ (medium–large scale)	+ (small scale)

In general, as summarized in Table 5.6, a hybrid system provides a flexible solution to smart parking applications where the functionality of the system may change in the future. Furthermore, the scalability of such systems for large application such as urban parking is high, and the deficiencies of a single-sensor approach in this scenario can be overcome. Hybrid systems can be the future of smart parking solutions, and the worldwide usage of these systems can be anticipated in the future prospect of IoT. A general concept of a hybrid smart parking application is shown in Figure 5.3. In this model, a series of parking sensors are installed throughout the parking lot. We also recommend the usage of UAVs to further enhance the detection and navigation. All systems are connected and exchange information using open standards with 5G or LPWAN communication modules. The gates are controlled by CCTV which can read the license plate of the vehicles. This can be used to estimate the total parking time and to automatically determine the fee using algorithms. The fee is also dynamically adjusted based on the traffic and the time of day the parking is utilized. Users have options to pay the fee in various ways. Many VMSs are also installed to indicate the useful information to the drivers. By introducing a cloud-based SPS, we ensure a seamless integration with many services such that users can reserve and find the best route to reach the parking lot with a guaranteed parking space at any time of the day. By incorporating open APIs and open standards with the power of open-source software and services, we can introduce a fast and safe SPS that can integrate with other elements in the smart city ecosystem.

Although there are challenges in these systems, such as the requirement to handle multiple data sources entailing complex algorithms and the requirement of having a safe and fast method of communication between multiple sensors to decrease the latency of data, but at the end, the outcome of these hybrid systems is both beneficial to the user and the parking lot owners. Because these systems are the most effective in providing useful services such as integrated smart parking reservation, smart payments, automatic gate control, and dynamic pricing strategies based on

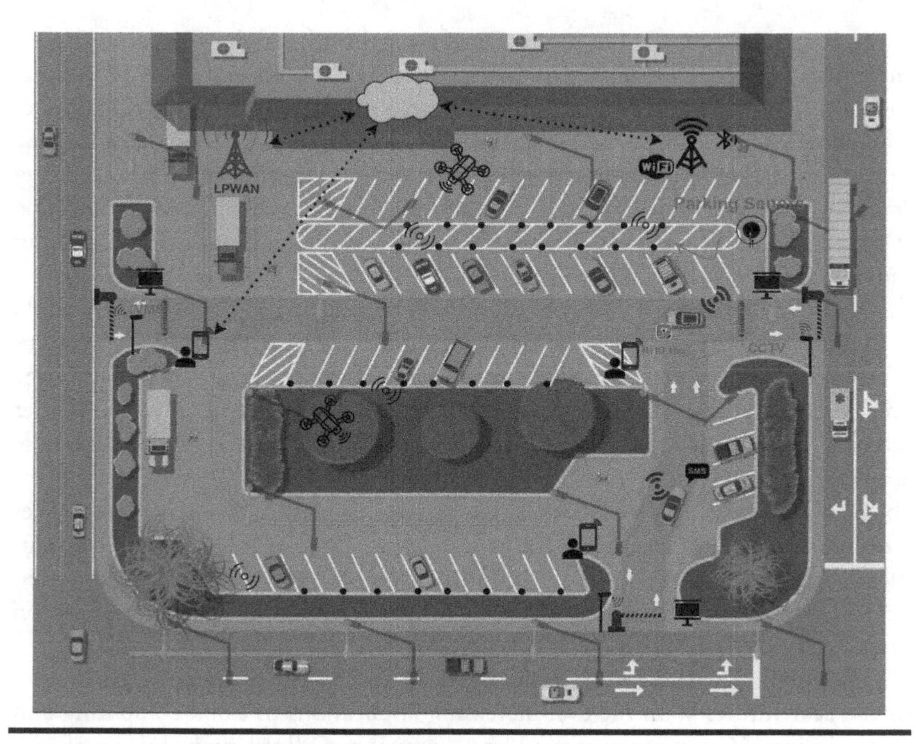

Figure 5.3 A model for a hybrid SPS.

the availability of the parking spot and other factors that maximize utilization of the parking lots during the peak and off-peak hours.

The cost-effectiveness of the hybrid approach is generally substantiated by the type and the number of sensors used. Single-sensor smart parking solutions as described earlier in this chapter may vary in terms of the accuracy of the collected data. Moreover, in these applications, in case of failure of the sensor, the whole application may fail, which results in discomfort and economical cost. This is where hybrid solutions can come into the play. Although the initial capital cost may deter the use of this approach, in the long term, the benefits of using such a system can easily be realized.

5.6 Open Research Issues

In this chapter, a complete survey of existing smart parking solutions was discussed where several types of sensors in terms of their functionalities and their strengths and weaknesses were introduced. As discussed in the chapter, we cannot find a single best sensor that is applicable for all the smart parking solutions. Some may fail in extreme weather conditions, or they are required to be placed inside pavements which have their own complications. Others may have limited use in certain applications due to privacy and security issues. Therefore, the efforts to diminish these shortcomings in the information collection are now put into mobile sensing devices where some of the limits or barriers of fixed sensors can be addressed; however, due to irregularities in spatiotemporal coverage and the issue of big data that are being transmitted, aggregation of such data and predicting the availability of the parking spots are the challenges that if tackled correctly, this could be the new future of mobile sensing devices.

Moreover, having a reliable, fast, and secure mode of communication in SPS is another problem. As discussed in this chapter, the new era of LPWAN communication protocol seems to be the future of commutations as they provide large area coverage, low power consumption, and high battery lifetime, as well as higher security measures compared to the legacy communication modules. Large-scale applications of LPWAN are still under study as there is still the issue of interoperability and coexisting with other WSN. A large amount of data and big packet sizes that are being transmitted require a steadfast, end-to-end encrypted communication. It was estimated that by 2023, there would be 20–40 billion connected devices, wherein this scenario could be a bottleneck in the existing communication infrastructures [122].

The 5G mobile network is expected to be the center of the emerging IoT devices in the near future. With the ever-increasing applications in cloud computing and smart devices, 5G can promise to address, some if not, all the current issues of telecommunications. Studies are now under way to integrate the existing devices with 5G wireless communication.

Connected vehicles are also another interesting option that is being used in several applications of smart parking solutions. Mobile sensors connected to the vehicles and smart mobile applications that could identify the parking spots are now introduced by car manufactures. The idea of connecting everything including our cars to each other may sound overwhelming as it can provide attracting services such as real-time navigation and crowdsourced information. However, the current technologies in connected vehicles may limit the full potential of these types of applications particularly in urban areas [123]. Varying speed, the need for a better routing protocol, specific bandwidth for communication, and the need for high-speed communication technology to send and receive information and in general the quality of service (QoS) including delays are the challenges in the current era of connected vehicles. The authors in Ref. [105] proposed a mathematical model for a new generation of forwarding QoS approaches that enables the allocation of the optimal path that satisfies the QoS parameters while supporting a wide range of communication-intensive IoT applications. The model is used to investigate the effects of multi-hop communication on a traffic system model designed with a Markov discrete-time M/M/1 queuing model, applicable to green deployment of duty-cycle sensor nodes. The authors in this chapter have presented analytical formulation for the bit error rate, and a critical path-loss model is defined to the specified level of trust among the most frequently used nodes. Additionally, the authors have addressed the degree of irregularity parameter for promoting adaptation to geographic switching with respect to two categories of transmission in distributed systems: hop-by-hop and end-to-end retransmission schemes.

Nowadays, most of the deployed SPSs rely on battery-powered sensors and wireless communication modules. Although there are possibilities of implementing energy-aware algorithms or routing protocols that could effectively reduce the overall energy consumption as discussed in Ref. [124]; however, at some point due to a size reduction of circuit boards and an increase in big data, we need to look at the potentials of energy-harvesting modules. These modules use power-generating elements such as solar cells, piezoelectric elements, and thermoelectric elements to generate electricity by converting different energy sources such as light, vibration, and heat, respectively. ParkHere [125] is a self-powered parking sensor that uses the weight of vehicles to power a micro generator which at the same time can send its information of the occupancy of the parking spot to the server via mobile radio.

As more and more things connect to each other, it is imperative to the IoT ecosystem for all systems in general to be able to communicate and exchange information to one another. Developing a multifaceted technology to improve the existing infrastructures while providing maximum security and privacy measures with a total and fully controlled management mechanism and with interoperability functionalities is vital for the future advancement of the ecosystem. Nowadays, more time and efforts are spent and devoted in artificial intelligence (AI), ML, and interoperability to provide solutions to the aforementioned obstacles. Many aspects of the ecosystem need to work coherently to overcome such difficulties. Such solutions would not only provide a better QoS to the users of the systems, but they also make sure the efficiency and the integrity of all the elements that represent the system.

5.7 Concluding Remarks

As the population of the urban area increases leading to traffic congestions and other problems, the need for parking spots is inevitable. In the age of IoT and smart city ecosystems, it is not difficult to believe why a smart and innovative solution is considered to be the paved way towards a more sustainable future for cities. Hence, to improve the current parking system and to address significant issues in overcrowded cities, the SPS was overviewed. Many parameters must be well studied and analyzed before implementing any SPS. Therefore, a comprehensive survey of the current state of SPSs including classification of the parking system, major vehicle detection technologies, and communication module was presented. The objective of this survey was to offer an insight into new researchers who seek to work in the ITS. We look at different elements of SPS and explain thoroughly the hardware and software aspects of this application. Software aspects of smart parking were presented, and several features such as parking prediction, path optimization, and parking assignment, and how the collected information can enhance the experience of parking operators as well as drivers were introduced. Different tables were generated which compared several key factors of each of the elements in SPS, sensors, and communication modules. Moreover, an overview of data security and privacy and the new trend in interoperability and data exchange were discussed. Next, emerging technologies in SPSs, namely, the V2X and UAVs, have been presented. Moreover, the concept of cloud-based hybrid models was suggested in order to solve the current problems in smart parking applications. As a future work, we plan to investigate our cloud-based hybrid concept with interoperability capabilities with the help of drones and provide a proof of concept of this new generation of the SPSs.

Notes

1 www.libelium.com/products/plug-sense/.
2 www.huawei.com/minisite/iot/en/smart-parking.html.
3 www.zte.com.cn/global/.
4 www.nwave.io/parking-technology/.
5 www.telensa.com/smart-parking/.
6 www.opengroup.org/iot/omi/.
7 www.opengroup.org/iot/odf/.

References

1. G. Cookson, Parking pain—INRIX offers a silver bullet, *INRIX—INRIX*. [Online]. Available: http://inrix.com/blog/2017/07/parkingsurvey/. [Accessed: 21-Nov-2017].
2. M. Gohar, M. Muzammal, and A. Ur Rahman, SMART TSS: Defining transportation system behavior using big data analytics in smart cities, *Sustainable Cities and Society*, vol. 41, 114–119, 2018. [Online]. Available: https://linkinghub.elsevier.com/retrieve/pii/S2210670717309757. [Accessed: 24-Feb-2019].
3. S. E. Bibri, The IoT for smart sustainable cities of the future: An analytical framework for sensor-based big data applications for environmental sustainability, *Sustainable Cities and Society*, vol. 38, 230–253, 2018. [Online]. Available: https://linkinghub.elsevier.com/retrieve/pii/S2210670717313677. [Accessed: 25-Feb-2019]
4. R. Lu, X. Lin, H. Zhu, and X. Shen, SPARK: A new VANET-based smart parking scheme for large parking lots, in *IEEE INFOCOM 2009, The 28th Conference on Computer Communications*, Rio De Janeiro, Brazil, 2009, pp. 1413–1421.
5. D. Uckelmann, M. Harrison, and F. Michahelles, An architectural approach towards the future Internet of Things, In *Architecting the Internet of Things*, D. Uckelmann, M. Harrison, and F. Michahelles, Eds. Berlin, Heidelberg: Springer, 2011, pp. 1–24. [Online]. Available: DOI: 10.1007/978-3-642-19157-2_1.
6. F. Al-Turjman, 5G-enabled devices and smart-spaces in social-IoT: An overview, *Future Genereration Computer System*, vol. 92, 732–744, 2017.

7. M. Chandrahasan, A. Mahadik, T. Lotlikar, M. Oke, and A. Yeole, Survey on different smart parking techniques, *International Journal of Computer Applications*, vol. 137, no. 13, 17–21, 2016. [Online]. Available: www.ijcaonline.org/research/volume137/number13/chandrahasan-2016-ijca-908920.pdf. [Accessed: 22-Feb-2019].
8. G. Revathi and V. R. S. Dhulipala, Smart parking systems and sensors: A survey, in *2012 International Conference on Computing, Communication and Applications*, Dindigul, Tamilnadu, India, 2012, pp. 1–5.
9. M. Y. I. Idris, Y. Y. Leng, E. M. Tamil, N. M. Noor, and Z. Razak, Car park system: A review of smart parking system and its technology, *Information Technology Journal*, vol. 8, no. 2, 101–113, 2009. [Online]. Available: www.scialert.net/abstract/?doi=itj.2009.101.113. [Accessed: 07-Nov-2017].
10. A. Hilmani, A. Maizate, and L. Hassouni, Designing and managing a smart parking system using wireless sensor networks, *Journal of Sensor and Actuator Network*, vol. 7, no. 2, 24, 2018. [Online]. Available: www.mdpi.com/2224-2708/7/2/24. [Accessed: 21-Feb-2019].
11. M. Faheem, S. A. Mahmud, G. M. Khan, M. Rahman, and H. Zafar, A survey of intelligent car parking system, *Journal of Applied Research and Technology*, vol. 11, no. 5, 714–726, 2013. [Online]. Available: www.sciencedirect.com/science/article/pii/S1665642313715803. [Accessed: 10-Dec-2017].
12. K. Hassoune, W. Dachry, F. Moutaouakkil, and H. Medromi, Smart parking systems: A survey, in *2016 11th International Conference on Intelligent Systems: Theories and Applications (SITA)*, Mohammedia, Morocco, 2016, pp. 1–6. [Online]. Available: http://ieeexplore.ieee.org/document/7772297/. [Accessed: 22-Feb-2019].
13. T. Lin, H. Rivano, and F. Le Mouel, A survey of smart parking solutions, *IEEE Transactions on Intelligent Transportation Systems*, vol. 18, no. 12, 3229–3253, 2017. [Online]. Available: http://ieeexplore.ieee.org/document/7895130/. [Accessed: 22-Feb-2019].
14. E. Polycarpou, L. Lambrinos, and E. Protopapadakis, Smart parking solutions for urban areas, in *2013 IEEE 14th International Symposium on "A World of Wireless, Mobile and Multimedia Networks" (WoWMoM)*, Madrid, 2013, pp. 1–6.
15. J. Chinrungrueng, U. Sunantachaikul, and S. Triamlumlerd, Smart parking: An application of optical wireless sensor network, in *2007 International Symposium on Applications and the Internet Workshops*, Hiroshima, 2007, pp. 66–66.
16. A. Bagula, L. Castelli, and M. Zennaro, On the design of smart parking networks in the smart cities: An optimal sensor placement model, *Sensors*, vol. 15, no. 7, 15443–15467, 2015. [Online]. Available: www.mdpi.com/1424-8220/15/7/15443. [Accessed: 28-Feb-2018].
17. Z. Hui-ling, X. Jian-min, T. Yu, H. Yu-cong, and S. Ji-feng, The research of parking guidance and information system based on dedicated short range communication, in *Proceedings of the 2003 IEEE International Conference on Intelligent Transportation Systems*, Shanghai, China, 2003, vol. 2, pp. 1183–1186.
18. Y. Qian and G. Hongyan, Study on parking guidance and information system based on intelligent mobile phone terminal, in *2015 8th International Conference on Intelligent Computation Technology and Automation (ICICTA)*, Nanchang, China, 2015, pp. 871–874.
19. M. Buntić, E. Ivanjko, and H. Gold, ITS supported parking lot management, in *International Conference on Traffic and Transport Engineering-Belgrade*, Belgrade, Serbia, 2012.

20. Y.-C. Shiue, J. Lin, and S.-C. Chen, A study of geographic information system combining with GPS and 3G for parking guidance and information system, *City*, vol. 65, no. 6, 9, 2010.
21. M. Chen and T. Chang, A parking guidance and information system based on wireless sensor network, In *2011 IEEE International Conference on Information and Automation*, Shenzhen, China, 2011, pp. 601–605.
22. M. Patil and V. N. Bhonge, Parking guidance and information system using RFID and Zigbee, *International Journal of Engineering Research and Technology*, vol. 2, no. 4, 2013, pp. 2490–2493.
23. E. Kokolaki, M. Karaliopoulos, and I. Stavrakakis, Opportunistically assisted parking service discovery: Now it helps, now it does not, *Pervasive and Mobile Computing*, vol. 8, no. 2, 210–227, 2012. [Online]. Available: www.sciencedirect.com/science/article/pii/S1574119211000782. [Accessed: 10-Dec-2017].
24. M. Ş. Kuran, A. C. Viana, L. Iannone, D. Kofman, G. Mermoud, and J. P. Vasseur, A smart parking lot management system for scheduling the recharging of electric vehicles, *IEEE Transactions on Smart Grid*, vol. 6, no. 6, 2942–2953, 2015. [Online]. Available: http://ieeexplore.ieee.org/document/7056538/. [Accessed: 21-Feb-2019].
25. K. Raichura and N. Padhariya, edPAS: Event-based dynamic parking allocation system in vehicular networks, in *2014 IEEE 15th International Conference on Mobile Data Management*, Brisbane, Australia, 2014, vol. 2, pp. 79–84.
26. D. Bong and K. C. Lai, Integrated approach in the design of car park occupancy information system (COINS), *IAENG International Journal of Computer Science*, vol. 35, no. 1, pp. 7–14, 2008.
27. M. Buntić, E. Ivanjko, and H. Gold, Its supported parking lot management, in *Presented at the International Conference on Traffic and Transport Engineering*, Belgrade, 2012.
28. K. Dieussaert, K. Aerts, S. Thérèse, S. Maerivoet, and K. Spitaels, SUSTAPARK: An agent-based model for simulating parking search, *URISA Journal*, vol. 24, pp. 63–76, 2012.
29. I. Benenson, K. Martens, and S. Birfir, PARKAGENT: An agent-based model of parking in the city, *Computers, Environment and Urban Systems*, vol. 32, no. 6, 431–439, 2008. [Online]. Available: www.sciencedirect.com/science/article/pii/S0198971508000689. [Accessed: 10-Dec-2017].
30. S.-Y. Chou, S.-W. Lin, and C.-C. Li, Dynamic parking negotiation and guidance using an agent-based platform, *Expert Systems With Applications*, vol. 35, no. 3, 805–817, 2008. [Online]. Available: www.sciencedirect.com/science/article/pii/S095741740700293X. [Accessed: 10-Dec-2017].
31. C. J. Rodier and S. A. Shaheen, Transit-based smart parking: An evaluation of the San Francisco Bay area field test, *Transportation Research Part C: Emerging Technologies*, vol. 18, no. 2, 225–233, 2010. [Online]. Available: www.sciencedirect.com/science/article/pii/S0968090X09001120. [Accessed: 09-Dec-2017].
32. S. Pal and V. Singh, GIS based transit information system for metropolitan cities in India, in *The Proceedings of Geospatial World Forum*, Hyderabad, India, 2011, pp. 18–21.
33. Z.-R. Peng, A methodology for design of a GIS-based automatic transit traveler information system, *Computers, Environment and Urban Systems*, vol. 21, no. 5, 359–372, 1997. [Online]. Available: www.sciencedirect.com/science/article/pii/S0198971598000064. [Accessed: 09-Dec-2017].

34. E. Kokolaki, G. Kollias, M. Papadaki, M. Karaliopoulos, and I. Stavrakakis, Opportunistically-assisted parking search: A story of free riders, selfish liars and bona fide mules, in *2013 10th Annual Conference on Wireless On-demand Network Systems and Services (WONS)*, Banff, AB, Canada, 2013, pp. 17–24.
35. D. Thierry, I. Sergio, L. Sylvain, and C. Nicolas, Sharing with caution: Managing parking spaces in vehicular networks, *Mobile Information System*, no. 1, 69–98, 2013. [Online]. Available: www.medra.org/servlet/aliasResolver?alias=iospress&genre=article&issn=1574-017X&volume=9&issue=1&spage=69&doi=10.3233/MIS-2012-0149. [Accessed: 10-Dec-2017].
36. M. Idris, E. M. Tamil, Z. Razak, N. M. Noor, and L. W. Kin, Smart parking system using image processing techniques in wireless sensor network environment, *Information Technology Journal*, vol. 8, no. 2, pp. 114–127, 2009.
37. N. V. Juliadotter, Hacking smart parking meters, in *2016 International Conference on Internet of Things and Applications (IOTA)*, Pune, India, 2016, pp. 191–196.
38. K. Mouskos, M. Boile, and N. A. Parker, Technical solutions to overcrowded park and ride facilities, New Jersey Department of Transportation, FHWA-NJ-2007–011, 2007.
39. N. H. H. M. Hanif, M. H. Badiozaman, and H. Daud, Smart parking reservation system using short message services (SMS), in *2010 International Conference on Intelligent and Advanced Systems*, Kuala Lumpur, Malaysia, 2010, pp. 1–5.
40. T. Yan, B. Hoh, D. Ganesan, K. Tracton, T. Iwuchukwu, and J.-S. Lee, CrowdPark: A crowdsourcing-based parking reservation system for mobile phones, 2012.
41. S. Noor, R. Hasan, and A. Arora, ParkBid: An incentive based crowdsourced bidding service for parking reservation, in *2017 IEEE International Conference on Services Computing (SCC)*, Honolulu, HI, USA, 2017, pp. 60–67.
42. P. Sauras-Perez, A. Gil, and J. Taiber, ParkinGain: Toward a smart parking application with value-added services integration, in *2014 International Conference on Connected Vehicles and Expo (ICCVE)*, Vienna, Austria, 2014, pp. 144–148.
43. Federal Highway Administration, Traffic control systems handbook: Chapter 6 Detectors—FHWA office of operations. [Online]. Available: https://ops.fhwa.dot.gov/publications/fhwahop06006/chapter_6.htm#t62fnb. [Accessed: 04-Jan-2018].
44. F. Al-Turjman, Mobile couriers' selection for the smart-grid in smart cities' pervasive sensing, *Elsevier Future Generation Computer Systems*, vol. 82, no. 1, 327–341, 2018.
45. A. Kianpisheh, N. Mustaffa, P. Limtrairut, and P. Keikhosrokiani, Smart parking system (SPS) architecture using ultrasonic detector, *International Journal of Software Engineering & Applications*, vol. 6, no. 3, pp. 55–58, 2012.
46. A. O. Kotb, Y. C. Shen, and Y. Huang, Smart parking guidance, monitoring and reservations: A review, *IEEE Intelligent Transportation Systems Magazine*, vol. 9, no. 2, 6–16, 2017.
47. P. T. Martin, Y. Feng, and X. Wang, Detector technology evaluation, Department of Civil and Environmental Engineering University of Utah Traffic Lab, MPC-03-154, Nov. 2003.
48. B. Song, H. Choi, and H. S. Lee, Surveillance tracking system using passive infrared motion sensors in wireless sensor network, in *2008 International Conference on Information Networking*, Busan, South Korea, 2008, pp. 1–5.
49. G. M. Someswar, R. B. Dayananda, S. Anupama, J. Priyadarshini, and A. A. Shariff, Design & development of an autonomic integrated car parking system, *Compusoft*, vol. 6, no. 3, pp. 2309–2312, 2017.

50. M. Bachani, U. M. Qureshi, and F. K. Shaikh, Performance analysis of proximity and light sensors for smart parking, *Procedia Computer Science*, vol. 83, 385–392, 2016. [Online]. Available: www.sciencedirect.com/science/article/pii/S1877050916302332. [Accessed: 28-Feb-2018].
51. M. Arab and T. Nadeem, MagnoPark—Locating on-street parking spaces using magnetometer-based pedestrians' smartphones, in *2017 14th Annual IEEE International Conference on Sensing, Communication, and Networking (SECON)*, San Diego, CA, USA, 2017, pp. 1–9.
52. Z. Pala and N. Inanc, Smart parking applications using RFID technology, in *2007 1st Annual RFID Eurasia*, Istanbul, Turkey, 2007, pp. 1–3.
53. E. Karbab, D. Djenouri, S. Boulkaboul, and A. Bagula, Car park management with networked wireless sensors and active RFID, in *2015 IEEE International Conference on Electro/Information Technology (EIT)*, Dekalb, IL, USA, 2015, pp. 373–378.
54. A. Khanna and R. Anand, IoT based smart parking system, in *2016 International Conference on Internet of Things and Applications (IOTA)*, Pune, India, 2016, pp. 266–270.
55. J. Rico, J. Sancho, B. Cendon, and M. Camus, Parking easier by using context information of a smart city: Enabling fast search and management of parking resources, in *2013 27th International Conference on Advanced Information Networking and Applications Workshops*, Barcelona, 2013, pp. 1380–1385. [Online]. Available: http://ieeexplore.ieee.org/document/6550588/. [Accessed: 21-Feb-2019].
56. H. Zhu, J. Liu, L. Peng, and H. Li, Real-Time parking guidance model based on stackelberg game, in *2017 IEEE International Conference on Information and Automation (ICIA)*, Macau SAR, China, 2017, pp. 888–893.
57. W. Liang, Y. Zhang, J. Hu, and X. Wang, A personalized route guidance approach for urban travelling and parking to a shopping mall, in *2017 4th International Conference on Transportation Information and Safety (ICTIS)*, Banff, AB, Canada, 2017, pp. 319–324.
58. K. Hantrakul, S. Sitti, and N. Tantitharanukul, Parking lot guidance software based on MQTT Protocol, in *2017 International Conference on Digital Arts, Media and Technology (ICDAMT)*, Chiang Mai, Thailand, 2017, pp. 75–78.
59. X. Zhang, L. Yu, Y. Wang, G. Xue, and Y. Xu, Intelligent travel and parking guidance system based on Internet of vehicle, in *2017 IEEE 2nd Advanced Information Technology, Electronic and Automation Control Conference (IAEAC)*, Chongqing, China, 2017, pp. 2626–2629.
60. L. Xie, J. Liu, C. Miao, and M. Liu, Study of method on parking guidance based on video, in *2016 IEEE 11th Conference on Industrial Electronics and Applications (ICIEA)*, Hefei, China, 2016, pp. 1394–1399.
61. I. Aydin, M. Karakose, and E. Karakose, A navigation and reservation based smart parking platform using genetic optimization for smart cities, in *2017 5th International Istanbul Smart Grid and Cities Congress and Fair (ICSG)*, Istanbul, Turkey, 2017, pp. 120–124.
62. A. Houissa, D. Barth, N. Faul, and T. Mautor, A learning algorithm to minimize the expectation time of finding a parking place in urban area, in *2017 IEEE Symposium on Computers and Communications (ISCC)*, Heraklion, Greece, 2017, pp. 29–34.
63. Y. Zheng, S. Rajasegarar, and C. Leckie, Parking availability prediction for sensor-enabled car parks in smart cities, in *2015 IEEE 10th International Conference on Intelligent Sensors, Sensor Networks and Information Processing (ISSNIP)*, 2015, pp. 1–6.

64. Y. Ji, D. Tang, P. Blythe, W. Guo, and W. Wang, Short-term forecasting of available parking space using wavelet neural network model, *IET Intelligent Transport Systems*, vol. 9, no. 2, 202–209, 2015.
65. F. Caicedo, C. Blazquez, and P. Miranda, Prediction of parking space availability in real time, *Expert Systems With Applications*, vol. 39, no. 8, 7281–7290, 2012. [Online]. Available: www.sciencedirect.com/science/article/pii/S0957417412001042. [Accessed: 20-Mar-2018].
66. J. S. Leu and Z. Y. Zhu, Regression-based parking space availability prediction for the Ubike system, *IET Intelligent Transport Systems*, vol. 9, no. 3, 323–332, 2015.
67. M. Maric, D. Gracanin, N. Zogovic, N. Ruskic, and B. Ivanovic, Parking search optimization in urban area, *International Journal of Simulation Modelling*, vol. 16, 195–206, 2017.
68. A. Moradkhany, P. Yi, I. Shatnawi, and K. Xu, Minimizing parking search time on urban university campuses through proactive class assignment, *Transportation Research Record Journal of the Transportation Research Board*, vol. 2537, 158–166, 2015. [Online]. Available: http://trrjournalonline.trb.org/doi/abs/10.3141/2537-17. [Accessed: 21-Mar-2018].
69. M. Rybarsch et al., Cooperative parking search: Reducing travel time by information exchange among searching vehicles, in *2017 IEEE 20th International Conference on Intelligent Transportation Systems (ITSC)*, Yokohama, 2017, pp. 1–6.
70. S. Banerjee and H. Al-Qaheri, An intelligent hybrid scheme for optimizing parking space: A Tabu metaphor and rough set based approach, *Egyptian Information Journal*, vol. 12, no. 1, 9–17, 2011. [Online]. Available: www.sciencedirect.com/science/article/pii/S1110866511000077. [Accessed: 21-Mar-2018].
71. Internet of Things outlook—Ericsson, *Ericsson.com*, 09-Nov-2017. [Online]. Available: www.ericsson.com/en/mobility-report/reports/november-2017/internet-of-things-outlook. [Accessed: 26-Dec-2017].
72. I. Chatzigiannakis, A. Vitaletti, and A. Pyrgelis, A privacy-preserving smart parking system using an IoT elliptic curve based security platform, *Computer Communications*, vol. 89–90, no. Supplement C, 165–177, 2016. [Online]. Available: www.sciencedirect.com/science/article/pii/S014036641630072X. [Accessed: 26-Dec-2017].
73. T. Braun, B. C. M. Fung, F. Iqbal, and B. Shah, Security and privacy challenges in smart cities, *Sustainable Cities and Society*, vol. 39, 499–507, 2018. [Online]. Available: https://linkinghub.elsevier.com/retrieve/pii/S2210670717310272. [Accessed: 24-Feb-2019].
74. J. Ni, K. Zhang, Y. Yu, X. Lin, and X. S. Shen, Privacy-preserving smart parking navigation supporting efficient driving guidance retrieval, *IEEE Transactions on Vehicular Technology*, vol. PP, no. 99, pp. 6504–6517,, 2018.
75. R. Lu, X. Lin, H. Zhu, and X. Shen, An intelligent secure and privacy-preserving parking scheme through vehicular communications, *IEEE Transactions on Vehicular Technology*, vol. 59, no. 6, 2772–2785, 2010.
76. H. Wang and W. He, A Reservation-based smart parking system, in *2011 IEEE Conference on Computer Communications Workshops (INFOCOM WKSHPS)*, Shanghai, China, 2011, pp. 690–695.
77. J. de C. Silva, J. J. P. C. Rodrigues, A. M. Alberti, P. Solic, and A. L. L. Aquino, LoRaWAN #x2014; A low power WAN protocol for Internet of Things: A review and opportunities, in *2017 2nd International Multidisciplinary Conference on Computer and Energy Science (SpliTech)*, Iasi, Romania, 2017, pp. 1–6.

78. M. Collotta, G. Pau, T. Talty, and O. K. Tonguz, Bluetooth 5: A concrete step forward towards the IoT," IEEE Commun. Mag., vol. 56, no. 7, pp. 125–131, Jul. 2018.
79. O. Wellnitz and L. Wolf, On latency in IEEE 802.11-based wireless ad-hoc networks, in *2010 IEEE 5th International Symposium on Wireless Pervasive Computing*, Modena, Italy, 2010, pp. 261–266.
80. R. S. Sinha, Y. Wei, and S.-H. Hwang, A survey on LPWA technology: LoRa and NB-IoT, *ICT Express*, vol. 3, no. 1, 14–21, 2017. [Online]. Available: www.sciencedirect.com/science/article/pii/S2405959517300061. [Accessed: 19-Dec-2017].
81. U. Raza, P. Kulkarni, and M. Sooriyabandara, Low power wide area networks: An overview, *IEEE Communications Surveys & Tutorials*, vol. 19, no. 2, 855–873, 2017.
82. S. Al-Sarawi, M. Anbar, K. Alieyan, and M. Alzubaidi, Internet of Things (IoT) communication protocols: Review, in *2017 8th International Conference on Information Technology (ICIT)*, Amman, Jordan, 2017, pp. 685–690.
83. A. Asaduzzaman, K. K. Chidella, and M. F. Mridha, A time and energy efficient parking system using Zigbee communication protocol, in *SoutheastCon 2015*, Fort Lauderdale, FL, USA, 2015, pp. 1–5.
84. M. Lauridsen, H. Nguyen, B. Vejlgaard, I. Z. Kovacs, P. Mogensen, and M. Sorensen, Coverage comparison of GPRS, NB-IoT, LoRa, and SigFox in a 7800 km #x000B2; Area, in *2017 IEEE 85th Vehicular Technology Conference (VTC Spring)*, Sydney, NSW, 2017, pp. 1–5.
85. J. Shi, L. Jin, J. Li, and Z. Fang, A smart parking system based on NB-IoT and third-party payment platform, in *2017 17th International Symposium on Communications and Information Technologies (ISCIT)*, Cairns, Australia, 2017, pp. 1–5.
86. A. Lavric and V. Popa, Internet of Things and LoRa #x2122; Low-power wide-area networks: A survey, in *2017 International Symposium on Signals, Circuits and Systems (ISSCS)*, Iasi, Romania, 2017, pp. 1–5.
87. M. Dalgleish, *Highway Traffic Monitoring and Data Quality*, 1 edition. Boston: Artech House Publishers, 2008.
88. N. Piovesan, L. Turi, E. Toigo, B. Martinez, and M. Rossi, Data analytics for smart parking applications, *Sensors*, vol. 16, no. 10, 2016.
89. A. Araújo, R. Kalebe, G. Girаõ, I. Filho, K. Gonçalves, and B. Neto, Reliability analysis of an IoT-based smart parking application for smart cities, in *2017 IEEE International Conference on Big Data (Big Data)*, Boston, MA, 2017, pp. 4086–4091.
90. S. Gupte and M. Younis, Participatory-sensing-enabled efficient Parking Management in modern cities, in *2015 IEEE 40th Conference on Local Computer Networks (LCN)*, Clearwater Beach, FL, 2015, pp. 241–244.
91. J. Villalobos, B. Kifle, D. Riley, and J. U. Quevedo-Torrero, Crowdsourcing automobile parking availability sensing using mobile phones, in *UWM Undergraduate Research Symposium*, 2015, pp. 1–7.
92. A. Broring et al., Enabling IoT ecosystems through platform interoperability, *IEEE Software*, vol. 34, no. 1, 54–61, 2017. [Online]. Available: http://ieeexplore.ieee.org/document/7819420/. [Accessed: 21-Feb-2019].
93. S. Kubler, J. Robert, A. Hefnawy, K. Framling, C. Cherifi, and A. Bouras, Open IoT ecosystem for sporting event management, *IEEE Access*, vol. 5, 7064–7079, 2017. [Online]. Available: http://ieeexplore.ieee.org/document/7898832/. [Accessed: 21-Feb-2019].
94. A. Karpenko et al., Data exchange interoperability in IoT ecosystem for smart parking and EV charging, *Sensors*, vol. 18, no. 12, 4404, 2018. [Online]. Available: www.mdpi.com/1424-8220/18/12/4404. [Accessed: 21-Feb-2019].

95. M. Elkhodr, S. Shahrestani, and H. Cheung, The Internet of Things: New interoperability, management and security challenges, *International Journal of Network Security & Its Applications*, vol. 8, no. 2, 85–102, 2016. [Online]. Available: www.aircconline.com/ijnsa/V8N2/8216ijnsa06.pdf. [Accessed: 22-Feb-2019].
96. M. Owayjan, B. Sleem, E. Saad, and A. Maroun, Parking management system using mobile application, in *2017 Sensors Networks Smart and Emerging Technologies (SENSET)*, Beirut, 2017, pp. 1–4.
97. S. Ravishankar and N. Theetharappan, Cloud connected smart car park, in *2017 International Conference on I-SMAC (IoT in Social, Mobile, Analytics and Cloud) (I-SMAC)*, Palladam, Tamilnadu, India 2017, pp. 71–74.
98. N. Larisis, L. Perlepes, G. Stamoulis, and P. Kikiras, Intelligent parking management system based on wireless sensor network technology, *Sensors & Transducers*, vol. 18, 100–112, 2013.
99. D. Kanteti, D. V. S. Srikar, and T. K. Ramesh, Smart parking system for commercial stretch in cities, in *2017 International Conference on Communication and Signal Processing (ICCSP)*, Chennai, 2017, pp. 1285–1289.
100. M. Y. I. Idris, E. M. Tamil, N. M. Noor, Z. Razak, and K. W. Fong, Parking guidance system utilizing wireless sensor network and ultrasonic sensor, *Information Technology Journal*, vol. 8, no. 2, 138–146, 2009. [Online]. Available: www.scialert.net/abstract/?doi=itj.2009.138.146. [Accessed: 24-Dec-2017].
101. S. Mathur et al., ParkNet: Drive-by sensing of road-side parking statistics, in *Proceedings of the 8th International Conference on Mobile Systems, Applications, and Services*, New York, USA, San Francisco, California, USA, 2010, pp. 123–136.
102. S. Y. Cheung, S. C. Ergen, and P. Varaiya, Traffic surveillance with wireless magnetic sensors, in *Proceedings of the 12th ITS world congress*, San Francisco, California, USA, 2005, vol. 1917, p. 173181.
103. S. Yoo et al., PGS: Parking Guidance System based on wireless sensor network, in *2008 3rd International Symposium on Wireless Pervasive Computing*, Santorini, Greece, 2008, pp. 218–222.
104. C. Trigona et al., Implementation and characterization of a smart parking system based on 3-axis magnetic sensors, in *2016 IEEE International Instrumentation and Measurement Technology Conference Proceedings*, Taipei, Taiwan, 2016, pp. 1–6.
105. J. Chinrungrueng, S. Dumnin, and R. Pongthornseri, iParking: A parking management framework, in *2011 11th International Conference on ITS Telecommunications*, St. Petersburg, Russia, 2011, pp. 63–68.
106. Z. Zhang, M. Tao, and H. Yuan, A parking occupancy detection algorithm based on AMR sensor, *IEEE Sensors Journal*, vol. 15, no. 2, 1261–1269, 2015. [Online]. Available: http://ieeexplore.ieee.org/document/6919252/. [Accessed: 21-Feb-2019].
107. S. Banerjee, P. Choudekar, and M. K. Muju, Real time car parking system using image processing, in *2011 3rd International Conference on Electronics Computer Technology*, Kanyakumari, India, 2011, vol. 2, pp. 99–103.
108. H.-C. Tan, J. Zhang, X.-C. Ye, H.-Z. Li, P. Zhu, and Q.-H. Zhao, Intelligent car-searching system for large park, in *2009 International Conference on Machine Learning and Cybernetics*, Baoding, China, 2009, vol. 6, pp. 3134–3138.
109. B. N. Silva, M. Khan, and K. Han, Towards sustainable smart cities: A review of trends, architectures, components, and open challenges in smart cities, *Sustainable Cities and Society*, vol. 38, 697–713, 2018. [Online]. Available: https://linkinghub.elsevier.com/retrieve/pii/S2210670717311125. [Accessed: 24-Feb-2019].

110. S. Kubler, J. Robert, A. Hefnawy, C. Cherifi, A. Bouras, and K. Främling, IoT-based smart parking system for sporting event management, in *Proceedings of the 13th International Conference on Mobile and Ubiquitous Systems: Computing, Networking and Services—MOBIQUITOUS 2016*, Hiroshima, Japan, 2016, pp. 104–114. [Online]. Available: http://dl.acm.org/citation.cfm?doid=2994374.2994390. [Accessed: 21-Feb-2019].
111. Z. Wei, Y. Li, Y. Zhang, and L. Cai, Intelligent parking garage ev charging scheduling considering battery charging characteristic, *IEEE Transactions on Industrial Electronics*, vol. 65, no. 3, 2806–2816, 2018. [Online]. Available: http://ieeexplore.ieee.org/document/8012480/. [Accessed: 21-Feb-2019].
112. S. M. Bagher Sadati, J. Moshtagh, M. Shafie-khah, A. Rastgou, and J. P. S. Catalão, Operational scheduling of a smart distribution system considering electric vehicles parking lot: A bi-level approach, *International Journal of Electrical Power & Energy Systems*, vol. 105, 159–178, 2019. [Online]. Available: https://linkinghub.elsevier.com/retrieve/pii/S0142061518304824. [Accessed: 21-Feb-2019].
113. H. Y. Chang, H. W. Lin, Z. H. Hong, and T. L. Lin, A novel algorithm for searching parking space in vehicle ad hoc networks, in *2014 10th International Conference on Intelligent Information Hiding and Multimedia Signal Processing*, Kitakyushu, Japan, 2014, pp. 686–689.
114. M. Santhiya, M. M. S. Karthick, and M. keerthika, Performance of various TCP in vehicular ad hoc network based on timer management, *International Journal of Advanced Research in Electrical, Electronics and Instrumentation Energy*, vol. 2, no. 12, 6160–6166, 2013. [Online]. Available: www.rroij.com/peer-reviewed/performance-of-various-tcp-invehicular-ad-hoc-networkbased-on-timer-management-42995.html. [Accessed: 26-Dec-2017].
115. W. Balzano and F. Vitale, DiG-Park: A smart parking availability searching method using V2V/V2I and DGP-class problem, in *2017 31st International Conference on Advanced Information Networking and Applications Workshops (WAINA)*, Taipei, Taiwan, 2017, pp. 698–703.
116. F. Al-Turjman, Fog-based caching in software-defined information-centric networks, *Elsevier Computers & Electrical Engineering Journal*, vol. 69, no. 1, 54–67, 2018.
117. C. Jeremiah and A. J. Nneka, Issues and possibilities in Vehicular Ad-Hoc Networks (VANETs), in *2015 International Conference on Computing, Control, Networking, Electronics and Embedded Systems Engineering (ICCNEEE)*, 2015, pp. 254–259.
118. X. Li, M. C. Chuah, and S. Bhattacharya, UAV assisted smart parking solution, In *2017 International Conference on Unmanned Aircraft Systems (ICUAS)*, Miami, FL, USA, 2017, pp. 1006–1013. [Online]. Available: http://ieeexplore.ieee.org/document/7991353/. [Accessed: 21-Feb-2019].
119. F. Al-Turjman, "Fog-based Caching in Software-Defined Information-Centric Networks", *Elsevier Computers & Electrical Engineering Journal*, vol. 69, no. 1, pp. 54–67, 2018.
120. Streetline, *Streetline*. [Online]. Available: www.streetline.com/. [Accessed: 26-Dec-2017].
121. C. H. Huang, H. S. Hsu, H. R. Wang, T. Y. Yang, and C. M. Huang, Design and management of an intelligent parking lot system by multiple camera platforms, in *2015 IEEE 12th International Conference on Networking, Sensing and Control*, Taipei, Taiwan, 2015, pp. 354–359.

122. F. Al-Turjman and S. Alturjman, 5G/IoT-enabled UAVs for multimedia delivery in industry-oriented applications, *Springer's Multimedia Tools and Applications Journal*, 2018. DOI: 10.1007/s11042-018-6288-7.
123. F. Al-Turjman, QoS–aware data delivery framework for safety-inspired multimedia in integrated vehicular-IoT, *Elsevier Computer Communications Journal*, vol. 121, 33–43, 2018.
124. P. Lee, H.-P. Tan, and M. Han, Demo: A solar-powered wireless parking guidance system for outdoor car parks, in *Proceedings of the 9th ACM Conference on Embedded Networked Sensor Systems*, New York, USA, 2011, pp. 423–424.
125. F. Al-Turjman and S. Alturjman, Context-sensitive access in industrial Internet of Things (IIoT) healthcare applications, *IEEE Transactions on Industrial Informatics*, vol. 14, no. 6, 2736–2744, 2018.

Chapter 6

Indecision Service Delivery in Cloud-Based IoT

Fadi Al-Turjman
Antalya Bilim University

Hadi Zahmatkesh
Middle East Technical University

Contents

6.1 Introduction

Cloud computing is a growing field where the Internet of Things (IoT) paradigm is influencing and changing the world. IoT allows people and things (e.g., sensors, actuators, and smart devices) to be connected anytime and anywhere, with anyone and anything. Most of these things are expected to communicate with each other, collect data, and provide services via the Cloud paradigm. Delivering data in a cost-effective and scalable manner is vital in order to support the continued growth of Cloud services [1]. According to Ref. [2], in a typical datacenter, 90% of bandwidth is consumed by only less than 1% flows with a size of around 100 MB or bigger each. Therefore, efficient data delivery is crucial in order to enhance the overall quality and performance of various Cloud services such as transport, traffic, and health care while reducing their overall cost of delivery and energy consumption. On the other hand, there would be huge cost for the Cloud providers if an unplanned data delivery is utilized without paying attention to the price and resource scarcity.

Furthermore, cost-effectiveness is one of the main characteristics defined in an IoT setting [3]. Therefore, an efficient IoT-specific pricing model would help Cloud providers to identify different factors affecting the level of service that aggregates data from various sources. This in turn causes enhancements in the reliability of services.

Nowadays, we are all connected with our computers, smartphones, and many other objects and devices that can send and receive data. This in turn raises a couple of questions such as how much does it require to send data over the network? and what happens if the pricing model of energy or the topological order of the nodes changes? Therefore, it is significant to investigate different approaches to optimize the pricing schemes in order to have cost-efficient data delivery in Cloud-based IoT infrastructures. However, since the IoT is growing fast, a problem that any service provider may face is the uncertainty problem. Uncertainty in Cloud-based IoT solutions arises mainly from the heterogeneity of the utilized devices and communication links towards the Cloud. This is an important issue that has an impact on computing efficiency and brings additional challenges to scheduling problems in the Cloud environment [4]. Therefore, uncertainty analysis should be an important part of design and service provisioning approaches in the IoT era.

In this chapter, we propose an optimized pricing framework covering a cost-effective IoT infrastructure to address the data delivery issue in the IoT era by considering uncertainty factors.

In this regard, two metaheuristic algorithms are employed to optimize the pricing model, dealing with uncertainty. The computational power needed to find the optimal pricing is already a computationally expensive task, while introducing the uncertainty problem makes this to a highly computational problem that would require more power to solve compared to a single computer. This introduces the IoT solution by itself where using distributed systems can actually help to solve the problem by taking all the uncertainty factors into consideration. This would allow companies not only to solve the uncertainty in their mathematical models but also to predict future pricing if used properly.

In order to assist the readers, a list of abbreviations along with their brief definitions used throughout the chapter is provided in Table 6.1. The rest of this chapter is organized as follows: Section 6.2 outlines the related works and

Table 6.1 List of Abbreviations

Abbreviation	*Definition*
3G	Third generation of mobile network
4G	Fourth generation of mobile network
CSB	Cloud services brokerage
GA	Genetic algorithm
HEED	Hybrid Energy-Efficient Distributed
IoT	Internet of Things
LTE	Long-Term Evolution
MANET	Mobile ad hoc network
RFID	Radio frequency identification
SAA	Simulated annealing algorithm
WCDMA	Wideband Code Division Multiple Access

background behind the pricing model and uncertainty factors which are introduced in Section 6.3. Moreover, the algorithms used in the study are presented in Section 6.4. Section 6.5 discusses the results and findings of the study. Finally, Section 6.6 concludes this chapter.

6.2 Background

Uncertainty features can be utilized to provide services in a way that the overall requirements of the communication systems are satisfied. Cloud-based IoT paradigms provide unique services that are possible only through smart devices such as smartphones, tablets, sensors, and tags connected via the Internet. These services can be classified into three categories: computing services, storage services, and communication services. In Cloud-based IoT systems, several configurations for the shared computing load among various devices/nodes can be considered, and the processing requirements may differ according to the actual work. Pricing the uncertainty in these configurations plays a key role in providing a real-time response and improves the reliability of the system.

Meanwhile, in IoT systems, a massive amount of data can be generated using billions of sensor devices in smart environments. These devices are not even capable of storing the generated data for 1 day. Moreover, it is not necessary to push all the data directly to the Cloud, if there is redundancy or irrelevance in data. Therefore, a few of these devices can volunteer to store partial data based on a pricing model temporarily. Together with computing services, storage services can filter, analyze, and compress data for efficient transmissions. On the other hand, the utilized wireless protocols control the communication services in the Cloud-based IoT systems. These protocols can adapt for narrow-band transmission, low-power operation, or longer range of coverage because of the constrained resources per smart device. An efficient pricing model can combine these protocols into a single globalized communication system utilized by the Cloud. This would help to manage subnetworks of smart devices such as sensors and actuators in addition to providing security and reliability in the IoT system.

Nowadays, these three categorizes of services are offered based on static/mobile devices in the IoT paradigm. However, the data delivery in these services varies as well based on the utilized device and technology. For example, 4G Long-Term Evolution (LTE) has lower costs since no radio network controller for 3G Wideband Code Division Multiple Access (WCDMA) and service continuity can be obtained by interfacing different systems such as WCDMA [5]. Moreover, costs are measured and distributed between different service layers in the production of Cloud services. For various communication technologies, costs of hardware, software, cooling, and Cloud utilization would be different, and therefore, the prices of services vary accordingly [6].

There have been a number of attempts for addressing data delivery problem by providing a pricing framework in the IoT-based Cloud environment. For example, Hybrid Energy-Efficient Distributed (HEED) clustering algorithm [7] tackles the computational power that is needed when it comes to distributed systems. The new simplified version of this algorithm reduces communication between the master node and the slave nodes [8]. In this study, the approach is based on scattering data into the computing nodes, since every node computes a completely independent uncertainty problem. The master node will only gather the data after it is completely processed a coupled analysis approach.

Addressing data delivery and exchange in the IoT-based Cloud by providing a pricing framework is considerably important nowadays [3]. It allows us to compute the required price for each node in the Cloud. Different incentives can take part in the pricing model that dictates the choice of a group of candidates for data processing. Recent results in incentive-based data exchange have been well studied. For example, in Ref. [9], the authors present a comprehensive investigation on routing and forwarding in MANETs by focusing on the gain per individual node.

The pricing model presented in this study is based on the laws of supply and demand [3]. Moreover, uncertainty is available in delivering data through the Cloud, but a proper research about the uncertainty in the Cloud has not been conducted yet. This allows us to address the problem of pricing model simultaneously with uncertainty factors in terms of the topological order, transmission and reception energy, nodal charge and nodal power as well as computation capacity in the pricing model.

Adaptive optimization algorithms have been applied in this study in order to overcome the pricing uncertainty in Cloud-based services. The genetic algorithm (GA) and simulated annealing algorithm (SAA) were used to find the suboptimal solutions using distributed and parallel systems.

6.3 System Model

In this section, we describe our main system models used in addressing the pricing uncertainty in the Cloud-based IoT infrastructure. The assumed pricing is as follows [3]:

$$p_i = \gamma_i * \left(\frac{E_{tx} + E_{rx}}{E_i} + \pi + U_i + S + \text{CSBs} \right) \tag{6.1}$$

where γ_i is the pricing factor for each node providing the same service in the IoT paradigm. This factor can be calculated based on the state of the current resources in the system. E_{tx} and E_{rx} are the transmission and reception energies, respectively. E_i, π, and U_i represent the nodal charge, nodal power, and computation capacity, respectively. Furthermore, S is the surety level and Cloud services brokerages (CSBs) represent the accuracy of the brokers in the pricing framework.

In this study, GA and SAA are used to solve the problem as we are trying to optimize the outcome of the mathematical model. Together with the uncertainty factors, it would be hard to actually compute on a single node since this would require too much time to compute if the uncertainty entries are significantly large.

Uncertainty factors are represented by the constants of the pricing model where γ_i, π, U_i, and E_i are assigned different values every time the algorithm works. This would allow us to simulate various conditions that might actually be present in the future based on a level of probability of uniform distribution. The constants are represented in an array of uncertainty numbers. For example, [0.6, 4, 55, 14] is an uncertainty array, where 0.6 represents the network interconnectivity. The numbers 4 and 55 would represent the amount of charge every node will need and the capability per node, respectively. The number 14 in the uncertainty array would represent the number of packets the node has to process and relay. Moreover, the size of the uncertainty array can vary ranging from 8 to 2,000 entries. The issue of uncertainty would be crucial when it comes to the whole networks of smart environments integrated with other complex IoT networks such as smart cities and smart energy systems. The aforementioned factors may influence the occurrence of the uncertainty in the Cloud-based IoT environments. Heterogeneity of devices and wide-scale use of wireless technology may cause uncertainty in transferring speeds and delay in delivery of data.

For example, uncertainty is one of the most important problems in most IoT systems based on the radio frequency identification (RFID) technology [10]. In this context, uncertainty can happen due to inconsistent data. Since RFID tags can be read using different readers simultaneously, it is possible to obtain inconsistent data about the precise location of the tags. Captured data may include redundant data with a significant amount of additional information, which may be another cause of uncertainty.

In this chapter, using a distributed computation approach, the uncertainty array is mapped to different nodes with the same algorithm that is ready to perform the computation. This would allow different nodes to take apart of the heavy process instead of one processor doing all the tasks. The array of a certain number of entries would be mapped to these different nodes, and this provides a speedup of 60% since both algorithms are distributed according to Amdahl's law [11].

6.4 Genetic and Annealing-Based Approaches

In this section, we propose two algorithms to solve the pricing framework in the Cloud-based IoT infrastructure. The main idea here is to optimize the abovementioned uncertain values in the pricing framework presented in Equation 6.1. First, GA is used to quantify the uncertainty in the pricing model.

Since it is a metaheuristic algorithm, its main purpose is to optimize the solution. The pseudocode for the GA is given in Algorithm 6.1.

Algorithm 6.1: Genetic Algorithm (GA)

1. Create initial population (uncertainty array)
2. Compute P_i in Eq. 6.1
3. **begin**
4. Select two individuals using tournament_selection
5. Apply crossing-over between the parents
6. Compute P_i in Eq. 6.1
7. Generate the new population
8. **Repeat** (if max_num_of_generations is not reached or price is decreasing)
9. **Return** best price in current generation

The centralized version of the algorithm would run on a single node, which is processing all the entries of the uncertainty array and running the algorithm on each entry. However, the distributed version of the GA utilizes the map function that will chunk the uncertainty array and will scatter the entries upon the number of nodes that are available.

SAAs are used to compare the performance of the GA in its centralized and distributed forms. The pseudocode for the SAA is given in Algorithm 6.2.

Algorithm 6.2: Simulated Annealing Algorithm (SAA)

1. Generate a random solution
2. Calculate its price using P_i in Eq. 6.1
3. Generate a random neighboring (close to) solution
4. Calculate the new solution's price P_{new}
5. Compare them:
   If $P_{new} < P_{old}$, choose the new solution
   If $P_{new} > P_{old}$, choose the new solution based on a predefined probability
6. Repeat steps 3–5 until an acceptable price is found.

We remark here that the centralized version of the SAA would have the same processing time as the centralized version of the GA. Meanwhile, the distributed version of the SAA uses all the nodes available to compute the uncertainty array by mapping each entry to a node. In Section 6.5, both algorithms are compared in both forms, centralized and distributed, in terms of various uncertainty factors in the pricing model.

6.5 Performance Evaluation

An i7-4770HQ processor and 16 GB of RAM are used to simulate the nodes of the distributed system. The number of entries used in the experiment creates the uncertainty array. Each entry represents a number of constants in the pricing model equation. The equation is considered as the objective function for the algorithms used in this experiment. The simulation code is written in Python and uses the multiprocessing library to map the entries of the uncertainty array to the nodes processing it.

6.5.1 *Simulation Metrics and Parameters*

The simulation tool randomly generates the uncertainty array based on the amount of entries required. The aforementioned algorithms are utilized to compute the optimal results for each entry. The metrics used in the simulation are as follows.

Network topology (γ_i): This is the topological order that would affect the pricing model. Companies try to reduce the cost and increase their profits because moving data from a source to a destination would have a cost for them. This would help them achieve that by representing the number of nodes the data package went through before reaching destination. This factor can be calculated based on the state of the current resources in the system.

Transmission energy (E_{tx}): It is the energy consumed per transmitted packet measured in joules. This factor represents the varying energy requirement per connected device to the Cloud.

Reception energy (E_{rx}): It is the energy consumed by the revived packet measured in joules.

Nodal charge (E_i): It represents the charge that is needed per node. Nodal charge varies from a node to another.

Nodal power (π): It represents the power consumption per node.

Computation capacity (U_i): It represents the capacity to compute and relay messages as a factor of utilization.

6.5.2 Results and Discussions

The flexibility of Cloud-based infrastructures requires a cost-effective pricing model to enhance the reliability of the system especially for the ones under uncertainty. Pricing the uncertainty in the Cloud-based IoT configurations plays a significant role in providing a real-time response and improving the reliability of the system. Moreover, understanding the pricing problem in a Cloud-based platform will provide a useful insight into the complexity of the problem in the presence of uncertainty as well as the Cloud monitoring value network [12]. In this chapter, during experiment, γ_i was set to have a uniform distribution between 0 and 1, whereas U_i was set to have a value between 5 and 20. Furthermore, π would have a uniform distributed value between 40 and 100 to illustrate the different power consumption that every node uses. In the pricing model problem, both the SAA and the GA have the same entries of data to show which one would be more accurate on distributed systems. The test is performed by creating uncertainty arrays that have the following sizes: 8, 16, 32, 64, 128, and 256. They are all computed by both algorithms to find the best algorithm on distributed systems.

As shown in Figure 6.1, GA and SAA perform well when they are distributed, but the GA shows its powerful computation power compared to the SAA when it

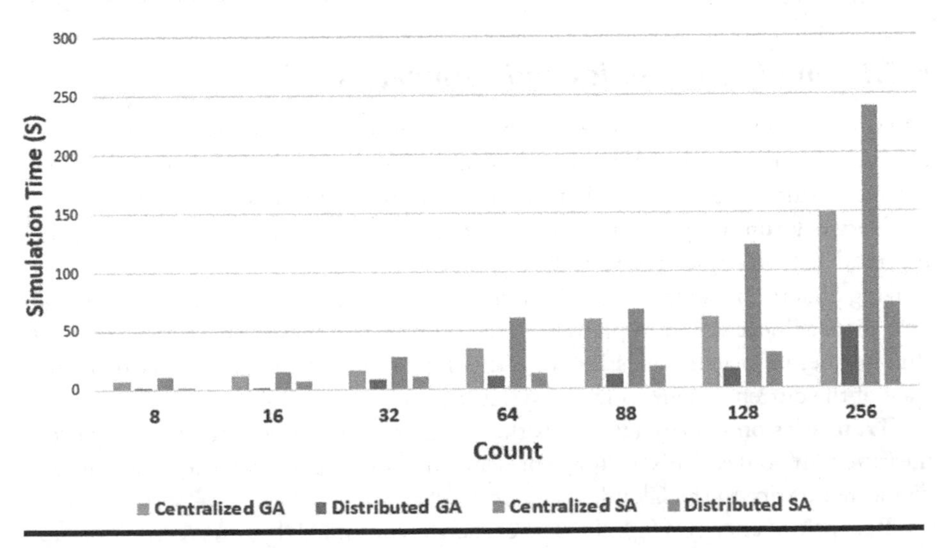

Figure 6.1 Uncertainty factors.

Table 6.2 Time Consumed while Varying Uncertainty Factors' Count under both GA and SAA (Centralized/Distributed)

	8	*16*	*32*	*64*	*88*	*128*	*256*
Centralized GA	5 s	9 s	18 s	39 s	1 m	1 m 15 s	2 m 33 s
Distributed GA	1 s	2 s	5 s	10 s	16 s	22 s	49 s
Centralized SAA	7 s	15 s	31 s	1 m 2 s	1 m 14 s	2 m 3 s	3 m 58 s
Distributed SAA	2 s	4 s	8 s	18 s	27 s	39 s	1 m 19 s

comes to a centralized or distributed system. The 256 entries reveal that the running time of the distributed GA is less than 50 s with a 4 s difference than the previous experiment, which would be counted as a computational timing error.

The results presented in Table 6.2 show that how the SAA grows exponentially when we take an uncertainty value of 2^n in its centralized form. Furthermore, its performance in the distributed form is reduced to half. However, the GA provides promising results when it runs in a distributed manner. Every array of uncertainty is mapped to a node with a GA ready to compute the variables.

Moreover, the SAA shows that it takes longer when calculating the 256 entries with a running time of 4 min, whereas it takes only 2 min and 41 s for computing the previous 256 entries.

Please note that the experiment was done using a distributed system that has four cores in its computational power.

6.5.3 *Number of Cores vs. Uncertainty Entries*

The pricing problem is solved using different core powers to represent the speedup that distributed systems would achieve when it solves/computes uncertainty. For this experiment, the test is performed by creating a matrix of uncertainty that held 64, 88, 128, and 256 entries of uncertainty. In addition, it considers various computational powers to see how it would affect the running time of the algorithms compared to the centralized approach.

Figure 6.2 shows how the distributed system helps in computing uncertainty problems when it comes to time and core efficiency. Considering the computation on two different nodes, the algorithm shows that it takes approximately 2 min compared to the 50% speedup. This shows that predicting the pricing model of any node is possible if the algorithm is efficient enough to be distributed.

The results presented in Table 6.3 show how the uncertainty factors are affected by the number of present nodes. The centralized GA is compared with the distributed approach considering two, three, and four nodes. With 256 entries, it is shown that the centralized algorithm takes 2 min to actually compute all the entries, whereas increasing the nodes would make the computational time lower.

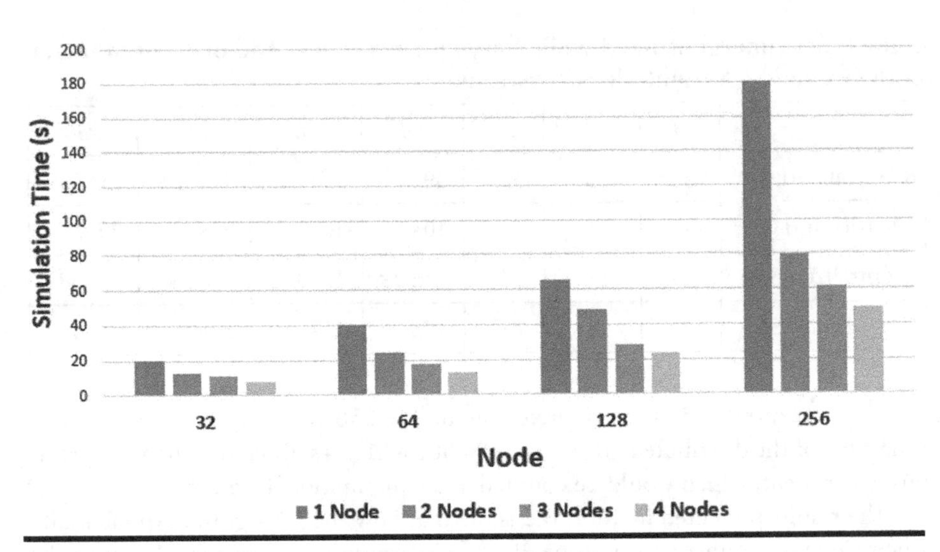

Figure 6.2 Uncertainty factors—GA.

Table 6.3 Time consumed while Varying Uncertainty Factors' Count under Centralized/Distributed GA with Varying Number of Nodes

	32	*64*	*128*	*256*
Centralized GA	18 s	39 s	1 m 15 s	2 m 35 s
Distributed GA 2	10 s	22 s	46 s	1 m 31 s
Distributed GA 3	8 s	18 s	29 s	1 m 09 s
Distributed GA 4	5 s	10 s	22 s	49 s

This approach can be implemented with more nodes and more uncertainty factors allowing us to compute more factors in the pricing model.

The same experiment was conducted for the SAA with a different number of distributed nodes to see the effect of uncertainty on the computational power of the algorithm. The uncertainty array that was computed in the previous test was used again to have a consistent metric.

As shown in Figure 6.3, the distributed system actually helps in computing uncertainty problems when it comes to time and core efficiency. The algorithm shows that running 256 entries in a single node is computationally expensive using the SAA. On the other hand, using distributed systems help to find the optimal solution most of the time.

The results presented in Table 6.4 compare the distributed SAA and the centralized SAA in terms of running time. With 32 entries, the centralized SAA surpasses the 1 min mark, in which the GA is able to compute the 256 entries with a lower

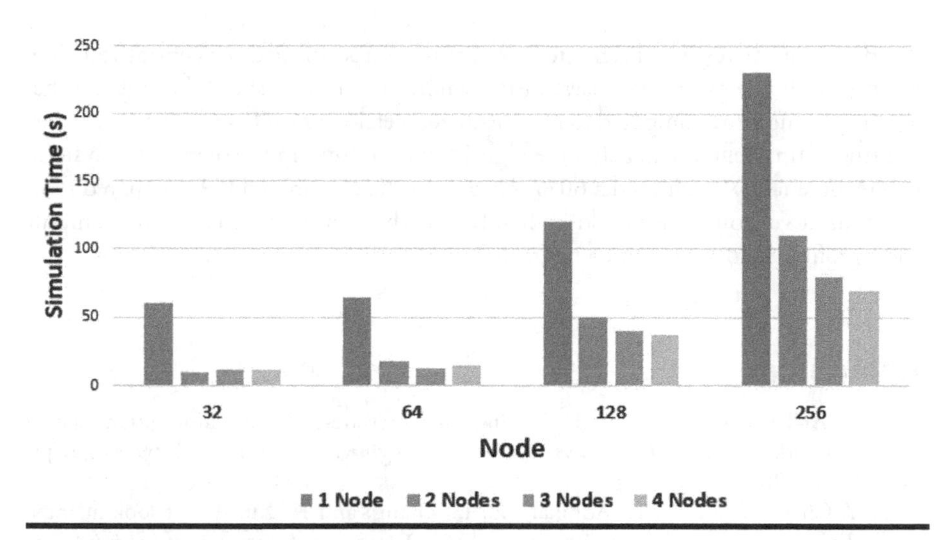

Figure 6.3 Uncertainty factors—SAA.

Table 6.4 Time Consumed while Varying Uncertainty Factors' Count under Centralized/Distributed SAA with Varying Number of Nodes

	32	*64*	*128*	*256*
Centralized SAA	1 m 2 s	1 m 14 s	2 m 3 s	3 m 58 s
Distributed SAA 2	13 s	27 s	54 s	1 m 49 s
Distributed SAA 3	18 s	21 s	44 s	1 m 25 s
Distributed SAA 4	18 s	27 s	39 s	1 m 19 s

running time. Moreover, the distributed one performs better because multiple nodes perform the heavy computation instead of a single node.

6.6 Conclusion

Uncertainty is an important issue that affects computing efficiency and brings additional challenges to Cloud providers in the IoT environment. In this chapter, we propose a pricing framework for a Cloud node in the IoT era by considering uncertainty factors such as network topology, transmission/reception energy, nodal charge and power, and computation capacity. With all the factors that might change in the network, the aim of this study is to use IoT itself to calculate these uncertainty variables for the pricing of the Cloud nodes. The results presented based on calculating the uncertainty on centralized and decentralized algorithms show a significant improvement in terms of calculating uncertainty in the Cloud-based IoT era.

Based on the results, distributed algorithms outperformed the centralized algorithms in all the given test cases. Furthermore, the results show that the GA has lower running time compared to its centralized version as well as to the SAA. When the uncertainty entries increase, the algorithms' speedup in the distributed systems is phenomenal. We achieved a 60% speedup in the distributed GA compared to its centralized version. This would be helpful for the companies to actually compute their profit and loss when it comes to IoT.

References

1. F. Al-Turjman, "Fog-based Caching in Software-Defined Information-Centric Networks", *Elsevier Computers & Electrical Engineering Journal*, vol. 69, no. 1, pp. 54–67, 2018.
2. Y. Chen, S. Jain, V. K. Adhikari, Z. L. Zhang, and K. Xu, A first look at inter-data center traffic characteristics via yahoo! Datasets, in *Proceedings of IEEE Infocom*, Shanghai, China, pp. 1620–1628, April 2011.
3. F. Al-Turjman, Price-based data delivery framework for dynamic and pervasive IoT, Elsevier Pervasive and Mobile Computing Journal, vol. 42, 299–316, 2017.
4. A. Tchernykh, U. Schwiegelsohn, V. Alexandrov, and E. G. Talbi, Towards understanding uncertainty in cloud computing resource provisioning, *Procedia Computer Science*, vol. 51, 1772–1781, 2015.
5. J. Byun, B. W. Kim, C. Y. Ko, and J. W. Byun, 4G LTE network access system and pricing model for IoT MVNOs: spreading smart tourism, *Multimedia Tools and Applications*, vol. 76, no. 19, 19665–19688, 2017.
6. F. Al-Turjman, "Price-based Data Delivery Framework for Dynamic and Pervasive IoT", *Elsevier Pervasive and Mobile Computing Journal*, vol. 42, pp. 299–316, 2017.
7. O. Younis and S. Fahmy, HEED: A hybrid, energy-efficient, distributed clustering approach for ad hoc sensor networks, *IEEE Transactions on Mobile Computing*, vol. 3, no. 4, 366–379, 2004.
8. R. Pawlak, B. Wojciechowski, and M. Nikodem, New simplified HEED algorithm for wireless sensor networks, in *International Conference on Computer Networks*, Zurich, Switzerland pp. 332–341, June 2010.
9. S. Zhong, L. E. Li, Y. G. Liu, and Y. R. Yang, On designing incentive-compatible routing and forwarding protocols in wireless ad-hoc networks, *Wireless Networks*, vol. 13, no. 6, 799–816, 2007.
10. A. Magruk, The most important aspects of uncertainty in the Internet of Things field–context of smart buildings, *Procedia Engineering*, vol. 122, 220–227, 2015.
11. C. Delimitrou and C. Kozyrakis, Amdahl's law for tail latency, *Communications of the ACM*, vol. 61, no. 8, 65–72, 2018.
12. C. Lai and L. Xu, Bilevel fee-setting optimization for cloud monitoring service under uncertainty, *IEEE ACCESS*, vol. 6, 9473–9483, 2018.

Chapter 7

Home Automation in Cloud-Based IoT

Fadi Al-Turjman and Mohamad Sanwal
Antalya Bilim University

Contents

7.1 Introduction

A smart home is an application enabled by ubiquitous computing in which the home environment is monitored by ambient intelligence to provide context-aware services and facilitate remote home control [1]. Furthermore, it is considered as a combination of several enabling technologies such as sensors, multimedia devices, communication protocols, and systems. From a different scope, a smart home is merely a residence equipped with different Internet-connected devices to remotely monitor and manage the appliances and systems installed in the home such as lighting and heating, to just mention a few. Such a smart residence would be useful in managing the daily lives of the inhabitants. With the recent developments in the Information and Communications Technology (ICT) and the reduction in the costs of low-powered electronics, a new paradigm, called smart cities, has drawn the attention of the research community.

The smart city is constructed by a number of buildings including the smart homes, which are designed and maintained by using high advanced integrated technologies. These technologies include advanced electronics, sensors, base stations, and different communication techniques which are computerized using databases and advanced algorithms [2]. With the increasing demands for good governance, health services, and energy savings, better utilization of the existing resources becomes more complex with the passage of time. Accordingly, smart cities and home paradigms form the best alternative to overcome these challenges in a smart way [3]. There are several applications that have already been deployed or proposed in these paradigms such as smart grid, smart meters, smart water control, health care, and surveillance. Smart homes are also forming the future of conventional homes we are experiencing nowadays. Smart homes with lots of sensors and wireless connectivity techniques have been anticipated for a long time. Research and commercial versions have been built for experiments as well as to check the positive and negative aspects of this paradigm [4]. The purpose of smart homes is to automate different appliances in an intelligent and optimized manner without the need for human interaction. For example, the smart home contains cameras, motion sensors, fire, and window alarms, which should work by themselves while collecting and sensing information and data from the home rooms and collaborate in what we call wireless sensor networks (WSNs). These networks are usually energy constrained and have a very limited communication range. Thus, it aims at delivering data in a multi-hop communication fashion in what we call the Internet of things (IoT).

IoT is a revolutionizing technology that tends to connect the entire world by connecting physical smart devices used for sensing, processing, and actuating [2–3]. By integrating the machine-to-machine (M2M) communication technologies with the smart devices, these devices can connect and interact without any human intervention. As a result, IoT is believed to enable a fully conductive environment that can influence the life of the society in different aspects such as everyday activities of the individuals, business and economy applications, health-care applications, energy applications, traffic and road controlling, and even political systems, to just mention a few. Moreover, the "things" are merely the devices and objects connected to a common interface with the ability to communicate with each other. By integrating the three core components of the IoT, namely, the Internet, the things, and the connectivity, the value of IoT is to close the gap between both the physical and digital worlds in the self-reinforcing and self-improving systems. The concept of smart homes is considered as an IoT-based application enabled by connecting the home appliances to the Internet. The home system's main goal is to provide security by monitoring and controlling all devices in homes over a cloud.

To achieve security, the system detects any threats in the home such as gas leaks, water leaks, and fires; then, it is alarming the residents to prevent any losses in lives or properties. In addition, the system provides instant detection for any robberies happening. The controller manages all the devices installed in the home, and it can remotely control these devices with the aid of smartphones. In addition, the system is compatible with all kinds of devices with the ability to manage their running time.

Amazon Web Server has built IoT-specific services, such as AWS Greengrass and AWS IoT. These services help people to collect and send data to the cloud, to load and analyze data, and to manage devices. AWS IoT is a managed cloud platform that allows the connected devices to easily and securely interact with each other and with cloud applications. AWS IoT is a managed service built for the purpose of connecting the devices to each other and to the cloud. Moreover, it can handle billions of devices and trillions of messages, with the ability to reliably and securely process and route these messages to the AWS endpoints and to other devices.

In order to secure homes, security issues in exchanging the gigantic data amounts of sensitive/personal data over the network via the multi-hop strategy shall be addressed first. Because it can make the exchanged data prone to several attacks due to attackers. The authors in Ref. [5] proposed a security approach related to the black hole attack while maintaining the active routes with nodal trust. However, there are still several attacks to be addressed and considered in the IoT/Big Data era.

In this chapter, we focus on smart home security issues in IoT-enabled smart cities and Big Data paradigms. We overview several related aspects such as network architecture, devices, wireless technologies, applications, and used machine learning (ML) techniques. Our main contributions can be summarized as follows:

- This chapter provides a critical overview of the IoT-based home automation considering the main application areas, architecture, limitations, and design factors.
- Related intelligence and learning techniques are discussed.
- Security measures and requirements are outlined for easy access.
- Expected and common security threats are overviewed and classified.
- Specific tools and assessment methods are reported as well.
- Potential enabling technologies and communication protocols as well as the interaction with the cloud-based infrastructures techniques are presented.
- Security threats and promising solutions in addition to several open issues to be addressed in the future are reported as well.

For more readability, a list of used abbreviations along with their definitions is provided in Table 7.1. The rest of this chapter is organized as follows. Section 7.2 provides some background about smart homes in smart city along with the objectives and constraints. Some of the existing and popular applications are discussed in Section 7.3. The architecture and basic components of smart homes along trustworthy computing models are discussed in Section 7.4. Section 7.5 outlines the most popular home automation techniques. The different devices used are categorized in Section 7.6. In Sections 7.7 and 7.8, secure deployment as well as wireless protocols used in smart homes are briefly explained. In Section 7.9, different security issues are discussed along with its countermeasures. In Section 7.10, we describe how we can deploy smart homes. Furthermore, in Section 7.11, we report and discuss open research issues. Finally, Section 7.12 is concluding this research paper.

Table 7.1 List of Abbreviations and Their Definitions

Abbreviation	*Definition*
IoT	Internet of things
LAN	Local area network
WAN	Wide area network
MAN	Metropolitan area network
DDoS	Distributed denial of service
PDoS	Permanent denial of service
Wi-Fi	Wireless fidelity
GSM	Global System for Mobile Communications
LTE	Long-Term Evolution
UMTS	Universal Mobile Telecommunications System
DNS	Domain Name System
NAN	Neighborhood area network
HAN	Human Area Network
PLC	Power line carrier
IEC	International Electrotechnical Commission
DSL	Digital subscriber line
NB-IoT	Narrow-band IoT
CT	Communication technology
ITU	International Telecommunication Union
WSDL	Web Services Description Language
SDN	Software-defined network
P2P	Peer-to-peer
SOAP	Simple Object Access Protocol

7.2 Background

Due to increasing population in urban areas, there has been noticeable reduction in resources, so we have to resolve this problem for the benefit of citizens. Smart cities are the best solution with a flexible infrastructure which can be sustainable in disasters. Smart cities using wireless technology will be more efficient regarding resources as well as in disasters to provide better solutions. A smart city and home require some basic components to operate such as smartphones, sensors, and networks to operate wirelessly and in mobile conditions. The connected sensors will support the infrastructure such as smart grids, smart homes, surveillance, vehicular movements, and quake detection in buildings [6]. Wireless networks connect different devices using radio waves in air as a medium. There are some advantages such as flexibility and low cost. There are many networks which are currently being used with some advantages over others such as local area network (LAN), wide area network (WAN), metropolitan area network (MAN), WiMAX, Wi-Fi, and ZigBee [7].

7.2.1 Objectives

The main objective of a smart home is to build and deploy cost- and performance-efficient homes. In Ref. [8], the main objective is to build effective and cost-efficient smart homes with maximum security. In Ref. [9], the main objective is to automate homes using different techniques and technologies such as through Web browser, cloud server, Global System for Mobile Communications (GSM), or Bluetooth. In Ref. [10], different security issues regarding smart homes are mentioned with their issues such as standards, integration, and authentication. Furthermore, how we can control different modules and equipment using IoT is also discussed. Moreover, in Ref. [11], the author described a mobile application in which we can remotely control multiple home appliances. It offers more flexibility for handling things at home automatically. Also, it assures the user about more security. It is also providing better solutions for water management and child security. In Ref. [12], the authors described methodologies for home automation.

7.2.2 Constraints

In Ref. [8], the author mentioned the main problems regarding home automation. The main concern is to build cost-efficient homes. Further, he described the price ranges of different pieces of equipment along with their installation cost, wiring cost, and development cost, and how can we decease them. Moreover, in the paper he described the ways to achieve power efficiency along security. In Ref. [9], the author highlighted the Internet connection issues in rural areas along with maintenance of cloud. Furthermore, issues such as the requirement of continuous Internet along with installation problems were mentioned. Moreover, in Ref. [10], the authors described the issue regarding controlling the multiple devices at the same time as well as efficiency and security. Authors in Ref. [12] described the challenges in different communication techniques along with their cost, efficiency, and data rate. In Ref. [13], the author briefly explained the security issues regarding smart home security and their effect on the system.

Table 7.2 mentioned the common constraints that smart homes are facing in the industry. Their challenges and issues are summarized in the same table.

Table 7.2 Comparison of Smart Homes Constraints for Some References

Reference	*Technology*	*Energy*	*Control*	*Cost*	*Efficiency*	*Security*	*Connectivity*	*Speed*	*Performance*
[7]	Yes	Yes	No	Yes	No	Yes	No	No	No
[8]	Yes	No	No	Yes	No	No	Yes	Yes	No
[9]	No	No	Yes	No	Yes	Yes	No	No	No
[10]	Yes	No	No	Yes	Yes	No	No	Yes	Yes
[12]	No	No	No	Yes	Yes	Yes	No	No	Yes

7.3 Applications

In the recent years, engineers and researchers are working on different applications of smart homes. Smart applications use different sensors such as motion, light, and fire sensors to detect activities and information. The information collected by these sensors is utilized to take appropriate actions. Some of the following applications are being used in the smart homes for a while.

Smart lightening: It requires smart bulbs, motion sensors, light sensors, and wireless connection to control them using mobile phone application. Smart bulbs can be controlled by the mobile application through an Internet or Bluetooth connection. On the other hand, motion and light sensors are used to control the lights automatically. Furthermore, the authors in Ref. [14] described the lightening control system in smart homes (Figure 7.1).

Smart garage: The basic functionality of a smart garage is the door opener. The garage door opener has rolling-code implementation. The most popular rolling-code implementation is a product called KeeLoq. It is a lightweight block cipher that generates codes based on the cryptographic key. Furthermore, in Ref. [15] the author has explained the security countermeasures and its functionality in detail.

Smart meters: Smart homes are the basic component of smart homes. The main purpose is to provide monitoring and control functions over homes units that consume energy. A smart meter consists of three sections: base station, appliance controller, and user interface. Base station has different components such as

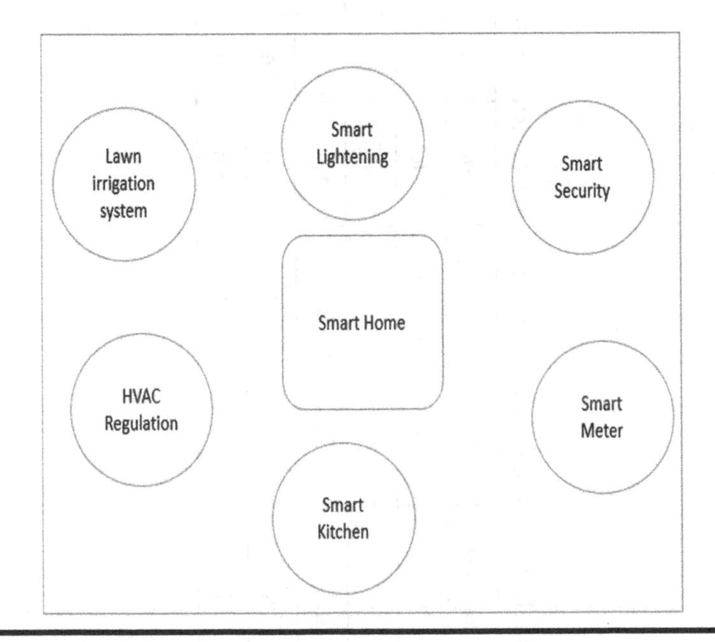

Figure 7.1 Smart home applications.

IoT gateway, appliances tracker, and data repository. Appliance controller contains sensors and relay modules [15]. User can monitor its total consumption by using user interface. In Ref. [16], the authors have explained in detail the working of smart meters.

Smart kitchen: Smart appliances are the major part in smart home automation. Smart kitchen appliances such as the smart refrigerator (LG's Smart ThinQ) can scan grocery receipts, check the missing items, and warn the user if an item is about to expire. Moreover, it suggests recipes based on the refrigerator content. Furthermore, by automating the kitchen appliances and accessing them with a smartphone, energy can be saved by monitoring the energy consumption of appliances [17].

HVAC (heating, ventilation, and air conditioning) regulation: Heating and cooling homes consumes 50% of the annual energy cost at homes. Recently, automated home systems have become more popular. With automated HVAC, the homeowner can remotely reduce/increase the room temperature when the room is not occupied and reprogram it over the air [17].

Lawn irrigation system: A lush and healthy lawn is the beauty of the house, but the weather does not cooperate always. In the past few decades, we have seen sprinkler systems, which were not efficient regarding the usage of water. According to the survey, half of the water is wasted due to the inefficiency. However, sprinkler control systems, such as Skydrop, provide real-time weather communication with real-time weather data. If rain deposit two inches of water, the automatic sprinkler will detect and disable the scheduled watering [17].

Smart security: Security is the main concern of everybody. Secure home security consists of Close Circuit Television (CCTV) cameras, alarms, motion sensors, thumb readers, and face recognition devices. Authentic person can use his or her face or fingerprint to enter the house. If someone tries to enter, the alarms will go off, and an automatic call will be dialed to the police as well as the owner (Figure 7.2).

In Section 7.4, we discuss the architecture and main components of smart homes.

7.4 Architecture and Main Components

The generally used components are sensors, actuators, wireless signal control devices, appliances, and monitoring devices. Sensors are responsible to sense an activity, for example, light sensing, motion sensing, and temperature sensing. An actuator is a component that is responsible for moving and controlling the mechanism. It requires a source of energy and controlled signal. The signal control devices such as modem are used to provide and receive wireless signals. Appliances can be refrigerators, air conditioners, washing machines, and so on. Moreover, surveillance devices such as cameras and monitoring screens are used for security purpose.

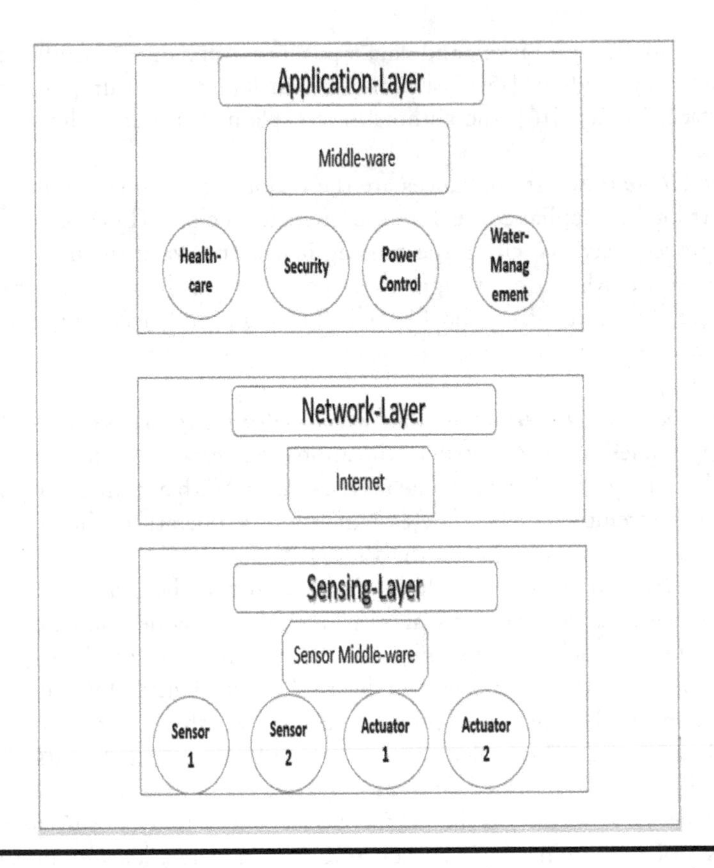

Figure 7.2 Architecture of the smart home network.

Different system and architecture models have been considered for implementation of smart homes in smart cities, but mostly researchers suggest and impose a three-layered architecture model for smart home automation [18]. This model consists of three layers: (1) sensing layer, (2) network layer, and (3) application layer.

7.4.1 Sensing Layer

The sensing layer is responsible for collecting data from different home appliances present in the house. This data is collected through different sensors that are installed in the appliances and doors and different things present in the home. The sensors that are responsible for the collection of data can be heat, temperature, light, motion, and so on. Data collection and data processing at the sensing layer are done by the microprocessors such as SAMSUNG S3C2440A. To transfer the collected data towards the network layer, it uses the wireless module, for example, ZigBee [10].

7.4.2 Network Layer

The network layer collects data from the sensing layer using wireless networks, for example, Wi-Fi and ZigBee, and transfers this data to the upper layer, which is called application layer. This layer works as a bridge between the two layers. It also uses different protocols to send the data optimally.

7.4.3 Application Layer

This is the upper most layer of this architecture. This layer collects data from the network layer and uses this data for different purposes. These applications are mostly installed in the smartphones to control the home appliances. For example, you can use the android application to control your home wirelessly. The applications to control smart homes are Nest and Samsung SmartThings [19].

7.4.4 Cloud-Based Architecture for Big Data

Since there is no cloud-based architecture so far that can provide the users with special services to the digital devices installed in smart homes, we recommend a potential cloud-based architecture, which relies on the current cloud industry with a modification of the service layer. This can result in providing efficient and stable services for the smart homes' owners. We also recommend two network architectures: Web and P2P (see Figure 7.3a). This way, the cloud server will be able to provide a better service for the higher quality audio/video signals by reducing the bandwidth pressure while transmitting them. In addition, the smart home gateway is responsible for both describing their services in Web Services Description Language (WSDL) and registering them in the cloud service directory. This way, the other homes can search for the service and benefit from it. Consequently, we can consider smart homes as service consumers and suppliers at the same time. P2P and Web are reasoned solutions that can be introduced to the cloud for combining both the cloud and smart homes. By using these two technologies, the cloud successfully provides a more special functionality to the smart homes. Recently, the home gateway has experienced too much research focusing on it. One of the initiatives resulted is the Open Services Gateway Initiative (OSGI). This initiative intends to create a platform that enables deploying services over both the wide-range network and the local network or device. In OSGI solutions, the home gateway has two main functions: connecting the appliances inside the home and linking them to the outside network. This way, homeowners can experience a better home life without the need to overpower the smart devices with spontaneous user interfaces.

7.4.4.1 Peer-to-Peer networks

In this section, we consider the smart home services that are classified into two main categories: residential entertainment and video streaming. A typical client–server

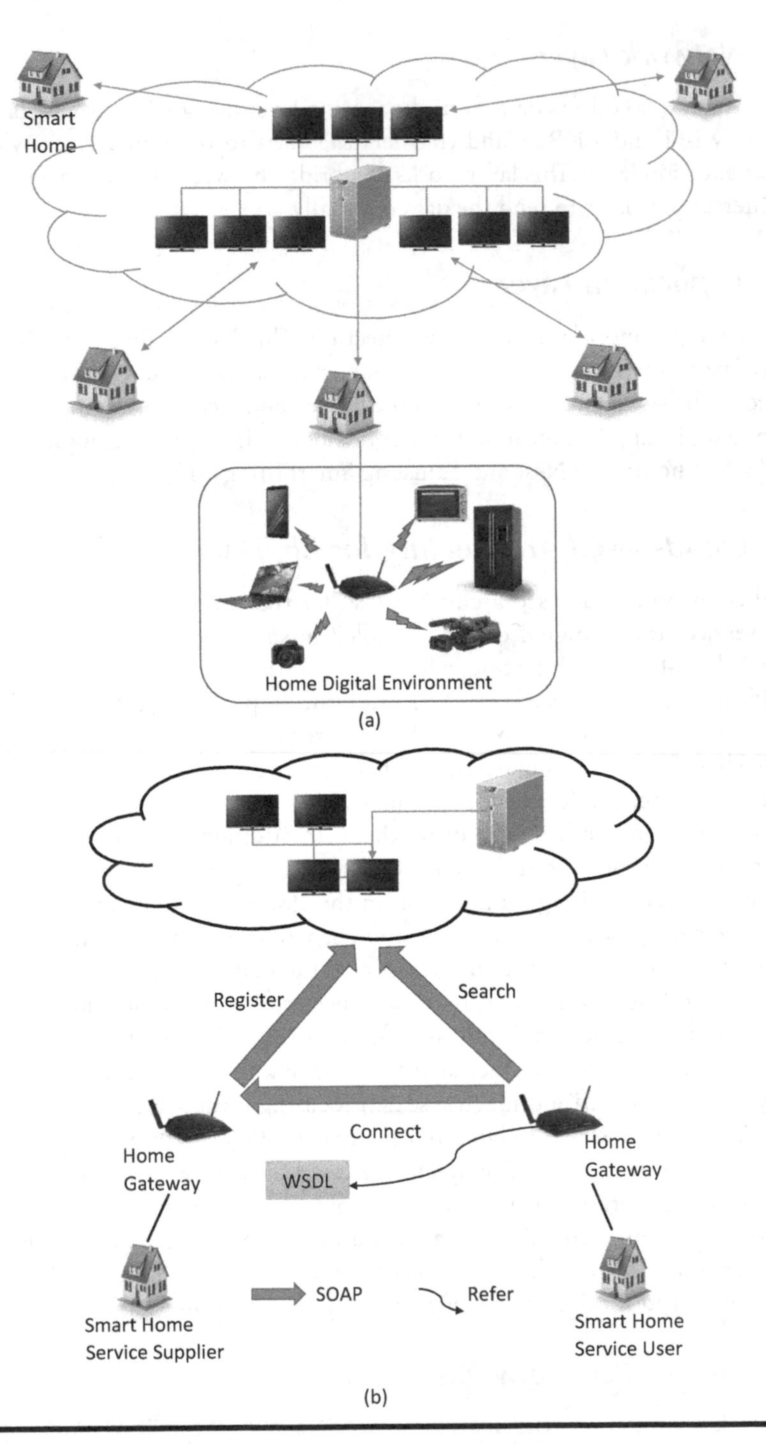

Figure 7.3 (a) The P2P network; (b) the Web-based diagram.

cloud architecture is constrained by the problem of limited bandwidth while transferring Big Data amounts. Solving such a problem required us to propose a mechanism where cloud server and smart home nodes can set a P2P network. This will decrease the bandwidth usage. With this model, each user will be responsible to register some information in the cloud such as his or her name and IP. Moreover, a specific software for the P2P communication should be installed in the gateways of each home. As a result, home entities can segment the bandwidth equally. For instance, a user can set a cloud service to discern online video streaming for the other clients who share the same cloud service with him or her. Further, the other users can benefit from such a service by watching this video. So, a participant who is broadcasting the video stream is a customer downloading the video and a seller who is uploading it to others at the same time.

7.4.4.2 Web Service

In the traditional client–server model, homes are considered as consumers of services only. In contrast, in our model, smart homes (peers) are consuming services and sharing them with others. So, we can consider them as both consumers and suppliers of a service at the same time. To understand this feature, a Web service technology is introduced [14]. It uses XML packets and follows the Simple Object Access Protocol (SOAP) standard. The residence gateway is responsible for connecting and managing various devices and networks installed within the house such as home automation network and PC network; as a result, this can create an integrated digital environment. Moreover, this model enables the users to use the home gateway to share some features provided by their appliances with the neighboring houses. Hence, the cloud is considered as a UDDI (Universal Description, Discovery, and Integration) server in which services can be offered. Further, it allows services and applications to interact with each other on the Internet using a predefined set of rules. Figure 7.3b shows the recommended Web architecture for home automation.

7.4.4.3 Architecture Layers

The cloud architecture can be separated into three parts: platform, infrastructure, and service layers as shown in Figure 7.4.

Service layer: This part includes three main components: interface, directory, and control. Service interface directly interacts with the residence owner. Furthermore, the owners can easily publish their service description using the service interface. Service control part is responsible for analyzing, processing, and responding to the owners' demands.

Platform layer: This part is considered as the essence of the cloud architecture presented above. It includes two main modules: specifically security manage module and resources management module. Further, the platform layer with the help

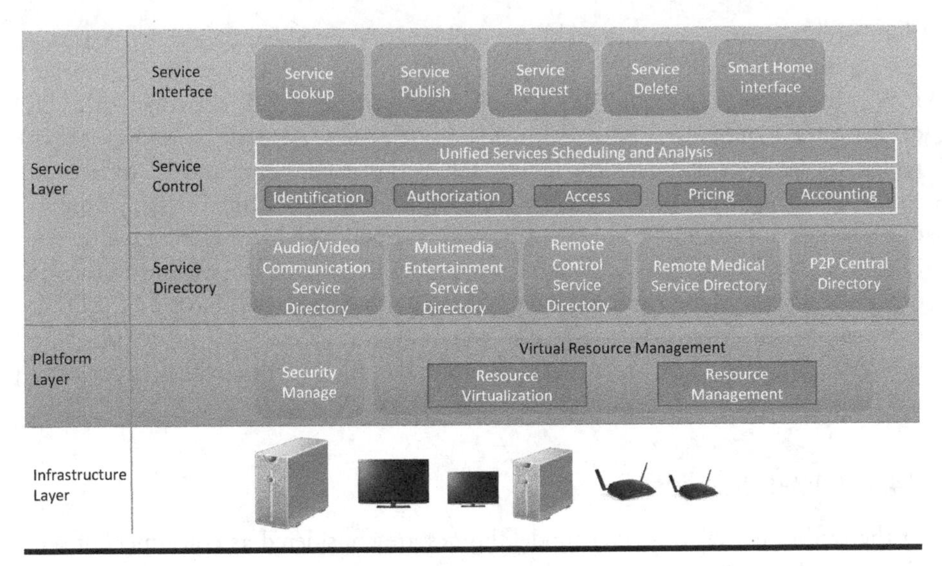

Figure 7.4 Three-layer cloud hierarchy for smart homes.

of the infrastructure layer can create the base for delivering Platform as a Service (PaaS) to smart homes.

Infrastructure layer: This part contains a huge amount of physical resources responsible for delivering the smart services. These resources are handled by a virtual unit, which is controlled by the cloud as a realization for Infrastructure as a Service (IaaS).

7.5 Trustworthy Computing and Assessment Models

There are various trustworthy computing models for smart homes. Every model has its own features and advantages over another. Different modeling techniques include feature modeling, variable modeling, etc. In Ref. [20], the author has described the object-based IoT management model. In Ref. [21], the authors described the modeling and analysis of ZigBee-based smart home systems. From the literature review, many researchers have considered the feature modeling technique because it has good tool support for variability reasoning. Feature modeling is widely used to specify the system functionalities, which vary according to the end-user need. In Ref. [22], the authors described the feature model in the case of smart homes. Features are hierarchically linked in a tree structure as shown in Figure 7.5. The blue boxes are current features, whereas the black boxes represent potential future variants.

The emerging IoT/Big Data project deals not only with the massive amount of manipulated data in smart homes but also with its usage. With the IoT, any item or device at home, even in the nanoscale, can be associated with the Internet and thus generate gigantic data amount. Therefore, sophisticated assessment methods and

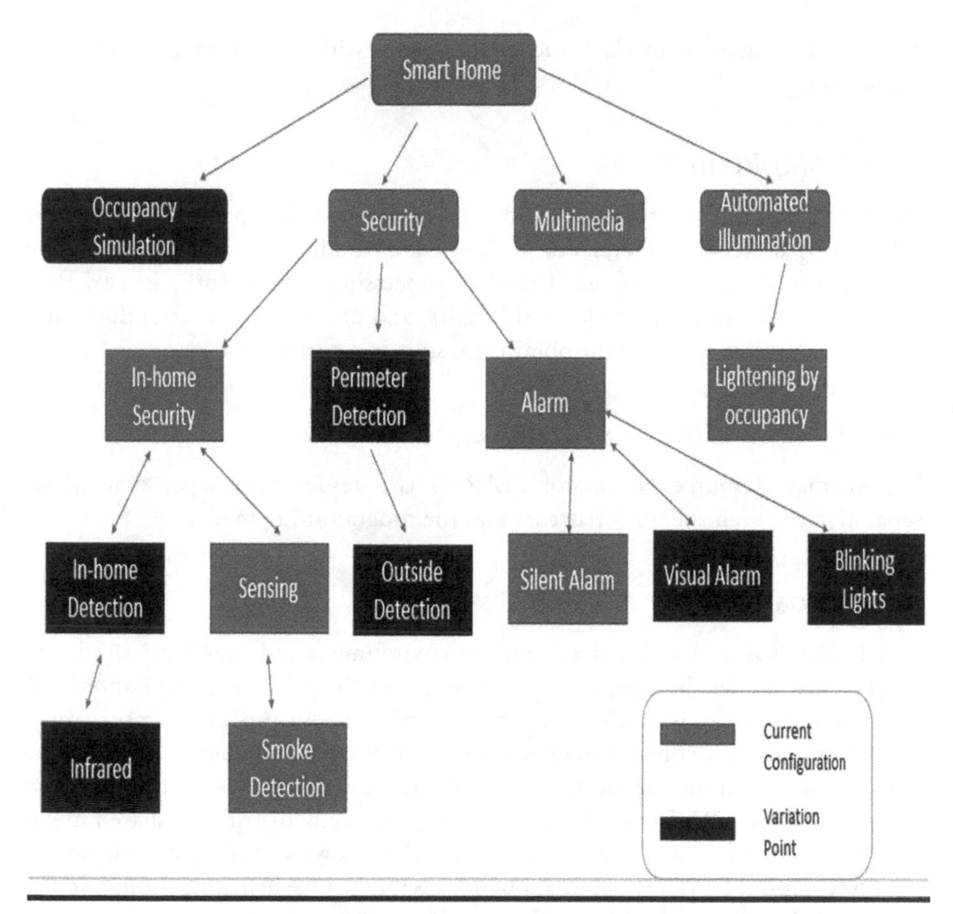

Figure 7.5 Computing model for the smart home paradigm.

benchmarking tools are of utmost importance in this field. We list the common ones in the following sections.

7.5.1 Hadoop

Hadoop platform is used for a wide assortment of techniques that have been created to record, arrange, and test massive amounts of data. It includes open-source devices, methods, and libraries for Big Data analysis and models. Majority of the other tools/techniques are connected with Apache Hadoop including the Hadoop Distributed File System (HDFS), MapReduce, Mahout, Spark, and Hive tools.

7.5.1.1 Hadoop Distributed File System

HDFS was developed for Big Data processing. It can support several users frequently and simultaneously. It is a file system that sits at the bottom of the Hadoop architecture made up of data and name nodes with a built-in fault-tolerance mechanism.

7.5.1.2 MapReduce

MapReduce is the most powerful tool in distributed and parallel applications functioning under the umbrella of 5G/Big Data paradigm. It depends mainly on conquer and divide techniques. This data processing engine works in two fold: 1) mapping the raw data into key/value pairs, and 2) reducing the data duplicates by combining/summarizing the obtained results in a parallel manner.

7.5.1.3 Yet Another Resource Negotiator

Yet Another Resource Negotiator (YARN) is a resource manager that allows separation between the infrastructure and the programming model.

7.5.1.4 Spark

Apache Spark is also a cluster computing environment and uses ideas similar to MapReduce model, but improves speed by using in-memory computations. Its response time is significantly faster than MapReduce in processing tasks stored in memory and Hadoop at disk operations. It stores data in memory and provides fault tolerance without replication with abstraction called Resilient Distributed Datasets (RDDs). RDD can be understood as read-only distributed shared memory. It was extended to include DataFrames. This allows grouping of collection of data by columns; hence, it can be thought as RDD with schema. Learning process is through in-memory caching of intermediate results. Spark is easy to program and supports integration with Java, Python, Scala, and R programming languages. It supports multiple data sources, including Cassandra, HBase, or any Hadoop data source. Besides its effective features, Spark has some inefficiencies in terms of stream processing and bottlenecks, which can occur because of data transfer across nodes using the network [23].

7.5.1.5 Mahout

Mahout offers a wide selection of robust algorithms. It is good for batch processing (not streaming). There is a lack of active user community and documentation. It is commonly claimed to be difficult to set up an existing Hadoop cluster. Configuration problems may occur. Algorithms focus on classification, clustering, and collaborative filtering. Extensibility is good, but strong Java knowledge is required. Mahout is best known for collaborative filtering (recommendation engines), which offers similarity measures such as Pearson correlation, Euclidean distance, Cosine similarity, Tanimoto coefficient, and log likelihood.

7.5.1.6 Hive

The storage layer includes data integration tools such as Hive, which allows running standard SQL queries on data stored in the HDFS and NoSQL databases using HiveQL, an extension of ANSI SQL. This is a powerful and simple way to query the system, which is then distributed across MapReduce/TEZ commands, and then runs on top of Hadoop YARN. Metadata for tables and partitions is kept in the Hive Metastore. HIVE provides an interactive way of working with Big Data on a cluster and is easier than writing MapReduce code in Java. It is highly optimized and extensible. Hive is good for online transaction processing and stores data de-normalized as flat text files. However, the record-level updates, inserts, and/or deletes are not allowed because of the absence of the underneath relational database.

7.5.1.7 Sqoop

Sqoop is another tool that is used to import and export data between relational databases and Hadoop ecosystems. This is useful when HDFS is used as an enterprise data warehouse preprocessing engine. The idea behind Sqoop is that it leverages map tasks of the MapReduce framework. It has the following features: (1) connections with all main Relational Database Management Systems (RDBMS), (2) Kerberos security integration, and (3) data transfer from RDBMS to Hive or HBase and vice versa.

7.5.1.8 Apache HBase

HBase is a column-oriented NoSQL database utilized in Hadoop, in which the client can store substantial quantities of code lines and sections. HBase performs write/read activities. It additionally supports record-level updates, which are not conceivable using HDFS. HBase gives parallel data capacity by means of the hidden data file frameworks over the cloud servers. It is an open-source code to handle data in petabytes in thousands of nodes. It has the following features: (1) compatible with Java API for client access, (2) bloom filters and block cache for real-time queries, (3) linear and modular scalability, (4) strictly consistent reads and writes, (5) extensible JIRB (Interactive JRuby) shell, (6) supports exporting metrics via Hadoop, and (7) convenient base classes for backing MapReduce tasks with HBase tables.

7.6 Intelligent Techniques in Home Automation

There are several attempts in the literature which have been used mainly for the smart home automation in order to improve the quality of connected devices in the vicinity of the home. For example, if a water pipe is broken in the house and water flows from it continuously for many hours, this will destroy the devices and appliances in the home. A common solution for this is to deploy a moisture detection sensor to alarm the house owner if there is a water leak in it. These sensors can be placed near water heaters, dishwashers, refrigerators, sinks, and sump pumps. Further, if the sensor detects any unwanted moisture, an intelligent home technique can be used to automatically send a notification to the owner about the problem and might suggest a quick solution. This intelligent technique can be categorized into (1) voice recognition-based home automation and IoT-based home automation [24].

Voice recognition-based home automation: A voice acknowledgment-based home robotization framework was proposed and executed in Ref. [24]. The equipment for this framework comprises Arduino UNO and a cell phone. The remote connection between Arduino and the cell phone is done through Bluetooth technology. A mobile application can function based on the homeowner's voice recognition [11]. Furthermore, Android and iOS nowadays are using Google Assistant and Siri for voice-based commands related to home automation applications.

IoT-based home automation: In Ref. [23], the authors exhibited a home control and checking framework dependent on the IoT innovation. They planned and executed it by utilizing a Web server, controlling in-home gadgets, and programming tools. The engineering which they utilized comprises home condition, home gateway, and remote condition (Table 7.3).

Remote condition enables the approved clients to control and screen the home machines from anywhere by using smartphones, which use Wi-Fi or 4G. Home environment consists of a hardware interface module and a home gateway. For the monitoring purpose, home environments support sensors such as humidity and temperature.

These techniques are usually online and apply ML algorithms for more efficient decisions. Online smart home applications in the Big Data era are supposed to be generated and processed almost instantaneously. The change in the structure of data, variety, or velocity creates challenges for the learning systems and needs to be addressed with advanced adaptive algorithms. Typically, developing and deploying a learning system takes time, but most of the time is spent on understanding and preprocessing data. Efficient learning can help a lot in this regard. ML techniques in this domain can be classified into three main categories: (1) supervised learning, (2) unsupervised learning, and (3) semi-supervised learning.

Table 7.3 Comparison between Home Automation Techniques

Technology	*Cost*	*Efficiency*	*Flexibility*	*Response*
Voice recognition	Moderate	Moderate	High	Moderate
IoT-based	High	High	High	High

Supervised learning techniques use the data to make accurate predictions and learn the mapping between the input and its corresponding output while receiving a feedback during the learning process to identify things based on similar features. Unlike supervised learning that uses labeled data, unsupervised learning has no labels/features and no feedback signal. This technique is mostly used to find the hidden structure of the data and move it to similar groups. Semi-supervised technique resembles to extend the way humans learn and navigate through their daily life tasks. Semi-supervised learning is neither fully supervised nor unsupervised; it's kind of a hybrid approach. Some issues such as data redundancy, inconsistency, noise, heterogeneity, transformation, labeling for learning, data imbalance, and feature selection need to be addressed during the data preprocessing stage.

7.7 Devices Used in Smart Homes

There are a wide number of devices that are being used in home automation to improve the quality of life [25,26]. We have categorized these devices into the following categories.

Sensing devices: Smart home has some basic sensors in its paradigm, which are responsible for sensing information. Some of the sensors are fire, moisture, window, and motion detectors. Other sensors are light, water, pressure, and gas sensors. Fire sensor is responsible for detecting the smoke produced by fire and releases water. Moisture detector detects the amount of the moisture present in the walls. Motion sensor detects some movement on the floor and controls the lightening of the house. Light sensor detects the amount of light from outside, and if it is not enough, it turns on the lights. Water sensor detects the level of water in the tanks. Pressure sensor detects the pressure of home and outside, and gives you the information. There are other sensors such as chemical and thermostats, which work according to their functionality.

Appliances: Smart home is incomplete without appliances. There are different appliances for kitchen, washroom, and living rooms. Appliances for kitchen include refrigerators, kettles, juicers, microwave, blenders, and cookers. Appliances for washroom are washing machines, hand dryers, toilet roll dispensers, and so on. Appliances for living room include air conditioners, fans, lights, and heaters.

Entertainment devices: Entertainment devices are divided into categories such as home theater devices, gaming devices, and music devices. Home theater devices include TVs, DVD boxes, and speakers. Gaming devices are computers, monitors, Xbox, PlayStation, and so on. Music devices include speakers and microphones.

Security devices: Security is the main feature of smart homes. Security devices include CCTV cameras, alarms, and monitoring screens to monitor the activities inside and outside of the house. Alarms are activated when someone breaks into the house (Table 7.4).

Table 7.4 Categories of Devices in Home Automation

Categories	*Devices*
Sensing devices	Fire, moisture, motion, light, water
Appliances	Kitchen: refrigerators, kettles, juicers, microwave Washroom: washing machines, hand dryers, toilet roll dispensers Living room: air conditioners, fans, lights, heaters
Entertainment devices	Home theater devices: TVs, DVD box, speakers Gaming devices: computers, monitors, Xbox, PlayStation Music devices: speakers, microphones
Security devices	CCTV cameras, alarms, monitoring screens

7.8 Security in Deployment

Secure deployment of smart home devices is the major concern nowadays. Due to increasing security attacks, it is necessary to deploy secure devices. There are some techniques and parameters which we should consider while deploying the smart home devices.

Reviews about device: Before purchasing the device, customer should review and research about the device. By reviewing, we can get some information about that device – for example, whether it is attacked by any hacker or not. Moreover, user can get information about its security and performance.

Connecting application: It is extremely important to check which application connects the smart devices in home and which company designs this application and its rating. If the company is good, you can trust that application.

Password protection: In this modern era, almost every manufacturing company supports the password protection of devices. Along with the password protection, it is very important to check that the specific device supports the changing of password. Some of the companies do not allow to change the default password, which deceases the security and increases the vulnerability of the device.

Authentication: It is an essential requirement of the emerging Big Data networks, and it is of utmost importance to allow only authentic users for service/data access in heterogeneous systems such as those found in the smart homes. Thus, IoT devices are supposed to intelligently verify the system user and their connected gadgets. However, current identification and authentication methods are not sufficient for the emerging 5G-oriented IoT applications where transparency and reliability are key elements.

Preparation: In this stage, the application gets a product_id by checking the QR code on the gadget or by choosing the correct item title recorded within the app and sends it to the cloud. A while later, the application broadcasts the PT_SCAN message demonstrating the product_id and the EC public key of the application. Meantime, the application inquires client to input Wi-Fi credentials and begins a Wi-Fi provisioning process [27].

Wi-Fi provisioning: A few schemes are commonly utilized when completing a Wi-Fi setup of a headless IoT gadget employing a portable application (e.g., Get to Point Mode, Wi-Fi direct, TI's SmartConfig). JoyLink takes the strategy in which the application encodes the Service Set Identifier (SSID) and password into an arrangement of IP addresses and sends each IP a null character. At the same time, the application proceeds broadcasting the PT_SCAN message. This procedure rehashes until the point when the contraption adequately watches the traffic design, separates the Wi-Fi certification, accesses the system, and finally sends a PT_SCAN reaction to the application. The reaction joins the contraption MAC address and device key, which is the EC open key of the device [27].

Device initialization: When the application gets the MAC address, it sends demand to the cloud containing the MAC id and other information such as product_id and customer account information. The cloud asserts the legitimate relationship and responds with a message containing feed_id and access_key. The application makes the neighborhood key out of access_key and sends it with the feed_id and the access_key to the device in a PT_WRITEACCESSKEY message, which is mixed with the key orchestrated out of ECDH (tmp_key) in the midst of PT_SCAN. The contraption saves the (feed_id, access_key, local_key) and sends the authentication request to the cloud. Once the cloud gets the demand, it delivers a session_key, scrambles it with the access_key, and sends the outcome inside the PT_AUTH response message [28].

Communication security: For the deployment of smart devices, the communication security is extremely important. Different secure communication protocols are used for the communication to make it secure and authentic. In Ref. [29], the author has described some techniques and methods for smart home devices security, for example, key management system.

7.9 Communication Protocols

It is of great importance to study the communication nature in very dense environments such as the IoT one while serving and exchanging Big Data. The most commonly used wireless communication protocols in home automation can be classified into either wired or wireless. Table 7.5 categorizes the communication technologies that can be used in IoT-based smart homes based on the type (wired vs. wireless), data rate, coverage distance, frequency, standard, and application area in smart homes' neighborhood area networks (NANs), human area networks (HANs), and WANs.

Table 7.5 Categorization of Communication Technologies for Home Automation

Type	*Technology*		*Data Rate*	*Coverage*	*Frequency*	*Standard*	*Application Area*
Wired	*Coaxial cable*		10 Gbps	0–28 km	2 GHz	ITU-T J.222	WAN
	DSL	ADSL	1–8 Mbps	0–5 km	4000 Hz–4 MHz	ITU-T G.992	HAN, NAN
		HDSL	0–2 Mbps	0–3.6 km			
		VDSL	15–100 Mbps	0–1.5 km			
	Fiber optic		10 Gbps	10–60 km	180–330 THz	ITU-T G.652, and IEC 60793	NAN, WAN
	PLC	HomePlug	14–200 Mbps	0–200 m	1.8–250 MHz	IEEE Std 1901	HAN
		Narrowband	10–500 Kbps	0–3 km	3–500 kHz	IEC 61131-3	NAN, WAN
	Ethernet		10 Mbps–10 Gbps	0–100 m	100–500 MHz	IEEE 802.3	HAN

(Continued)

Table 7.5 (*Continued*) Categorization of Communication Technologies for Home Automation

Type	*Technology*	*Data Rate*	*Coverage*	*Frequency*	*Standard*	*Application Area*
Wireless	ZigBee	250 Kbps	0–100 m	2.4 GHz	IEEE 802.15.4	HAN
	Wi-Fi	54 Mbps	0–250 m	2.4 and 5 GHz	IEEE 802.11b/g/n	
	Bluetooth	721 Kbps	0–100 m	2.4 GHz	IEEE 802.15.1	
	Z-Wave	40 Kbps	0–30 m	900 MHz	ITU-T G. 9959	
	NB-IoT	14.4 Kbps (2G) 100 Mbps (4G)	10–100 km	824 and 1,900 MHz	GSM/GPRS/EDGE, (2G), UMTS/HSPA (3G), LTE (4G)	NAN, WAN
	Wi-SUN	300 Kbps	500 m–5 km	900 GHz	IEEE 802.15.4g	
	Sigfox	100 bps	30–50 km (rural) 3–10 km (urban)	868 MHz	Sigfox	
	LoRaWAN	50 Kbps	10–15 km (rural) 2–5 km (urban)	900 MHz	LoRaWAN	

7.9.1 Wired Communication

In the following, we list the common wired communication methods used in smart homes.

7.9.1.1 Coaxial Cable

CT Coaxial cable relies on Data Over Cable Service Interface Specification (DOCSIS) principles that are approved by the International Telecommunication Union (ITU), under ITU-T J.222 recommendation. This technology comes with six versions, which was first initiated in 1997 with version 1.0. The most recent version is DOCSIS 3.1 Full Duplex, completed in 2017, which supports a transmission rate up to 10 Gbps at a frequency range of 0 MHz to 2 GHz. The technology can cover a range of approximately 28 km. Although earlier versions of this technology provided quality of service (QoS) features such as capacity enhancement and channel bonding where the channel can be coupled with an adjacent channel that has the same frequency band to enhance data transmission rate, DOCSIS 3.1 Full Duplex provides symmetrical streaming and enhanced uploading speeds. Considering its high data rate, the technology can make a good communication medium in smart homes' NANs and HANs where most of data congestions occur.

7.9.1.2 Digital Subscriber Line

Digital subscriber line (DSL) technology is a wired communication medium that transfers data through the standard telephone lines at a frequency range of 4,000 Hz to 4 MHz depending on the type. The technology is approved by ITU G.992 recommendations. Three different types of this technology are commonly used: Asymmetric DSL (ADSL) offers a data rate of 1–8 Mbps with a maximum coverage of 5 km, high bit-rate DSL (HDSL) provides a data rate of 2 Mbps for a distance of 3.6 km, and very high-speed DSL (VDSL) can cover a maximum of 1.5 km distance but with relatively higher data rates that range between 15 and 100 Mbps. Unlike ADSL, the key feature associated with HDSL type is the symmetricity of downstream and upstream where it processes an equal data packet in both directions. DSL technologies can be integrated with IoT-based smart homes' HANs and NANs. They can also provide access to rural areas and distribution and transmission substations.

7.9.1.3 Fiber Optic

Fiber optic technology uses a glass medium where data packets are transmitted as a form of light pulses standardized by various international organizations such as ITU-T G.652 and International Electrotechnical Commission (IEC), IEC 60793. The technology is known for consistent transmission speeds and long range with a frequency band of 180–330 THz. Fiber optic can provide a data rate up to 10 Gbps for a coverage range of 10–60 km. In fiber optic communication, transmission reliability is proved where link failure is less likely to occur relative to copper-based wired transmission. The high data transmission rate and low costs make the technology a possible alternative to handle network requirements in NANs and WANs. However, fiber optic wires are fragile, and thus, difficult to be integrated in dynamic regions and/or above the ground through electric power transmission towers.

7.9.1.4 Power Line Carrier

Power line carrier (PLC) technology makes use of the electric power transmission infrastructure to provide bidirectional data communication. The technology comes in two types: HomePlug and Narrowband PLC and standardized under IEC 61131-3. The frequency band of HomePlug PLC is 1.8–250 MHz, whereas for Narrowband PLC, it is 3–500 kHz. HomePlug PLC has a data rate between 14 and 200 Mbps, whereas the data rate for Narrowband PLC ranges from 10 to 500 Kbps. Coverage range differs due to the difference in data rates. Unlike HomePlug that can cover a distance of 200 m, Narrowband can reach up to 3 km in coverage distance, which makes it a good alternative for NAN or WAN applications.

7.9.1.5 Ethernet

Ethernet communication is a conventional communication protocol that targets connecting HAN devices. The technology is standardized under IEEE 802.3. Data rate in Ethernet can range from 10 Mbps to 10 Gbps but with a short communication distance of 0–100 m with a frequency band of 100–500 MHz. Considering this high data rate, Ethernet can be used in smart homes' HANs.

7.9.2 Wireless Communication

The wireless communication techniques can be classified into LAN protocols or NAN/WAN protocols.

7.9.2.1 LAN Protocols

We list the most commonly used communication LAN protocols as follows:

Wi-Fi: It is the most popular wireless technology used in home automation. It is based on IEEE 802.11 standards and can operate in 2.4, 3.5, and 5 GHz frequency bands. It can provide a data rate up to 300 Mbps. It provides secure and high-speed communication. Its range is up to 100 m [30]. Further, in Ref. [31], the authors described the way to enhance the Wi-Fi network by using small cells, which improves the spectrum as well as the communication.

ZigBee: It uses different frequency bands, but mostly, it operates on 2.4 MHz. Recently, ZigBee technology has been considered as an open environment for developers to design many applications that can work with it. As a result, currently, there are more than 1,200 products certified and compatible with a ZigBee hub. Early on, interoperability of the connected devices, which are made by several manufacturers, was a big challenge facing ZigBee. However, recently released versions tend to have a better operability, regardless of their manufacturer or version.

Z-Wave: It operates on 2.4 GHz frequency. As a result, it is not affected by interference, and there are no traffic jams in this band. One significant advantage of Z-Wave is the interoperability of the devices connected to it, which means that the devices communicate to all other Z-Wave devices, regardless of the type, version, or brand. Further, the interoperability is backward and forward compatible in the Z-Wave ecosystem. In order to set up and manage the home automation network, Z-Wave uses one central hub. In addition, the user can easily add any smart home devices and manage them using the Z-Wave protocol once the network is installed [8]. With more than 1,700 certified Z-Wave devices available in the world, it is easy for the homeowners to select from a plenty of options for automating their homes. In addition, Z-Wave devices can be easily set up and use and consume less energy.

UPB: It stands for universal power line bus. It is a home automation protocol that uses the wired communication technology. UPB was released in 1999 and is considered as one of the most technically advanced protocols. Based on the X10 standard, the UPB home automation protocol uses the power lines for enabling communication between different devices. UPB devices can connect to the network using two enabling devices, a central home controller which is manually set up by the user, and links for each device connected to the network. Further, UPB protocol doesn't require too much technical savvy for the users to set up and run. Compared to the other protocols, UPB has fewer home automation products available nowadays. One reason for this is that UPB is difficult to be combined with any wireless protocols. However, there are around 150 commercial UPB-compatible products existing in the market. UPB is one of the most reliable protocols. This is because it is hard-wired into the power lines, so a limited interference is experienced compared to wireless home automation technologies.

Thread: It is another remote convention for intelligent home gadgets. More than 250 gadgets can be associated with a single Thread arrangement. Further, the majority of the gadgets associated with the system are battery-worked in order to be economical in power consumption. Utilizing indistinguishable recurrence and radio chips from ZigBee, Thread is expected to give a dependable, low-controlled, self-recuperating, and secure system that makes it basic for individuals to interface in excess of 250 gadgets in the home with one another. Besides, they can be associated with the cloud for omnipresent access. Home Learning Thermostat and Nest Protect are utilizing a rendition of Thread, and more items should enter the market soon.

Insteon: Insteon protocol is a hybrid technology that uses both wireless and wired communication technologies. This makes it unique to be added to the field of home automation. Insteon home automation runs on a patented dual-mesh network that uses both wired and wireless communication technologies to overwhelm general difficulties facing each type while operating solely. An Insteon hub connects with Insteon-compatible devices. This creates the ability to control the user's smart home via smartphones. With more than 200 Insteon-compatible home automation devices available commercially, Insteon tends to rely more on them, unlike many other protocols. This creates a restricted compatibility with smart devices made by other manufacturers. On the other hand, Insteon-compatible devices have a feature to be forward and backward compatible, meaning that they can work well with both new and old versions of devices. With this technology, the user needs only to know how a smartphone is used. Moreover, as the user turns any Insteon-compatible devices, they can be automatically added to the network. This makes them speedy-setup devices and the user faces no troubles with connecting them. In addition, the Insteon protocol shows a notable feature regarding the network size, the network is not limited, and it can be as large as the user desires. This is because that the network is a dual-mesh, and one Insteon hub is able to work with hundreds of devices across a broad range, without facing any problems.

Bluetooth: It is another short-range communication alternative that can be utilized for in-door systems since it has a reasonable scope of 1–100 m. It follows the IEEE 802.15.1 standard to exchange information at a rate of 700 Kbps/2.4 GHz. It is well known by its low power usage. The far reaching of this innovation among the edge devices, particularly cell phones, gives it another preferred standpoint to be incorporated in home networks [31]. It is one of the quickest developing conventions for home mechanization, and it is incorporated into numerous related gadgets.

EnOcean: EnOcean technology is an energy-harvesting technology. It means that it produces its energy from natural resources, that is, ambient light. It is cost-efficient and can be used in battery-less equipment. Its maintenance cost is very low. It can provide the data up to 125 Kbps, and it operates on 902 and 315 MHz [30].

Wave2M: It is designed for ultra-low-power transmission. It is used for long-range transmission for a small amount of data. Its range can be up to 1,000 m. It can provide the data rate up to 100 Kbps, and it operates on 2.4 GHz frequency.

Near-field communication (NFC): It is also a short-range (~10 cm–20 m) communication method, and it consists of a reader and a tag. The tag can vary and come in different alternatives such as stickers, cards, or unpowered labels appended to the house structure. Furthermore, NFC enables distributed correspondence in which multiple tags ought to be controlled. It provides an electronic check of the surrounding appliances, which thus enables giving better administrations to home residents. It gives low power and ease remote availability inside a short scope, which makes it appropriate for integration with WSNs, M2M, and IoT-empowered paradigms in the Big Data era.

ONE-NET: It is an open-source standard for remote system administration. It is intended for minimal effort and low-energy control systems, for example, home robotization, security checking, gadget control, and sensor systems. It can cover up to 100 m and provide a communication rate that can reach 38.4 Kbps at a frequency of 915 MHz.

7.9.2.2 Neighborhood/Wide Area Network

We list the most commonly used communication NAN/WAN protocols as follows.

Narrowband IoT **(NB-IoT):** The NB-IoT technology plays an important role in home automation. This technology consists of 3G, 4G, LTE, and LTE-A modes that provide the capability of Big Data transmission. It rates from 14.4 Kbps (for 3G) to 100 Mbps (for 4G) at a licensed frequency band (824 and 1,900 MHz). This technology targets a large area because of its long domain that ranges between 10 and 100 km. With that being said, the NB-IoT technology consumes high power for the transmission process.

Sigfox: It is a new emerging long-range communication method that operates at a frequency band of 868 MHz covering 30–50 km. Its transmission rate is equal to 100 bps. Leverage with this innovation, is recognizable in its low power utilization for information transmission. Sigfox gives a mid-arrangement considering scope of inclusion and power utilization against the NB-IoT innovation.

LoRaWAN: It is an ongoing non-gainful association that is chiefly settled to be coordinated in IoT WAN applications. This technology works at 900 MHz frequency to transmit at a rate of 50 Kbps for 10–15 km in provincial regions and 2–5 km in urban zones [29]. With respect to Sigfox, LoRaWAN offers a superior transmission rate with diminished power utilization.

Wi-SUN: This technology follows the IEEE 802.15.4g standard in which it works at a low frequency of 900 MHz to transmit the gathered information at a rate of 300 Kbps for a zone area of roughly 500–5 km. It gives a higher information transmission rate in contrast to LoRaWAN and Sigfox technologies. The value of this innovation is found in low inertness correspondences that settle on it a decent decision for home robotization applications.

In Table 7.6, our comparison is summarized among these different wireless communication protocols.

7.10 Challenges and Design Issues

Integrated IoT/Big Data solutions for home automation are expected to be available very soon. One of the main objectives specified is to have ubiquitous communication anytime and anywhere between all home devices/people. In this section, we overview significant challenges and design factors towards realizing this vision. These issues can significantly affect energy and spectral efficiency, QoS, and availability.

Table 7.6 Comparison of the Used Wireless Communication Protocols

Technology	*Cost*	*Power*	*Speed*	*Operating Frequency*	*Operating Range (up to Meters)*
WIFI	High	High	300 Mbps	2.4 GHz	100
ZigBee	Low	Low	250 Kbps	2.4 GHz	100
Z-Wave	Low	Low	30 Kbps	2.4 GHz	30
Bluetooth	Low	Low	21 Kbps	2.4 GHz	100
IEEE 802.15.3a	High	High	20 Mbps–1.3 Gbps	3.1–10 GHz	10
EnOcean	Low	Moderate	125 Kbps	315 and 902 MHz	30
Wave2M	Low	High	100 Kbps	2.4 GHz	1000
RFID	Low	Low	4 Mbps	120 KHz–10 GHz	10 cm–200 m
ONE-NET	Low	Low	38.4 Kbps	915 MHz	100
LTE	High	High	50–100 Mbps	450–2600 MHz	Mobile
NB-IoT	Low	Moderate	14.4 Kbps (2G) 100 Mbps (4G)	824 and 1900 MHz	10–100 km
Wi-SUN	Low	Moderate	300 Kbps	900 GHz	500 m–5 km
Sigfox	Low	Low	100 bps	868 MHz	30–50 km (rural) 3–10 km (urban)
LoRaWAN	Low	Low	50 Kbps	900 MHz	10–15 km (rural) 2–5 km (urban)
LoRa	High	Moderate	0.3–37.5 Kbps	900 MHz	10–15 km (rural) 3–5 km (urban)

Interoperability: It is defined as the ability of system, applications, and services to work together in a predictive fashion. It is a primary concern because consumer must have easy-to-connect and easy-to-use devices. In smart homes, devices come from different vendors with different network interfaces. So, they have to interoperate to achieve the join execution of tasks. Recently, a large number of devices have been working on Wi-Fi and ZigBee because they allow a wide range of devices to interoperate. Different standards have been taken into account, but still there are some areas that need improvement. In Ref. [32], the author has described briefly about the interoperability and connectivity issues.

Self-management: Intelligent systems can monitor their own health and notify the user about the potential issues before a critical situation. It is the main requirement of the sensor node that it should adopt itself according to the changes in the environment. It should completely independent of human intervention. It should collaborate with other devices independently. Further, in Ref. [32], the author explained briefly about the self-management issue.

Maintainability: It is a fundamental necessity in a system that reflects how dependable and tough the savvy home system is. Unexpected events like falling in flat hubs, tired batteries, and new errands can happen. So, the system should monitor its own strength and change operational parameters, such as providing low quality with limited energy resources, accordingly [32].

Bandwidth: It is another challenge for IoT connectivity in Big Data. Managing bandwidth in a network is crucial because a huge amount of data is created with increasing number of devices. Specially, video streaming is the popular application that requires the highest amount of bandwidth. So, the system should be capable of connecting and transferring data without any delay and data loss [32].

Power consumption: Many IoT devices connected in smart homes send and receive signals, and CPUs process the data, which causes the consumption of power. An efficient IoT network needs minimal energy to give better performance. It is a trade-off between power and data. If your system needs to send and receive more data, then it will consume more power [32].

Integration: By the passage of time, many companies are manufacturing smart equipment for smart homes and cities. Users have a variety of choices to purchase equipment, but the problems occur when one brand is working on a different frequency than other. Then, integration problems occur, which lead to complexity.

Data storage: As applications are increasing, the amount of data getting collected is huge. To store more data, storage is required, which increases the cost.

High cost of ownership: Smart homes need plenty of new devices such as sensors, relays, smart appliances, and embedded systems. Therefore, these things cost money, which increases the high cost of ownership. Industry is still facing problems to generate and install smart equipment in affordable prices. Using cloud technology in the IoT era can somehow lower the cost for network deployment and complex data processing while facilitating a quick setup and integration of newly connected devices.

7.11 Security and Privacy Requirements

Secure smart homes are the need of today's world. Along with all features, security is the main concern for people living in smart homes. To secure the homes, there are several requirements for the services. Many researches have considered different requirements for the security of smart homes. In Ref. [33], the authors considered some basic security and privacy requirements. These requirements are data confidentiality, data integrity, device-to-device authentication, and so on. Further, the security and privacy are divided into several main categories, for example, confidentiality, integrity, and availability. By fulfilling these requirements, we can secure smart homes at the moderate level.

In Table 7.7, we have summarized these basic security requirements in home automation.

Table 7.7 Security and Privacy Requirements for Smart Home

Category	*Security Requirement*
User and device authentication	In a shrewd home, numerous gadgets are associated with the Web. These gadgets require programming refreshes, security spackle, and information trade. Just-approved clients can play out every one of these procedures; however, a solid verification component or key administration system is required for unapproved gadgets and clients to spare the shrewd home from enemy [33,34].
Monitor the network	In intelligent home distinctive substance (home machines, Electronically Stored Information (ESI), and Renewable Energy Sources (RES) associated with a system, the foe can target shrewd home system through DoS. Thus, new methods shall be introduced for checking any interruption in location instrument without observing any apparatus.
Integrity	It guarantees that data can't be changed by an unapproved client amid of any procedure [13]. Integrity guarantees that data is transmitted in a steady and right way. However, attackers break the integrity through malignant programming assaults [35].
Availability	It guarantees that all the system administrations and assets are securely accessible [36]. In intelligent home, pernicious and DoS attack the system administrations and assets. Calamity recuperation arrangement is connected in the targeted home system to guarantee accessibility [35].
Confidentiality	It guarantees the protection of clients. In secrecy, clients' private data is kept and just-approved clients can access it. Cryptography and key agreements are utilized to accomplish confidentiality in smart homes [21].

7.12 Security Threats and Solutions

Security challenges in home automation can be classified into layer and attack security issues. A typical IoT architecture in Big Data is shown in Figure 7.6. IoT consists of three main layers [37]: perception/physical layer, network/transport layer, and application layer. The perception layer is the hardware layer. It consists of sensors and actuators that send/receive data. The network layer deals with data routing. It uses the aforementioned communication protocols in providing different security levels. The application layer is the top layer that provides systems with the business flavor and offers the interfaces to users as illustrated in Figure 7.4 [38]. Exchanged data in these layers are exposed to security hazards. Meanwhile, malicious and common security attacks can be a threat as well. We list these attacks as follows.

Authentication: It is defined as the verification of communicating parties or users, that is, who they are and what they claim and what data should be sent to specific users. Authentication is the major feature of security. The server, which has desired information of users, should give access to authentic users. Solutions for this type of attacks are hard passwords and generating random captcha [39].

Man-in-the-middle: It means an attacker or hacker breaches or interrupts communication between the two systems. For example, fake temperature data generated by temperature sensors can be imitated and forwarded to the servers and the Cloud. Reasonable solutions for this kind of attack are strong cryptography, secure authentication, and data integrity verifications.

Data and identity theft: Data that is generated by unprotected or less secured wearables and appliances can be used by cyber attackers. For example, personal information can be exploited for fake or fraudulent transactions and identity theft. Some of the solutions to prevent these attacks are strengthening passwords, limiting how much information you share, protecting your mobile devices with passwords, and so on [40].

Device hijacking: It means the hacker hijacks the device and uses it according to his or her own purpose and will. These attackers are hard to detect because they do not change the basic functionality of the device. So, users cannot notice these

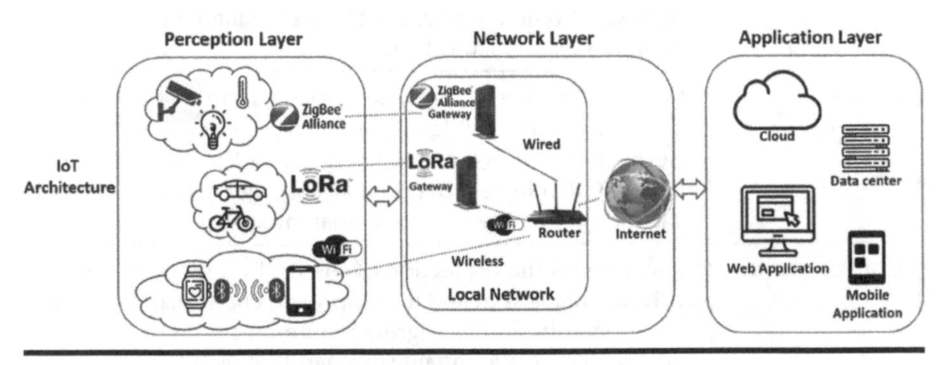

Figure 7.6 The IoT-based big data layers.

changes. Moreover, it takes only one device to infect the whole home devices. For example, an attacker who initially compromises a thermostat can gain access to the doors or other appliances. He or she can change the PIN to restrict the original authenticated person. To prevent device hijacking, we need to strengthen your passwords and regularly update your devices [41].

Denial of service (DoS): The attacker breaks down the network by using malicious nodes to forge a significant number of bogus identities, such as IP addresses, with the objective of disrupting the proper functioning of data and information transformation.

Distributed DoS (DDoS): A DoS means an attacker can render any device or network resource unavailable for authenticated user temporarily. This DoS can be power, Internet, and so on. DDoS attack originates from multiple sources, making it difficult to stop, simply by blocking a single source. For example, a compromised smart sensor can infect the same device which runs the same software. To prevent these attacks, we can deploy anti-DDoS hardware and software modules, and protect the Domain Name System (DNS) server.

Permanent DoS (PDoS): In PDoS, the damage of the attack is much severe, and it damages the device so badly that it requires replacement or reinstallation. For example, BrickerBot is designed to exploit hard-coded passwords in IoT devices and cause PDoS. Reasonable solutions for these types of problems are to protect DNS servers and improve firewall security.

Sybil: The perpetrator creates multiple identities in an effort to simulate multiple nodes. This type of attack is significantly hazardous in the constraint dense IoT networks because a member can also potentially claim to be in a different spot at the same time causing substantial security risks and chaos in the network.

Masquerade: In this type of attack, the member is actively masking its own identity to appear like another vehicle by using false identities, such as public keys. This technique is usually employed in conjunction with other types of attacks. This type of attack can be prevented by establishing a multi-person approval process.

Eavesdropping: This is a type of passive attack in which an unauthorized interception occurs on an ongoing communication without the consent of communication parties. Various forms of communication include phone calls, instant messages, or any other Internet service. To prevent this attack, encryption is the best possible solution.

False information injection: Attacker transmits erroneous information and data in the network, which might affect the behavior of other drivers. It can be both intentional and unintentional.

Black hole: is a network member that has some nodes (gadgets) that refuse to broadcast or forward data packets to the next hop. This attack can be avoided by keeping redundant paths between the sender and the destination.

Security countermeasures: Connected gadgets ought to be ensured by a complete IoT security arrangement that does not disturb the service provider, OEMs gainfulness, and the advertisement time. A thorough IoT security arrangement ought to incorporate the accompanying capacities: secure booting, mutual authentication, secure communication (encryption), security monitoring and analysis, and security life cycle management.

Meanwhile, security threats can be solved in different ways while considering the smart home projects. We overview some security and privacy solutions in the sections that follow.

7.12.1 Cryptography

Cryptography is derived from the Greek word *kryptos* that means "hidden," and it is the process of securing messages [24]. In cryptography, a simple text is converted into a complex form that allows only the sender and intended person to view its content. Cryptography concerns with confidentiality, integrity, and authentication of data. The aim of cryptography is to build protocols that can complete tasks even in the certain adverse environment. The first task of cryptography is the secure communication over an insecure channel in the home to guarantee the user's authentication, transmission, and privacy. Cryptographic algorithms are the backbone of security and privacy in smart homes nowadays because they are more secure and can avoid attacks during the data storage, transmission, and processing. However, typical encryption standards are not fully suitable for energy-constrained devices and their limited computational power.

7.12.2 Blockchain

Blockchain is an emerging technology that provides a distributed ledger or database, which is shared among all participants in the home network based on the consensus mechanism. This creates completely decentralized, secure, time-stamped and shared tamper-proof ledgers [42]. Due to its robust and decentralized infrastructure, blockchain technology is applied to handle issues related to trust, efficiency, privacy, and data sharing [43]. This technology eliminates the requirement of a third-party transaction authority by leveraging the potential of cryptography algorithms to provide trustworthy solutions for the entities participating in the chain.

7.12.3 Machine Learning

On the basis of current security and privacy issues, ML is another very efficient solution to overcome these security threats. ML technologies have improved the efficiency of intrusion detection system, which is the most commonly used security infrastructure to protect the networks from attacks. In smart homes, WSNs have gained lots of interest. To secure these WSNs, many ML technologies have been adopted. Authors in [43] proposed an ML-based technique to secure data and fusion in WSNs. To detect the attacks in Wi-Fi networks, a recent study developed a new extraction and selection model, which has high detection rate [44].

7.12.4 Adaptive Key Agreement

IoT architecture consists of constrained nodes that can communicate securely by a shared key. Key agreement is the critical part of this security approach. Symmetric keys are preferred usually for their simplicity and efficiency. In smart homes, smart meters, for instance, are used to measure the real-time power consumption. Data of these meters are aggregated based on a key agreement policy and sent to a control center to optimize the energy consumption. These meters contain sensitive data, and thus, there are several security threats. Moreover, the increase in data causes a new challenge of managing secrets keys for every device in smart homes. Keys must be generated, stored, communicated, and used in a secure way. Key distribution is a critical problem for symmetric keys in the distributed mobile system at homes. In Ref. [44], the authors propose a lightweight framework to strengthen the security of IoT networks. They introduce a cloud-supported mobile-sink authentication and an elliptic curve-based seamless secure authentication and key agreement (S-SAKA). However, in Ref. [45], the authors propose a hash and a global assertion value-based authentication scheme for the evolving 5G technology. Their proposal considers a context-sensitive seamless identity provisioning (CSIP) framework for futuristic IIoT.

7.12.5 Trusted Software-Defined Network Controller

Software-defined network (SDN) is a new paradigm of networking that offers many opportunities to protect the network in a more efficient and flexible way. It uses a special node that is called SDN controller, which is used for creating a communication medium with forwarding devices through diverse communication protocols. The SDN controller maintaining an IoT network topology view and a sleep/wake timing view can predetermine the packet's route and synchronize the hubs for safer routing. Another approach using the SDN controller is to create a random, dynamic route for each hop and route data packets through awake intermediate nodes.

7.13 Open Research Issues

Many attempts have been considered for the smart homes. Many problems and issues have been highlighted, but still there are some fields in which improvement is required. Following are some open research issues, although there are efforts that have been done, but they are not enough. Some are very critical and some have enough amount of fault finding. Following are some issues that need more attention.

Transformation of conventional homes to smart homes: The major problem is that how we can convert conventional homes to smart homes with minimum cost and design change. For example, if you want to transform your house to smart home, you have to modify lot of things from sensors to appliances. As we know, smart appliances cost money which everybody cannot afford. In the literature, researchers are not proposing a reasonable solution to this problem. Mostly, researchers suggest that smart homes should be built from scratch because of huge physical changes.

Interoperability between different brands: Different manufacturer companies manufacture their devices with different standards. Nowadays, consumers have many options to purchase smart products for their house. Although there are some standards that have been manipulated worldwide, still it is the main concern. Consumers face problems when they install different products manufactured by different companies. These factors generate problems such as interoperability and integration.

Inflexibility: Some of the equipment and system have preinstalled functionalities. According to studies, people want more flexible applications in which they need personal control over the application. For example, in a family every person wants to watch his or her favorite show at the same time. Perhaps, the structural changes are the main hurdle for broad adaptation of home automation.

Energy efficiency: It is the main concern in the communication devices. Different devices consume energy according to their need, and performance is directly proportional to energy consumption. Different communication techniques and protocols are being used in the smart homes, but this department needs improvement. There are different communication protocols such as ZigBee, Wi-Fi, and Bluetooth, but each has its own advantage and disadvantage over other.

Security: As the wireless technology is evolving day by day, along with the growth challenges in the security area are also increasing. Enough efforts have been done, but still these are not adequate. Security issues include authentication, DoS, and data and identity theft. Moreover, handheld gadgets are usually lightweight physical gadgets that are hard to anchor. They can be stolen or broken. This may affect secrecy, as well as information accessibility. Besides, many hazard partners, life partners, and relatives can be physically accessed to their connected sensors and/or related accounts.

7.14 Conclusion

In this chapter, we have surveyed different home automation systems along with their methodologies, techniques, architecture, and challenges. We have overviewed a precise description for smart homes in the smart cities and IoT-enabled environments. Moreover, we have discussed the efforts that have been done to promote smart homes along the existing challenges. We have described what have been done in the past few decades, the technologies that have been used, and the challenges that researchers are still facing. Further, we have discussed the main objectives and constraints to accomplish and overcome. We have overviewed different applications of smart homes along with implementation issues in practice. We have briefly described the technical details of applications along with required devices and functionality, and reported the main components such as sensors, actuators, and appliances in addition to their functionality in the overall system. We have explained the architecture of smart homes in detail. Although researchers suggest varying architectures, most of the studies suggest and implement a three-layered architecture as reported above. We have described the related trustworthy computing models for secure smart home applications and architectures. We have listed and discussed the samples of the different automation techniques. We have categorized the used devices into sensing, appliances, entertainment, and security devices; discussed security deployment issues at home; and investigated different communication protocols along with their details. We have overviewed each technology along with its cost, performance, and other design factors. Different security threats have been reported and categorized. Conventional security challenges, for example, authentication, man-in-the-middle, and identity theft, have also been included. Finally, open research issues were mentioned. These issues are very critical. Although some efforts have been done, still it needs further attention. Among the most common issues, which need special care, are the security and interoperability issues. From our survey on smart homes, we did not find any uniform concept for the secure smart homes in the smart city paradigm.

References

1. F. Al-Turjman, 5G-enabled devices and smart-spaces in social-IoT: An overview, *Elsevier Future Generation Computer Systems*, vol. 92, no. 1, 732–744, 2019.
2. F. Al-Turjman and M. AbuJubbeh, IoT-enabled smart grid via SM: An overview, *Elsevier Future Generation Computer Systems*, 2019. DOI: 10.1016/j.future.2019.02.012.
3. S. Ijaz, M. A. Shah, A. Khan, and M. Ahmed, Smart cities: A survey on security concerns, *(IJACSA) International Journal of Advanced Computer Science and Applications*, vol. 07, 2016.
4. B. Brush, B. Lee, R. Mahajan, and S. Agarwal, Home automation in the wild: Challenges and opportunities, in *Proceedings of the International Conference on Human Factors in Computing Systems, CHI 2011*, Vancouver, BC, Canada, May 7–12, 2011.
5. L. Yuxin, D. Mianxiong, O. Kaoru, and L. Anfeng, ActiveTrust: Secure and trustable routing in wireless sensor networks, *IEEE Transactions on Information Forensics and Security*, vol. 11, 2013, 2016.
6. F. Al-Turjman, "The Road Towards Plant Phenotyping via WSNs: An Overview", *Elsevier Computers & Electronics in Agriculture*, vol. 161, pp. 4–13, 2019.
7. A. G. Fragkiadakis, I. G. Askoxylakis, E. Z. Tragos, and C. V. Verikoukis, Ubiquitous robust communications for emergency response using multi-operator heterogeneous networks, *EURASIP Journal on Wireless Communications and Networking*, vol. 2011, 1–16, 2011.
8. S. M. Shaheed, Effective smart home system based on flexible cost in Pakistan, in *The 4th HCT Information Technology Trends (ITT 2017)*, Dubai, 2017.
9. P. Waghmare, Survey on: Home automation systems, in *International Conference on Trends in Electronics and Informatics ICEI*, Tirunelveli, India, 2017.
10. P. P. Gaikwad, J. P. Gabhane, and S. S. Golait, A Survey based on smart homes system using internet-of-things, in *International Conference on Computation of Power, Energy, Information and Communication*, Chennai, India, 2015, pp. 330–470.

11. V. Yadav, Smart home automation using virtue of IoT, in *2nd International Conference for Convergence in Technology (I2CT)*, Mumbai, India, 2017, pp. 313–318.
12. F. Al-Turjman, "5G-enabled Devices and Smart-Spaces in Social-IoT: An Overview", *Elsevier Future Generation Computer Systems*, vol. 92, no. 1, pp. 732–744, 2019.
13. N. Komninos, Survey in smart grid and smart home security: Issues, challenges and countermeasures, *IEEE Communication Surveys & Tutorials*, vol. 16, no. 2014, 1933, 2014.
14. C. L. Hu, T. K. Chan, Y. C. Wen, T. Tantidham, S. Sanghlao, B. Yimwadsana, and P. Mongkolwat, IoT-based LED lighting control in smart home, in *2018 IEEE International Conference on Applied System Invention (ICASI)*, Tokyo, Japan, 2018.
15. J. Margulies, Garage door openers: An internet of things case study, *IEEE Security Privacy*, vol. 13, 80–83, 2015.
16. T. Balikhina, A. Al Maqousi, A. AlBanna, and F. Shhadeh, System architecture for smart home meter, in *2017 2nd International Conference on the Applications of Information Technology in Developing Renewable Energy Processes Systems (IT-DREPS)*, Amman, Jordan, 2017.
17. Applications of smart homes, 17-Mar-2015. [Online]. Available: www.link-labs.com/blog/applications-of-home-automation. [Accessed 16-Dec-2018].
18. K. Bing, L. Fu, Y. Zhuo, and L. Yanlei, Design of an Internet of Things-based smart home system, in *The 2nd International Conference on Intelligent Control and Information Processing*, northeastern China, 2011.
19. N. Jose, The best home automation apps to make your life easier. [Online]. Available: www.thinklions.com/blog/best-home-automation-apps/#6_Samsung_SmartThings_A_Central_Hub_For_Home_Automation. [Accessed 04-Dec-2018].
20. J. Y. Kim, Smart home web of objects-based IoT management model and methods for home data mining, in *2015 17th Asia-Pacific Network Operations and Management Symposium (APNOMS)*, Busan, Korea, 2015, pp. 327–331.
21. H. Zheng, H. Zhang, and L. Pan, Modeling and analysis of ZigBee based smart home system, in *2014 5th International Conference on Digital Home*, Guangzhou, China 2014, pp. 242–245.
22. C. Cetina, P. Giner, J. Fons, and V. Pelechano, Autonomic computing through reuse of variability models at runtime: The case of smart homes, *Computer*, vol. 42, 37–43, 2009.
23. S. Sen, S. Chakrabarty, R. Toshniwal, and A. Bhaumik, Design of an intelligent voice controlled home automation system, *International Journal of Computer Applications*, vol. 121, 39–42, 2015.
24. H. AlShu'eili, G. S. Gupta, and S. Mukhopadhyay, Voice recognition based wireless home automation system, in *4th International Conference on Mechatronics*, Kuala Lumpur, 2011.
25. Smart home sensors, IBM, 2016. [Online]. Available: www.ibm.com/blogs/internet-of-things/sensors-smart-home/.
26. R. Sharma, Top 15 sensor types being used in IoT. [Online]. Available: www.finoit.com/blog/top-15-sensor-types-used-iot/.
27. H. Liu, Smart solution, poor protection, in IoT S&P'1, Dallas, TX, USA, 2017.
28. F. Al-Turjman, *Smart-cities: Performability, Cognition & Security*, Switzerland: Springer, 2019. ISBN 978-3-030-14717-4.
29. F. Al-Turjman, *Intelligence in IoT-Enabled Smart-Cities*, New York: Taylor & Francis, 2019. ISBN: 9781138316843.

30. M. Kuzlu, M. Pipattanasomporn, and S. Rahman, Review of communication technologies for smart homes/building applications, in *2015 IEEE Innovative Smart Grid Technologies-Asia (ISGT ASIA)*, Columbus, OH, USA, 2015.
31. F. Al-Turjman, E. Ever, and H. Zahmatkesh, Small cells in the forthcoming 5G/IoT: Traffic modelling and deployment overview, *IEEE Communications Surveys and Tutorials*, vol. 21, no. 1, 28–65, 2019.
32. S. S. Samuel, A review of connectivity challenges in IoT-smart home, in *3rd MEC International Conference on Big Data and Smart City (ICBDSC)*, Muscat, Oman, 2016.
33. J. H. Han, Y. Jeon, and J. Kim, Security considerations for secure and trustworthy smart home system in the IoT environment, in *2015 International Conference on Information and Communication Technology Convergence (ICTC)*, Jeju Island, Korea, 2015.
34. A. Dorri, S. S. Kanhere, R. Jurdak, and P. Gauravaram, Blockchain for IoT security andrrivacy: The case study of a smart home, in *2017 IEEE International Conference on Pervasive Computing and Communications Workshops (perCom Workshops)*, Hawaii, USA, 2017.
35. R. Roman, J. Zhou, and J. Lopez, On the features and challenges of security and privacy in distributed internet of things, *Computer Networks*, vol. 57, no. 2013, 2266–2279, 2013.
36. M. Gaboardi and C. J. Skinner, Special issue on the theory and practice of differential privacy, *Journal of Privacy and Confidentiality*, vol. 7, no. 2, 2017.
37. S. Krushang and H. Upadhyay, A survey: DDOS attack on internet of things, *International Journal of Engineering Research and Development*, vol. 10, no. 11, 58–63, 2014.
38. I. B. Ida, A. Jemai, and A. Loukil, A survey on security of IoT in the context of eHealth and clouds, in *2016 11th International Design Test Symposium (IDT)*, Hammamet, Tunisia, December 2016, pp. 25–30
39. F. Al-Turjman, H. Zahmatkesh, "An Overview of Security and Privacy in Smart Cities' IoT Communications", *Wiley Transactions on Emerging Telecommunications Technologies*, 2019. DOI. 10.1002/ett.3677.
40. B. O'Shea, How to prevent identity theft, nerdwallet, 04-Oct-2018. [Online]. Available: www.nerdwallet.com/blog/finance/how-to-prevent-identity-theft/. [Accessed 20-Dec-2018].
41. Here is how to fend off a hijacking of home devices. [Online]. Available: www.nytimes.com/2017/02/01/technology/personaltech/stop-hijacking-home-devices.html. [Accessed 20-Dec-2018].
42. S. Nakamoto, Bitcoin: A peer-to-peer electronic cash system, Whitepaper, 2008.
43. J. Leon Zhao, S. Fan, and J. Yan. Overview of business innovations and research opportunities in blockchain and introduction to the special issue in financial innovation, vol. 2, no. 1, 1, 2016.
44. F. Al-Turjman, Y. K. Ever, E. Ever, H. X. Nguyen, and D. B. David, Seamless key agreement framework for mobile-sink in IoT based cloud-centric secured public safety sensor networks, *IEEE Access*, vol. 5, 24617–24631, 2017.
45. F. Al-Turjman and S. Alturjman, Context-sensitive access in industrial internet of things (IIoT) healthcare applications, *IEEE Transactions on Industrial Informatics*, vol. 14, no. 6, 2736–2744, 2018.

Chapter 8

Learning in Cities' Cloud-Based IoT

Fadi Al-Turjman and Aissa Houdjedj

Antalya Bilim University

Contents

8.1 Introduction

We are living in the age of information technology (IT) revolution, where information can be used to improve our lives and speed up the process of decision-making for more competitive businesses. In order to have all decision makers connected, ubiquitous and intelligent communications have become a must. It employs mobility with context awareness, adaptability, scalability, and localization to create an environment where devices are smarter and take actions by predicting user behavior. Hence, a large variety of areas including energy conservation, manufacturing, health care, banking, education, and telecommunications are potential applications. A smart city (SC) is a principle of using the IT revolution and existing infrastructure in management for different purposes, such as planning, analysis, and improving the quality of new/existing services. In this sense, SCs include smart management, smart transportation, smart technologies, smart economy, and smart health [1].

Meanwhile, machine learning (ML) is a technique for artificial intelligence. Over time, its focus evolved and shifted more to algorithms that are computationally viable and robust. In the last decade, ML techniques have been used extensively for a wide range of tasks including classification, regression, and density estimation in a variety of IT applications such as bioinformatics, speech recognition, spam detection, computer vision, fraud detection, and advertising networks. These algorithms and techniques come from diverse fields including statistics, mathematics, neuroscience, and computer science. It has been used even boarder, and most areas are related to ML nowadays [2].

Recently, several countries have focused on new projects for SCs in metropolitan areas, and huge budgets have been allocated for this purpose [1]. In SC applications, it is necessary to collect, evaluate, and analyze the data to be obtained from varying sources of information and sensors and interpret the results, and the actions needed to be performed [3]. The action to be performed can be reducing the cost, taking new managerial decisions, improving the quality of service, and so on. For instance, smart traffic application gives real-time road information for the drivers by means of mobile applications. This information may indicate real-time situations such as instant traffic density on the route, maintenance work, and accidents reporting. The input data of the system to be created for this purpose consist of traffic lights, city surveillance, and cameras. The system has outputs that can be used to inform drivers, perform dynamic traffic light operations, and report driving directions [4].

Smart grid is another example for the use of ML in SCs. Many sensors are added to the smart grid, and the massive information collected from the end nodes is valuable for researchers and smart grid operators. In this case, advanced analytics on the smart grid are needed, a combination of ML algorithms and data mining techniques [5]. By exploiting the emerging smart grid collected data, we can develop data-driven solutions for the most pressing issues, such as prediction of electricity demands per region in an SC, residential photovoltaic system detection, electrical vehicle charging demand modeling, and time-variant load modeling.

SC applications generally require algorithms with different types of data inputs and outputs that will perform real-time learning on this big data. Traditional ML methods are not sufficient for SC applications in terms of processing power and memory consumption. Accordingly, several attempts have been introduced in the literature to improve these existing ML techniques. For example, deep learning (DL) is successfully applied in image recognition, object tracking, and analysis and interpretation applications on big and multilayer data [6]. Deep belief networks, convolutional neural networks (CNNs), and deep Boltzmann machines are widely used for DL applications in the literature. We can summarize our main contributions in this work as follows:

1. We overview all suitable ML techniques in the SC paradigm.
2. We overview the applications of ML algorithm in SCs.
3. We summarize the main components and basic layers of the SC architecture.
4. We summarize the key design factors that are required when selecting an ML algorithm for application in an SC.

5. We categorize the different ML algorithms used in SC use cases.
6. We highlight the wireless technologies used in SC that will enable an effective deployment of ML models in such network.
7. We summarize the main open research issues and challenges.

The organization of this chapter is as follows. Section 8.1 reviews the related academic surveys presented in the literature and outlines the contributions of this chapter. Section 8.2 overviews the ML techniques used in SCs. Section 8.3 presents the architecture and components of an SC considering different SC use cases. Section 8.4 provides some application of ML in SCs. Section 8.5 demonstrates the main design factors to be considered in developing and deploying an ML solution in SC. Section 8.6 defines the communication technologies in SC. Section 8.7 discusses the commercial products provides IoT Platform as a Service in SC. Section 8.8 presents the used metrics in the evaluation of ML algorithm. Sections 8.9 reveals the open research issues in order to direct future research trends. Section 8.10 concludes our thoughts introduced in this chapter. Table 8.1 provides the definitions of used abbreviations in this chapter for more readability.

Table 8.1 Abbreviations

ML	Machine learning
AI	Artificial intelligence
IoT	Internet of things
SC	Smart city
CNN	Convolutional neural network
DL	Deep learning
MRI	Magnetic resonance imaging
fMRI	Functional magnetic resonance imaging
DRL	Deep reinforcement learning
DIVA	Deep Intermodal Video Analytics
FDI	False data injection
ADGTS	Adaptive double-fitness genetic task scheduling
OppNets	Opportunistic networks
MLProph	Machine learning PROPHET
QS	Quality of service
VO	Virtual object

8.2 ML Overview

ML deals with algorithms that give computers the ability to learn, in much the same way as humans. This means that given a set of data, an algorithm infers information about the properties of the data, allowing it to make predictions about other data it may see in the future. The main focus of ML is the design of algorithms that recognize patterns and make decisions based on the input data. ML has found uses in areas such as biotechnology, fraud detection, wireless networks, stock market analysis, and national security.

A learning algorithm takes a set of samples as an input named a training set. In general, there exist three main categories of learning: supervised, unsupervised, and semi-supervised (reinforcement) learning [7]. Our chapter focuses on supervised and unsupervised learning because they have been and are still widely applied in IoT smart data analysis in SCs.

8.2.1 Supervised Learning

These algorithms use training data to generate a function that maps the inputs to desired outputs (also called labels). For example, in a classification problem, the system looks at sample data and uses it to derive a function that maps the input data to different classes. Artificial neural networks, radial basis function networks, and decision trees are the different forms of the supervised learning [8]. As an example, it is used for smarter health care in SCs [9]. The authors in Ref. [10] use CNNs to help create a new network architecture with the aim of multichannel data acquisition and also for supervised feature learning. Extracting features from brain images (e.g., magnetic resonance imaging (MRI), functional MRI (fMRI)) can help in early diagnosis and prognosis of severe diseases such as glioma. Moreover, the authors in Ref. [11] use a Deep Belief Network (DBN) for the classification of mammography images in a bid to detect calcifications that may be the indicators of breast cancer.

8.2.2 Unsupervised Learning

This set of algorithms works without previously labeled data. The main purpose of these algorithms is to find the common patterns in previously unseen data. Clustering is the most popular form of unsupervised learning. Hidden Markov models and self-organizing maps are other forms of unsupervised learning. It is used, for example, for controlling air and traffic and monitoring public places; to define the objective of the unsupervised learning. One of the major objectives is to identify sensible clusters of similar samples within the input data, known as clustering. Moreover, the objective may be the discovery of a useful internal representation of the input data by preprocessing the original input variable in order to transfer it into a new variable space. This preprocessing stage can significantly improve the result of the subsequent ML algorithm and is named feature extraction [12].

8.2.3 Semi-Supervised Learning

As the name indicates, these algorithms combine labeled and unlabeled data to generate an appropriate mapping function or classifier. Several studies have proven that using a combination of supervised and unsupervised techniques instead of a single type can lead to much better results. For example, it's used in self-driving cars. The frequently applied ML algorithms in smart data analysis and SC use cases are shown in Tables 8.2 and 8.3 accordingly.

Table 8.2 Overview of the Applied ML Algorithms in SC Use Cases

ML Algorithm	*SC Use Cases*	*Metric to Optimize*	*References*
K-nearest neighbors	Smart citizen	Passengers' travel pattern, efficiency of the learned metric	[13,14]
Naive Bayes	Smart agriculture, smart citizen	Food safety, passengers travel pattern, estimate the number of nodes	[13,15]
Support vector machine	All use cases	Classify data, real-time prediction	[16]
Linear regression	Economics, market analysis, energy usage	Real-time prediction, reducing amount of data	[16,17]
Support vector regression	Smart weather prediction	Forecasting	[8]
Classification and regression trees	Smart citizens	Real-time prediction, passengers travel pattern	[13,16]
K-means	SC, smart home, smart citizen, controlling air and traffic	Outlier detection, fraud detection, analyze small dataset, forecasting energy consumption, passengers travel pattern, analyze stream data	[13,18–21]
Density-based spatial clustering of applications with noise	Smart citizen	Labeling data, fraud detection, passengers travel pattern	[13,22]

(Continued)

Table 8.2 (*Continued*) Overview of the Applied ML Algorithms in SC Use Cases

ML Algorithm	*SC Use Cases*	*Metric to Optimize*	*References*
Principal component analysis	Monitoring public places	Fault detection	[23]
Canonical correlation analysis	Monitoring public places	Fault detection	[17]
Feedforward neural network	Smart health	Reduce energy consumption, forecast the states of elements, overcome the redundant data and information	[17,24]

Table 8.3 Overview of Frequently Used ML Algorithms for Smart Data Analysis

ML Algorithm	*Data Processing Tasks*	*ML Type*	*References*
K-nearest neighbors	Classification	Supervised learning	[25,26]
Naive Bayes	Classification	Supervised learning	[27]
Support vector machine	Classification	Supervised learning	[28]
Linear regression	Regression	Supervised learning	
Support vector regression	Regression	Supervised learning	[29]
Classification and regression trees	Classification/ regression	Supervised learning	[30]
Random forests	Classification/ regression	Supervised learning	[31]
Bagging	Classification/ regression	Supervised learning	[32]
K-means	Clustering	Unsupervised learning	[33]

(*Continued*)

Table 8.3 (*Continued*) Overview of Frequently Used ML Algorithms for Smart Data Analysis

ML Algorithm	*Data Processing Tasks*	*ML Type*	*References*
Density-based spatial clustering of applications with noise	Clustering	Unsupervised learning	[34]
Principal component analysis	Feature extraction	Unsupervised learning	[35]
Canonical correlation analysis	Feature extraction	Unsupervised learning	[36]
Feedforward neural network	Regression/ classification/ clustering	Semi-supervised learning	[37]

8.3 Architectures and Components

Figure 8.1 illustrates the overall position of ML approaches within the hierarchy of the SC infrastructure where each component of the SC system is controlled by an intelligent software agent which is deployed in the edge or the Cloud depending on the characteristics of the required analytics. The first layer consists of the SC infrastructures, where a network of sensors, cameras, wireless devices, and data centers forms the key component that allows civil authorities to provide essential services in a faster and more efficient manner. The second layer is called the edge computing layer. In the early version of the Cloud paradigm, all collected data samples from the connected devices used to be transported to the centralized Cloud layer. However, sending all data items captured by a smart object to the Cloud without aggregation is not efficient [38]. Such an approach would lead to resource wastage in terms of the network bandwidth, and storage. The edge is proposed to counterpart this weakness by pushing the processes of knowledge discovery using data analytics to the edges of the Cloud. However, edge devices have limited computational capabilities. To address the weaknesses inherited by the Cloud computing paradigm, a notion called "edge computing" has been proposed. Edge computing encourages a shift of handling everything in the core of Cloud computing to handling at the edges of a network. To be specific, instead of sending all the data collected to the Cloud, edge computing suggests processing the data at leaf nodes or at the edges.

In the edge layer, for example, the massive data generated from the end nodes in the IoT infrastructure for smart traffic, weather, and human activity control prediction are done in the edge layer rather than the Cloud layer [39–41].

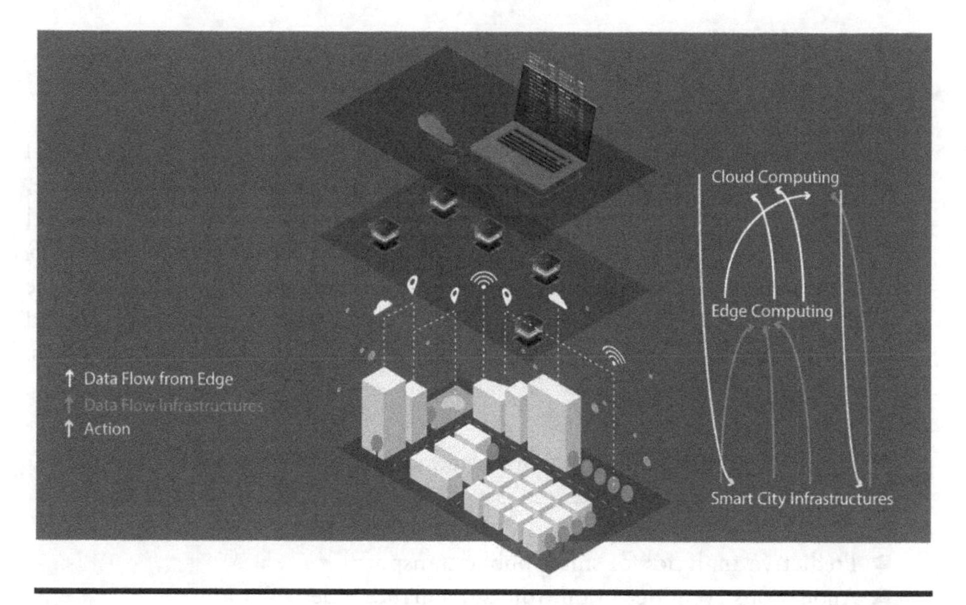

Figure 8.1 Architecture and components of the Cloud-based SC paradigm [38].

The edge-based analytics support local actions in predefined contexts, whereas the Cloud-based analytics are capable of covering larger geographical regions with various contexts. The limitation of the devices in the SC infrastructure is the deployment of complex and large learning models. Instead, several shallow ML approaches are required. However, to bring analytics and intelligence closer to the source of data (e.g., end users, IoT resource-constrained devices), there is a need to utilize modern and advanced learning models such as DL. A nascent research path is to overcome the resource limitation of these devices to allow them to utilize deeper neural network models. In recent years, several approaches have been proposed to compress or prune deep neural networks so that they can be loaded into IoT resource-constrained devices, wearable electronics, and smartphones [42]. Using such compressed neural networks, it is possible to integrate deep reinforcement learning (DRL) with these devices. At the edge computing level, the raw data is aggregated and transmitted to the Cloud computing level. Compressed DL models, DRL, and semi-supervised methods can be used at this level as the resources at this level have lesser constraints compared to the IoT resources. Also, at this level, lightweight intelligence needs to be brought to the IoT gateways and proxies to enable the efficient realization of horizontal integration of services in support of SC applications [43]. For the Cloud use case, the massive data generated from smart environment, smart home, and smart public place monitoring is processed in the Cloud layer [44].

8.4 Applications

8.4.1 ML in Transportation

Intelligent transport systems could help ease congestion, reduce pollution, and improve customer experiences on public transport. In one vision of the future, transport options could be intelligent; smartphones could review a range of travel options and make personalized suggestions, using ML to account for personal preferences such as lifestyle choices. ML also supports driverless cars, which could have a range of benefits. For example, Elon Musk has suggested that humans could be banned from driving due to the enhanced safety benefits of driverless cars. The benefits of ML are as follows:

- Monitoring and managing transportation system performance
- Freight transportation operations
- Air traffic control
- Predictive analytics for smart public transport
- Anomalous event detection from surveillance video
- Mobility services for data-driven transit planning, operations, and reporting
- Vehicle safety monitoring
- Passenger safety monitoring
- Efficient carpooling and ride sharing
- Object detection and traffic sign recognition
- Analysis of traveler's behavior.

8.4.2 ML in Captured Videos for Surveillance

The proliferation of surveillance and security cameras in cities is not only going to continue, but also NVIDIA predicts that there would be about one billion of them in use around the world by the year 2020 [45]. While this has raised concerns over privacy and possible police-state tactics, the presence of cameras has produced documented improvements in public safety, reduced crime rates in some areas, and catching criminals—security cameras played a key role in tracking the Austin, Texas, bomber. However, all those cameras produce far more video than human monitors can watch, let alone process, and analyze. Video can be checked forensically to identify people or actions after the fact, but monitoring them in real time has been out of reach. Government agencies and artificial intelligence (AI) developers want to change that. The Defense Department Project Maven is training AI systems to identify specific objects and activities in imagery, and the Intelligence Advanced Research Projects Activity is looking to use AI for real-time monitoring of multiple video feeds through its Deep Intermodal Video Analytics (DIVA) project. Industry is also developing tools for better video monitoring, such as NVIDIA's Metropolis platform [46], which uses DL AI to speed up analysis.

8.4.3 ML in Water and Power Monitoring/Automation

A clear advantage to AI is in streamlining power and water and usage. Google has said its AI has cut power requirements in its data centers by 40%. Cities are employing smart grids to better manage power use. Chattanooga, Tennessee, for instance, has been a pioneer in smart grid technology and is now building its partnership with Oak Ridge National Laboratory and the University of Tennessee–Knoxville on a solar-powered microgrid at the city's municipal airport. AI is also being applied to water metering to curb excess water and find leaks.

8.4.4 ML in Public Safety

AI is being applied to help solve—or, at least, ease—one of the banes of city existence: finding a place to park. Redwood City, California, for instance, uses a predictive modeling and AI framework from VIMOC Technologies to identify the patterns of use in parking garages and other areas to make parking more efficient.

License plate reader technology is used by police to find stolen cars, identify expired registrations, and otherwise run checks against criminal databases. Attached to the lower part of the police cruisers, automated License Plate Recognitions (LPRs) can scan up to 1,800 license plates in a minute across four lanes of traffic and almost immediately alert the police officer if a plate is flagged for some violation (see Table 8.4).

Table 8.4 Summary of Related Work in SC Use Cases

SC Uses Cases	*Data Processed In*		
Application	*Cloud*	*Edge*	*References*
Smart traffic		✓	[41,47]
Smart health	✓	✓	[48]
Smart environment	✓		[49]
Smart weather prediction		✓	[40]
Smart citizen	✓		[13,16]
Smart agriculture	✓	✓	[15]
Smart home	✓		[18]
Smart air controlling	✓		[50]
Smart public place monitoring	✓		[23]
Smart human activity control	✓	✓	[39,51]

Law enforcement agencies are also applying predictive policing in various other areas, including monitoring and managing sex offenders, and combating gang violence. Predictive systems are used in crime prevention, criminal sentencing, and parole recommendations.

Most of these innovations have raised some objections, from privacy concerns to claims that predictive policing systems reflect biases built into their programming. While civil rights concerns will need to be addressed and some kinks in the systems will undoubtedly have to be worked out, AI will be used more and more in in city systems. For no other reason, it has proven to be economical, with a Deloitte University Press study saying that AI can save the government $41 billion.

8.4.5 ML in Military

Nowadays, adopting ML in military application enhances the processing and utilization of data, which in turn improves the speed of decision-making on the battlefield. Many countries are spending billions of dollars to capitalize on AI as the next best thing in warfare. Drones, motherships, protective exoskeletons for troops, and unmanned vessels that can carry out missions and combat apps that can execute commands are some of the military-based AI applications.

ML can help to improve the computational military reasoning and also the intelligence in the autonomous unmanned weapon systems. It is used in cyber defense and cyber warfare, where the cyberattacks can be detected and curbed through pattern matching, statistical analysis, ML, and big data analysis. As an example, the Google's AI is being used by U.S. military drone.

8.4.6 ML in Food and Agriculture

ML algorithms provide new opportunities for data-intensive science in the multidisciplinary agri-technologies domain. ML is used in various areas of agriculture such as the usage of algorithms based on weather and historical yield data, image recognition algorithms to detect pests and diseases in plants, and robotics to harvest different types of specialty crops. So, it's a growing technology that has increasingly important applications to improve agricultural processes (Figure 8.2).

Yield prediction, one of the most significant topics in precision agriculture, is of high importance for yield mapping, yield estimation, matching of crop supply with demand, and crop management to increase productivity. The authors in Ref. [52] provided an efficient, low-cost, and nondestructive method that automatically counted coffee fruits on a branch. The method calculates the coffee fruits in three categories: harvestable, not harvestable, and fruits with disregarded maturation stage. In addition, the method estimated the weight and the maturation percentage of the coffee fruits. The authors in Ref. [53] developed a machine vision system for automating shaking and catching cherries during harvest. The system segments and detects occluded cherry branches with full foliage even when these

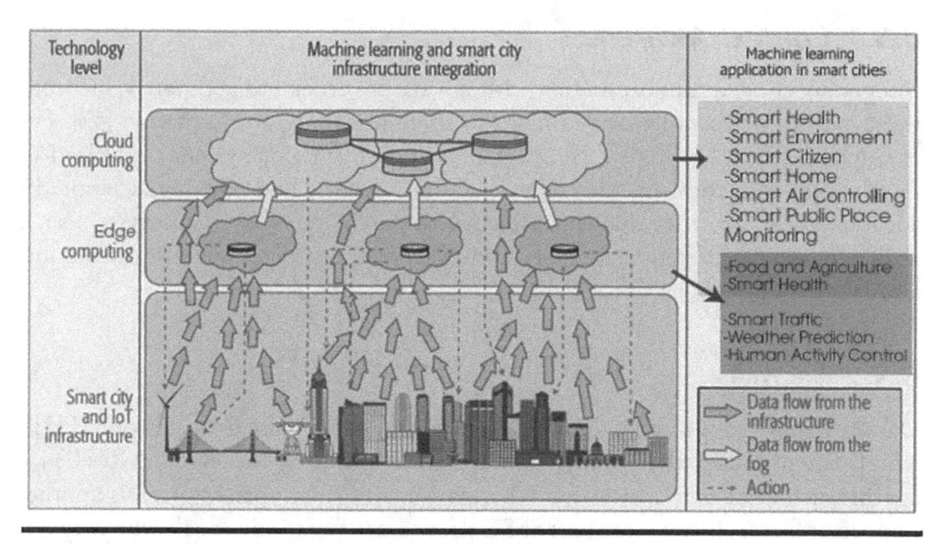

Figure 8.2 SC architecture and related use cases [38].

are inconspicuous. The main aim of the system was to reduce labor requirements in manual harvesting and handling operations.

Also, the ML techniques are used in disease detection. The authors in Ref. [54] developed a new method based on image processing procedure for the classification of parasites and the automatic detection of thrips in the strawberry greenhouse environment for real-time control. The author of Ref. [55] presented a CNN-based method for the disease detection diagnosis based on the images of simple leaves with sufficient accuracy to classify between healthy and diseased leaves in various plants. Finally, ML is also used in the stock management, which consists of two main topics: animal welfare and livestock production. Animal welfare deals with the health and well-being of animals, with the main application of ML in monitoring animal behavior for the early detection of diseases. On the other hand, livestock production deals with issues in the production system, where the main scope of ML applications is the accurate estimation of economic balances for the producers based on production line monitoring. The authors in Ref. [56] presented a system for the automatic identification and classification of chewing patterns in calves. The authors created a system based on ML applying data from chewing signals of dietary supplements, such as hay and ryegrass, combined with behavior data, such as rumination and idleness.

8.5 Design Factors for ML

Development of SC applications supported by big data analytics is subject to several challenges that need to be addressed to achieve a reliable and accurate system.

8.5.1 Context Awareness

Integrating contextual information with raw data is crucial to get more value from the data, and perform faster and more accurate reasoning and actuation [57]. For example, detecting a sleepy face in a human pose detection system could lead to totally different actions in the contexts of driving a car and relaxing at home. In addition, there are other challenges that affect the design of an SC ecosystem such as integration of different analytic frameworks, distribution of analytic operations, and lack of comprehensive testbeds.

8.5.2 Security

Data-driven ML approaches (e.g., DL) can be attacked by false data injection (FDI), which compromises the validity and trustworthiness of the system. Resilience against such attacks is a must for ML algorithms. Privacy preservation is another important factor since a large part of SC data comes from individuals who may not prefer their data to be publicly available [58]. ML algorithms should address these concerns to enable the wide acceptance of SC systems by organizations and citizens.

8.5.3 On-Device Intelligence

SC applications also call for lightweight ML algorithms deployable on resource-constrained devices for hard real-time intelligence. As intelligence is moving towards edge devices, increased computing power and sensor data along with improved AI algorithms are driving the trend towards ML run on the end device, such as smartphones or automobiles, rather than in the Cloud. This is also in line with the security and privacy preservation requirement because data is not transferred to the edge or Cloud.

8.5.4 Energy

To test the efficacy of an ML method, we need to check the following behaviors, which critically influence any smart system nowadays. These behaviors are mainly the energy usage, and the energy production. Energy usage/production can be predicted based on a holistic set of metrics such as time of the day, vehicular traffic patterns, human presence density, weather conditions, commercial and industrial activities, and the history of energy usage/generation [59].

8.5.5 Scheduling

Sensing, collecting, merging, transmitting, or even reverse controlling the data in the SC are all possible tasks to be performed by an IoT device/node. Nevertheless, the devices in IoT are fundamentally different in their processing abilities. Combined with the edge computing concept, a few super nodes in IoT that have better computing capabilities can be considered as the edge nodes in order to help other light/weak nodes in their task processing via data analysis and computing. Consequently, a new problem is raised as how to schedule the tasks among the different edge nodes while maintaining the minimum task makespan and communication cost. The authors in Ref. [60] proposed an adaptive double-fitness genetic task scheduling (ADGTS) algorithm. By the collaborative scheduling of tasks and edge computing resources, ADGTS can optimize the task makespan and communication cost simultaneously, as well as flexibly adapt to different importance of makespan and communication cost in practical applications.

There are some studies on edge computing in the context of IoT and SC. The authors in Ref. [61] proposed a resource allocation and task optimization method, which can reduce the delay by 39%. In addition, the authors in Ref. [62] presented early experiences of an orchestration scenario in terms of edge orchestration for IoT services and demonstrated the feasibility and initial results using a distributed genetic algorithm in the context. Finally, the authors in Ref. [63] proposed a resource allocation strategy for edge computing considering the price cost and time cost to complete a task.

8.5.6 Integrating Big and Fast/Streaming Data Analytics

In an SC context, many time-sensitive applications (e.g., smart and connected vehicles) need real-time or near-real-time analytics of the stream of data. Such applications call for new analytic frameworks that support big data analytics in conjunction with fast/streaming data analytics.

8.5.7 Storage

Development and evaluation of SC applications need real-world datasets, which are not readily available for many application domains. It is necessary to confirm the results based on simulated big data.

8.5.8 Scalability

Scalability is a central issue in the development and deployment of an ML technique. The scalability footprint of the ML technique can be defined by two core criteria: load and complexity. Load is the more critical factor. In order to isolate it, complexity must be minimized if possible. Load scalability encompasses memory, communication, and CPU and GPU loads. Thus, it is suggested to implement an ML solution with a hierarchical-based structure in order to ease/relax the scalability issues.

8.5.9 Cost

Data routing and process cost is another important factor that is essential when designing an ML technique. Several routing techniques are designed to reduce the cost of data delivery in SC applications.

As an example, the opportunistic networks (OppNets) [64] are a type of challenged networks where link performance is highly variable and the network contacts are intermittent. As such, routing in OppNets is a challenge because the nodes are required to buffer their packets till they find suitable forwarders that can eventually carry these packets to destination with minimal delay.

The authors in Ref. [65] proposes a novel routing protocol for OppNets called MLProph, which uses ML to determine the probability of successful deliveries. The ML model is trained by using various factors such as the predictability value inherited from the PROPHET routing scheme, node popularity, node's power consumption, speed, and location.

The main challenge is to reduce the complexity of the CNN model while keeping it accurately and reduce the cost of data delivery in SC applications.

8.5.10 Reliability

Reliability in SC-based ML has something to deal with failures when processing and/or exchanging data from the source to the destination, or even while exchanging data among the architecture layers. Heterogeneous sensor infrastructures of the individual information providers and data portal vendors tend to offer a hardly revisable information quality. Thus, reliability here comes in twofold: node and link reliability. Reliability of the link in the wireless physical network is defined as the ability to overcome the wireless signal degradation factors, and it significantly affects the QS in the SC system. On the other hand, node (resource) reliability is mainly defined by the computing power of the sensing/computing node itself. The assessment of quality of data can basically be evaluated in five common dimensions: completeness, correctness, concordance, plausibility, and currency [66]. Many protocols were designed to evaluate the data resources. The authors in Ref. [66] proposed a correlation model-based monitoring approach to evaluate the plausibility of SC data sources.

8.5.11 Deployment

SC applications require a framework capable of structuring and modeling its basic components, its interactions, and flow of data. In most of the SC applications, we are dealing with three levels: the physical components, the communications, and the application level [66]. Our main focus in this chapter is the application level, which includes all ML systems and developments conceived for the city.

8.6 Communication Technologies

In this section, we highlight the emerging proprietary technologies and the technical aspects of Sigfox, LoRa, narrowband-IoT (NB-IoT), Zwave, ZigBee, and Wi-Sun as summarized below.

8.6.1 ZigBee

ZigBee is a communication technology developed according to the IEEE 802.15.4 standard that transmits data at a rate of approximately 250 Kbps on 2.4 GHz frequency. The technology targets applications that require a short range of communication, 0–100 m. Despite the short range of coverage, ZigBee proves itself in its low power consumption characteristic as well as system scalability at the cost of data transmission rate. Hence, this technology can show a good performance in intra-ML information exchange.

8.6.2 Wi-Fi

Wireless fidelity (Wi-Fi) is another communication technology that is largely used nowadays in homes and business areas. It is designed according to the IEEE 802.11b/g/n standard with a capability of data transmission at frequencies 2.4 and 5 GHz in the domain of 0–250 m at a rate of 54 Mbps. It provides a relatively higher rate of data transmission at the cost of scalability. Wi-Fi technology can also be integrated into the SC system. The security in this technology is also an advantage because it has a robust authentication procedure, which can guarantee a secure SC application.

8.6.3 LoRaWAN

Everyday municipal operations are made more efficient with LoRa technology's long-range, low-power, secure, and GPS-free geolocation features. By connecting city services such as lighting, parking, and waste removal, cities can optimize the use of utilities and personnel to save time and money. LoRaWAN is designed for wide-range, low-power communications, and SC applications benefit from wireless battery-operated things and secure bidirectional communication, mobility, and localization service.

8.6.4 NB-IoT

NB-IoT is a standard-based low-power wide area (LPWA) technology developed to enable a wide range of new IoT devices and services. Where it is supported by all major mobile equipment, chipset, and module manufacturers, NB-IoT can coexist with 2G, 3G, and 4G mobile networks. It significantly improves the power consumption of user devices, system capacity, and spectrum efficiency, especially in deep coverage. A battery life of more than 10 years can be supported for a wide range of use cases.

8.6.5 Wi-SUN

The Wi-SUN Alliance has introduced a new certification program for large-scale outdoor IoT networks. Wi-SUN is an association dedicated to ensuring seamless connectivity and promotion of certified products, based on open standards for wireless systems, specified data rates, modulations, and frequency bands.

8.6.6 Sigfox

Sigfox is a narrowband (or ultra-narrowband) technology. It uses a standard radio transmission method called binary phase-shift keying (BPSK), and it takes very narrow chunks of spectrum and changes the phase of the carrier radio wave to encode the data. This allows the receiver to only listen in a tiny slice of spectrum, which mitigates the effect of noise. It requires an inexpensive endpoint radio and a more sophisticated base station to manage the network.

Sigfox communication tends to be better if it's headed up from the endpoint to the base station. It has bidirectional functionality, but its capacity from the base station back to the endpoint is constrained, and you'll have less link budget going down than going up. This is because the receiver sensitivity on the endpoint is not as good as on the expensive base station.

8.6.7 Z-Wave

Z-Wave is a wireless networking protocol primarily designed for use in SCs. With two-way communication through mesh networking and message acknowledgment, the Z-Wave protocol helps alleviate power issues and brings low-cost wireless connectivity to SC applications, offering a lower-power alternative to Wi-Fi and a longer-range alternative to Bluetooth. Also, it offers transmission rates of small data packets using a throughput rate of 9.6, 40, or 100 Kbps. The Z-Wave protocol operates on the low-frequency 908.42 band in the United States and the 868.42 MHz band in Europe. Although interference with other home electronics, such as cordless phones, is possible, the protocol avoids interference with the 2.4 GHz band where Wi-Fi and Bluetooth operate.

8.7 Tools and Commercial Products

A wide spectrum of commercial products provides IoT Platform as a Service (PaaS), which can reliably host a diverse range of SC services while ensuring their security, scalability, and interoperability. For example, Google's PaaS, Google Cloud IoT service [67], offers various application program interfaces (APIs) for device management, access control, data storage, and delivery, as well as tools for data analytics, machine intelligence, and data visualization. Alternatively, Microsoft Azure IoT Suite [68] makes available a rich collection of APIs and Software Development Kits (SDKs) for device management and configuration to developers. Other notable existing PaaS products include IBM Watson IoT [69], GE Predix [70], and Amazon AWS IoT [71].

8.8 Performability

One of the most important subjects in the implementation of ML in the SC is the evaluation of the algorithm, which is the essential part of any project. The used model may give a satisfying result when evaluated using a metric such as the accuracy score but may give poor results when evaluated against other metrics such as logarithmic loss or any other such metric. Most of the times we use classification accuracy to measure the performance of our model; however, it is not enough to truly judge our model. In this section, we will cover different types of evaluation metrics.

8.8.1 From an ML Perspective

8.8.1.1 Classification Accuracy

Accuracy of the used classification approach is what we usually mean, when we use the term accuracy. It is the ratio of the number of correct predictions to the total number of input samples.

$$\text{Accuracy} = \frac{\text{Number of correct predictions}}{\text{Total number of predictions made}} \tag{8.1}$$

It works well only if there are an equal number of samples belonging to each class. The real problem arises, when the cost of misclassification of the minor class samples is very high. If we deal with a rare but fatal disease, the cost of failing to diagnose the disease of a sick person is much higher than the cost of sending a healthy person to more tests.

8.8.1.2 Logarithmic Loss

Logarithmic loss or log loss works by penalizing the false classifications. It works well for multiclass classification. When working with log loss, the classifier must assign the probability to each class for all the samples. Suppose there are N samples belonging to M classes, the log loss is calculated as follows:

$$\text{Log loss} = \frac{-1}{N} \sum_{i=1}^{N} \sum_{j=1}^{M} y_{ij} \times \log\left(p_{ij}\right) \tag{8.2}$$

where y_{ij} indicates whether sample i belongs to class j or not, and p_{ij} indicates the probability of sample i belonging to class j. The Log loss has no upper bound, and it exists on the range [0, ∞). Log loss nearer to 0 indicates higher accuracy, whereas log loss away from 0 indicates lower accuracy. In general, minimizing log loss gives greater accuracy for the classifier.

8.8.1.3 Area Under the Curve

The area under the curve (AUC) is a performance metric for a binary classifier. By comparing the receiver operating characteristic (ROC) curves with the AUC, it captures the extent to which the curve is up in the northwest corner. A higher AUC is better of course.

8.8.1.4 Precision

Precision is used when the accuracy of the ML model is not enough to decide whether the network is valid or not. The idea of Precision is to calculate how many samples are predicted as class j and how many are really from class i. It is defined as follows:

$$\frac{\mathrm{TP}}{\mathrm{TP}+\mathrm{FP}} \tag{8.3}$$

Where TP represents the count of the truly predicted samples, and FP represents the count of false predictions.

8.8.1.5 Recall

The Recall factor also refers to sensitivity of the classification method. Our goal in ML training is to maximize both Recall and Precision. The idea of Recall is to calculate the percentage of the true prediction class over all samples that are predicted as follows:

$$\frac{\mathrm{TP}}{\mathrm{TP}+\mathrm{FN}} \tag{8.4}$$

Where TP indicates the true positive observations (samples), and FN represents the false negative observations.

8.8.1.6 F1 Score

The F1 score is the combination of Recall and Precision. F1 score is needed when you want to seek a balance between Precision and Recall; hence, it is referred to as the harmonic mean. It is defined as follows:

$$2\times\frac{\mathrm{Recall}\times\mathrm{Precision}}{\mathrm{Recall}\ +\ \mathrm{Precision}} \tag{8.5}$$

8.8.1.7 Mean Absolute Error

Mean absolute error (MAE) measures the average magnitude of the errors in a set of predictions, without considering their direction. It's the average of the absolute differences of the test sample between prediction and actual observation where all individual differences have an equal weight.

$$\mathrm{MAE} = \frac{1}{n} + \sum_{j=1}^{n} \left| y_j - \hat{y}_j \right| \tag{8.6}$$

8.8.1.8 Root Mean-Squared Error

Root mean-squared error (RMSE) is a quadratic scoring rule that measures the average magnitude of the error. It's the square root of the average of squared differences between prediction and actual observation.

$$\mathrm{RMSE} = \sqrt{\frac{1}{n} + \sum_{j=1}^{n} \left(y_j - \hat{y}_j \right)^2} \tag{8.7}$$

8.9 Open Research Issues

An SC project is basically dealing with the improvement of smartness of a city's systems and applications, and some of its requirements and characteristics are as follows:

- A robust and scalable framework combined with secure and open access
- A user-centric or citizen-oriented architectural approach
- A huge volume of storable, findable, sharable, tagged, mobile, and wearable public and private data enabling citizens to access information from anywhere and when needed
- An application level with analytically and integrated capabilities
- A smart physical and network infrastructure allowing the transfer of a huge volume of heterogeneous data and the support of complex and distributed services and applications.

The development of ML-SC applications supported by big data analytics is subject to several challenges that need to be addressed to achieve a reliable and accurate Cloud-based system. Majority of these challenges were discussed in the section of ML design factors (Section 8.5). The conventional analytic approach for IoT is to send raw data to the Cloud for processing. Components and architectures of the Cloud solutions have to be further developed for more decentralized analytics. The authors in Ref. [29,42] presented a new trend that aims to bring analytics closer to the Cloud edge and SC infrastructure devices. However, a key challenge was that a trained ML model works well when the same set of features and distribution models are used. By changing the distribution of these features, the trained model needs to be retrained based on a new training dataset. Thus, the transfer learning is a field of research that can help in such scenarios. Also, the integration with semantic technologies is a need for the development of Cloud-based SC applications. The need stems from the interaction of those systems with citizens and the use of social media data. Moreover, intelligent virtual objects can be used in SC services combined with DRL algorithms, considering each physical object has a virtual representation in the SC, and these virtual objects can learn, decide, and act autonomously.

8.10 Concluding Remarks

There are many ML algorithms that are used in SCs nowadays to learn from the big data collected through an SC infrastructure. The SC infrastructure consists of a vast number of varying devices that are connected to each other and transmit huge amounts of data. ML algorithm can be used in various applications of SC, such as health, urban, transportation, food, military, and surveillance. Every application presents a complex new set of requirements with multiple objectives that involve multiple layers using heterogeneous network technologies for a huge number of users. Choosing the efficient ML algorithm for SC use cases is not easy. Many performability functions are considered to measure the efficiency of the ML network. In this chapter, we overviewed the applications of ML in the SC and the design factors that are necessary to deploy an ML solution in an intelligent city-scale application.

References

1. T. Nam, and T. A. Pardo. 2011. Conceptualizing smart city with dimensions of technology, people, and institutions, in *Proceedings of the 12th Annual International Digital Government Research Conference: Digital Government Innovation in Challenging Times* (dg.o '11). ACM, New York, NY, USA, 282–291.
2. G. Singh, and F. Al-Turjman, Learning data delivery paths in QoI-aware information-centric sensor networks, *IEEE Internet of Things Journal*, vol. 3, no. 4, 572–580, 2016.
3. K. Su, J. Li, and H. Fu, Smart city and the applications, in *2011 International Conference on Electronics, Communications and Control (ICECC)*, 2011, pp. 1028–1031.
4. C. T. Barba, M. A. Mateos, P. R. Soto, A. M. Mezher, and M. A. Igartua, Smart city for VANETs using warning messages, traffic statistics and intelligent traffic lights, in *2012 IEEE Intelligent Vehicles Symposium*, 2012, pp. 902–907.
5. H. Xu, H. Huang, R. S. Khalid, and H. Yu, Distributed machine learning based smart-grid energy management with occupant cognition, in *2016 IEEE International Conference on Smart Grid Communications (SmartGridComm)*, 2016, pp. 491–496.

6. Y. LeCun, Y. Bengio, and G. Hinton, Deep learning, *Nature*, vol. 521, no. 7553, 436–444, 2015.
7. D. Barber, *Bayesian Reasoning and Machine Learning*. Cambridge: Cambridge University Press, 2012.
8. M. S. Mahdavinejad, M. Rezvan, M. Barekatain, P. Adibi, P. Barnaghi, and A. P. Sheth, Machine learning for internet of things data analysis: A survey, *Digitital Communication Networks*, vol. 4, no. 3, 161–175, 2018.
9. A. Obinikpo, B. Kantarci, A. A. Obinikpo, and B. Kantarci, Big sensed data meets deep learning for smarter health care in smart cities, *Journal of Sensor and Actuator Networks*, vol. 6, no. 4, 26, 2017.
10. D. C. Rose, I. Arel, T. P. Karnowski, and V. C. Paquit, Applying deep-layered clustering to mammography image analytics, in *2010 Biomedical Sciences and Engineering Conference*, 2010, pp. 1–4.
11. D. Kuang and L. He, Classification on ADHD with deep learning, in *2014 International Conference on Cloud Computing and Big Data*, 2014, pp. 27–32.
12. C. M. Bishop, *Pattern Recognition and Machine Learning*. Springer, New York, 2006.
13. X. Ma, Y.-J. Wu, Y. Wang, F. Chen, and J. Liu, Mining smart card data for transit riders' travel patterns, *Transportation Research Part C: Emerging Technologies*, vol. 36, 1–12, 2013.
14. C.-T. Do, A. Douzal-Chouakria, S. Marié, and M. Rombaut, Multiple metric learning for large margin kNN classification of time series, in *2015 23rd European Signal Processing Conference (EUSIPCO)*, Nice, France, 2015.
15. W. Han, Y. Gu, Y. Zhang, and L. Zheng, Data driven quantitative trust model for the internet of agricultural things, in *2014 International Conference on the Internet of Things (IOT)*, 2014, pp. 31–36.
16. W. Derguech, E. Bruke, and E. Curry, An autonomic approach to real-time predictive analytics using open data and internet of things, in *2014 IEEE 11th International Conference on Ubiquitous Intelligence and Computing and 2014 IEEE 11th International Conference on Autonomic and Trusted Computing and 2014 IEEE 14th International Conference on Scalable Computing and Communications and Its Associated Workshops*, Bali, Indonesia, 2014.
17. S. Hu, Research on data fusion of the internet of things, in *2015 International Conference on Logistics, Informatics and Service Sciences (LISS)*, 2015, pp. 1–5.
18. A. M. C. Souza and J. R. A. Amazonas, An outlier detect algorithm using big data processing and internet of things architecture, *Procedia Computer Science*, vol. 52, 1010–1015, 2015.
19. M. S. Shukla, Y. P. Kosta, and P. Chauhan, Analysis and evaluation of outlier detection algorithms in data streams, undefined, 2015.
20. H. Hromic et al., Real time analysis of sensor data for the internet of things by means of clustering and event processing, in *2015 IEEE International Conference on Communications (ICC)*, 2015, pp. 685–691.
21. X. Tao and C. Ji, Clustering massive small data for IOT, in *The 2014 2nd International Conference on Systems and Informatics (ICSAI 2014)*, 2014, pp. 974–978.
22. M. A. Khan, A. Khan, M. N. Khan, and S. Anwar, A novel learning method to classify data streams in the internet of things, in *2014 National Software Engineering Conference*, 2014, pp. 61–66.
23. D. N. Monekosso and P. Remagnino, Data reconciliation in a smart home sensor network, *Expert Systems with Applications*, vol. 40, no. 8, 3248–3255, 2013.

24. M. M. E. Mahmoud et al., Enabling technologies on cloud of things for smart healthcare, *IEEE Access*, vol. 6, 31950–31967, 2018.
25. H. V. Jagadish, B. C. Ooi, K.-L. Tan, C. Yu, and R. Zhang, iDistance: An adaptive B+-tree based indexing method for nearest neighbor search, *ACM Transactions on Database Systems*, vol. 30, no. 2, 364–397, 2005.
26. T. Cover and P. Hart, Nearest neighbor pattern classification, *IEEE Transactions on Information Theory*, vol. 13, no. 1, 21–27, 1967.
27. A. Mccallum and K. Nigam 2001. A comparison of event models for naive Bayes text classification. Work Learn Text Categ. 752.
28. B. Tang et al., Incorporating intelligence in fog computing for big data analysis in smart cities, *IEEE Transactions on Industrial Informatics*, vol. 13, no. 5, 2140–2150, 2017.
29. H. Drucker, C. J. C. Burges, L. Kaufman, A. J. Smola, and V. Vapnik, Support Vector Regression Machines. pp. 155–161, 1997.
30. W.-Y. Loh, Classification and regression trees, *Classification Trees, C*, vol. 1, 14–23, 2011.
31. L. Breiman, Random forests, *Machine Learning*, vol. 45, no. 1, 5–32, 2001.
32. L. Breiman, Bagging Predictors, 1994.
33. A. Likas, N. Vlassis, and J. Verbeek, The global k-means clustering algorithm, *Pattern Recognition*, vol. 36, 451–461, 2003.
34. M. Ester, H.-P. Kriegel, J. Sander, and X. Xu, A density-based algorithm for discovering clusters in large spatial databases with noise, in *KDD'96 Proceedings of the Second International Conference on Knowledge Discovery and Data Mining*, Portland, Oregon, 1996.
35. H. Hotelling, Relations between two sets of variates, *Biometrika*, vol. 28, no. 3–4, 321–377, 1936.
36. Y. Lecun, L. Bottou, Y. Bengio, and P. Haffner, Gradient-based learning applied to document recognition, *Proceedings of the IEEE*, vol. 86, no. 11, 2278–2324, 1998.
37. B. Schölkopf, J. C. Platt, J. Shawe-Taylor, A. J. Smola, and R. C. Williamson, Estimating the support of a high-dimensional distribution, *Neural Computation*, vol. 13, no. 7, 1443–1471, 2001.
38. C. Perera, Y. Qin, J. C. Estrella, S. Reiff-Marganiec, and A. V. Vasilakos, Fog computing for sustainable smart cities, *ACM Computing Surveys*, vol. 50, no. 3, 1–43, 2017.
39. M. Shukla, Y. P. Kosta, and P. Chauhan, Analysis and evaluation of outlier detection algorithms in data streams, in *2015 International Conference on Computer, Communication and Control (IC4)*, 2015, pp. 1–8.
40. P. Ni, C. Zhang, and Y. Ji, A hybrid method for short-term sensor data forecasting in internet of things, in *2014 11th International Conference on Fuzzy Systems and Knowledge Discovery (FSKD)*, 2014, pp. 369–373.
41. Y. Qin, Q. Z. Sheng, N. J. G. Falkner, S. Dustdar, H. Wang, and A. V. Vasilakos, When things matter: A survey on data-centric internet of things, *Journal of Network and Computer Applications*, vol. 64, 137–153, 2016.
42. S. Han, J. Pool, J. Tran, and W. J. Dally, Learning both Weights and Connections for Efficient Neural Networks, 2015.
43. A. Al-Fuqaha, A. Khreishah, M. Guizani, A. Rayes, and M. Mohammadi, Toward better horizontal integration among IoT services, *IEEE Communications Magazine*, vol. 53, no. 9, 72–79, 2015.

44. M. Mohammadi and A. Al-Fuqaha, Enabling cognitive smart cities using big data and machine learning: Approaches and challenges, *IEEE Communications Magazine*, vol. 56, 94–101, 2018.
45. NVIDIA, Computing for the Most Demanding Users NVIDIA, 2017.
46. NVIDIA Metropolis, Smart City Video Analytics & Applications | NVIDIA Metropolis. [Online]. Available: www.nvidia.com/en-us/autonomous-machines/intelligent-video-analytics-platform/. [Accessed: 12-Dec-2018].
47. M. A. Kafi, Y. Challal, D. Djenouri, M. Doudou, A. Bouabdallah, and N. Badache, A study of wireless sensor networks for urban traffic monitoring: Applications and architectures, *Procedia Computer Science*, vol. 19, 617–626, 2013.
48. Y. Yogita and D. Toshniwal, Clustering techniques for streaming data-a survey, undefined, 2013.
49. V. Jakkula and D. Cook, Outlier detection in smart environment structured power datasets, in *2010 6th International Conference on Intelligent Environments*, 2010, pp. 29–33.
50. C. Costa and M. Y. Santos, Improving cities sustainability through the use of data mining in a context of big city data, in *The 2015 International Conference of Data Mining and Knowledge Engineering*, 2015.
51. A. Shilton, S. Rajasegarar, C. Leckie, and M. Palaniswami, DP1SVM: A dynamic planar one-class support vector machine for internet of things environment, in *2015 International Conference on Recent Advances in Internet of Things (RIoT)*, 2015, pp. 1–6.
52. P. J. Ramos, F. A. Prieto, E. C. Montoya, and C. E. Oliveros, Automatic fruit count on coffee branches using computer vision, *Computers and Electronics in Agriculture*, vol. 137, 9–22, 2017.
53. S. Amatya, M. Karkee, A. Gongal, Q. Zhang, and M. D. Whiting, Detection of cherry tree branches with full foliage in planar architecture for automated sweet-cherry harvesting, *Biosystems Engineering*, vol. 146, 3–15, 2016.
54. M. A. Ebrahimi, M. H. Khoshtaghaza, S. Minaei, and B. Jamshidi, Vision-based pest detection based on SVM classification method, *Computers and Electronics in Agriculture*, vol. 137, 52–58, 2017.
55. K. P. Ferentinos, Deep learning models for plant disease detection and diagnosis, *Computers and Electronics in Agriculture*, vol. 145, 311–318, 2018.
56. V. Pegorini et al., In vivo pattern classification of ingestive behavior in ruminants using FBG sensors and machine learning, *Sensors*, vol. 15, no. 11, 28456–28471, 2015.
57. C. Perera, A. Zaslavsky, P. Christen, and D. Georgakopoulos, Context aware computing for the internet of things: A survey, *IEEE Communications Surveys & Tutorials*, vol. 16, no. 1, 414–454, 2014.
58. T. Braun, B. C. M. Fung, F. Iqbal, and B. Shah, Security and privacy challenges in smart cities, *Sustainable Cities and Society*, vol. 39, 499–507, 2018.
59. S. Khan, D. Paul, P. Momtahan, and M. Aloqaily, Artificial intelligence framework for smart city microgrids: State of the art, challenges, and opportunities, in *2018 3rd International Conference on Fog and Mobile Edge Computing (FMEC)*, 2018, pp. 283–288.
60. Q. Liu, Y. Wei, S. Leng, and Y. Chen, Task scheduling in fog enabled internet of things for smart cities, in *2017 IEEE 17th International Conference on Communication Technology (ICCT)*, 2017, pp. 975–980.

61. O. Skarlat, S. Schulte, M. Borkowski, and P. Leitner, Resource provisioning for IoT services in the fog, in *2016 IEEE 9th International Conference on Service-Oriented Computing and Applications (SOCA)*, 2016, pp. 32–39.
62. Z. Wen, R. Yang, P. Garraghan, T. Lin, J. Xu, and M. Rovatsos, Fog orchestration for internet of things services, *IEEE Internet Computing*, vol. 21, no. 2, 16–24, 2017.
63. L. Ni, J. Zhang, C. Jiang, C. Yan, and K. Yu, Resource allocation strategy in fog computing based on priced timed petri nets, *IEEE Internet Things Journal*, vol. 4, no. 5, 1216–1228, 2017.
64. L. Lilien, Z. H. Kamal, V. Bhuse, and A. Gupta, The concept of opportunistic networks and their research challenges in privacy and security, in *Mobile and Wireless Network Security and Privacy*, Boston, MA: Springer US, pp. 85–117.
65. D. K. Sharma, S. K. Dhurandher, I. Woungang, R. K. Srivastava, A. Mohananey, and J. J. P. C. Rodrigues, A machine learning-based protocol for efficient routing in opportunistic networks, *IEEE System Journal*, vol. 12, no. 3, 2207–2213, 2018.
66. J. S. B. Martins, Towards smart city innovation under the perspective of software-defined networking, *Artificial Intelligence and Big Data*, 2018. DOI: 10.5281/zenodo.1467770.
67. Google LLC, Cloud IoT core, Google Cloud Platform. [Online]. Available: https://cloud.google.com/iot-core/. [Accessed: 23-Dec-2018].
68. Microsoft Corp., Microsoft Azure cloud computing platform and services. [Online]. Available: https://azure.microsoft.com/en-us/. [Accessed: 23-Dec-2018].
69. IBM Corp., IBM Watson Internet of Things (IoT). [Online]. Available: www.ibm.com/internet-of-things. [Accessed: 23-Dec-2018].
70. F. Al-Turjman, M. Z. Hasan, and H. Al-Rizzo, Task scheduling in cloud-based survivability applications using swarm optimization in IoT, *Transactions on Emerging Telecommunications*, 2018. doi: 10.1002/ett.3539.
71. Amazon Inc., Amazon Web Services (AES)—Cloud computing services. [Online]. Available: https://aws.amazon.com/fr/. [Accessed: 23-Dec-2018].

Chapter 9

Data Delivery Pricing in Cloud-Based IoT[1]

Fadi Al-Turjman
Antalya Bilim University

Contents

---

[1] **F. Al-Turjman**, "Price-based Data Delivery Framework for Dynamic and Pervasive IoT", *Elsevier Pervasive and Mobile Computing Journal*, vol. 42, pp. 299-316.

9.1 Introduction

Internet of Things (IoT) is a pervasive technology for applications ranging from smart grid to vehicular networking and smart homes to smart workplace. IoT is growing as a framework to encompass all identifiable things, in a dynamic and interacting network. The promise of clever approaches and dynamic systems that could benefit from the aggregation and analysis of information over the IoT infrastructure is quite pervasive. Scientists in networking, R&D divisions and many businesses are in the race to develop an achievable and robust architecture to realize the IoT paradigm [1][3].

Yet, many hindrances render the IoT framework mostly a challenge. To date, much has been proposed on the promise and benefits of IoT, yet far less has covered the routing protocols to actually operate such a dynamic and large-scale paradigm [5][6]. The vision, however sparse, promises a robust and dynamic framework to integrate many enablers that are already outshined in research and development.

Obviously, Wireless Sensor Networks (WSNs) are envisioned to play a dominant role in IoT paradigms. The resilience, autonomous and energy efficient traits of WSNs render them a vital candidate for dominating the information collection task of an IoT framework [9][10]. Equally vital, the use of RFID technologies for non-LOS and seamless identification of objects is gaining much prominence as a key player in IoT frameworks [11]. The low cost associated with deploying RFID tags (passive, or active) is an important motivation. In fact, some argue that RFIDs have been a main motivator for the IoT framework [11][12].

The integration of these enablers, along with Internet based and context aware services facilitate a dynamic platform for the IoT. Nevertheless, much of current research has focused on developing these enablers in segregation, and optimizing their performance under local constraints and objectives. One of the most important tasks to be carried out, in such a large scale and dynamic environment, is relaying information from a source to a destination, given the new emerging characteristics in IoT. Typical routing approaches typically consider that all components belong to the same owner/provider, hence routing costs and link weights are directly proportional to their local provider characteristics. Though, IoT routing becomes inherently complex by multiple factors. An intrinsic design factor in IoT is delay-tolerance [2].

In reality, an IoT node has only partial knowledge regarding the full path to the destinations assigned to the packets it delivers. Due to splitting, which is mainly caused by nodal mobility, connectivity may occur on an irregular basis. In such circumstances, nodes are required to store and carry data packets until an appropriate forwarding chance ascends in a Store-Carry-Forward fashion [4]. Typical sensor networks' routing approaches are unfortunately mobility-intolerant since most of the WSN network architectures assume stationary sensor nodes [7][8]. As stated earlier, we adopt an expanded notion of sensor networks that incorporates MANET nodes. An abundance of routing-layer protocols have been proposed to accommodate the dynamic topology in MANETs and WSNs [7][8].

Yet, for all these protocols, it is implicitly assumed that the network is connected and there is a contemporaneous end-to-end path between any source/destination pair. In other words, the topology in the standard dynamic routing problem is assumed to be always connected and the objective of the routing algorithm, hence, is confined to finding the best currently available full path to move traffic from one end to the other. Unfortunately, none of these assumptions stand in a delay-tolerant setup. An IoT data delivery scheme must be delay-tolerant to cope with intermittent connectivity, in addition to providing faster delivery alternatives for other delay-sensitive types of data that demand minimal delays.

Furthermore, most entities participating in sensing, identification and relaying in IoT belong to different networks with multiple owners. It is not in the best interest of such networks to allow its resources to be utilized for relaying data across the network, without compensation. For example, an intermediate relay node, belonging to a WSN for surveillance, would not freely take part in relaying information of nearby RFID readers or other WSNs. Thus, price and trading, in addition to all of the routing metrics that govern a mesh ad-hoc network, need to be considered before a suitable routing protocols is presented to relay packets across an inherently diverse IoT.

To this end, we define an *IoT Setting* by the following four main characteristics: 1) Cost-effectiveness, 2) Seamless integration, 3) Reliability and trust, and 4) Delay-tolerance. Hence, we provide a framework, encompassing a cost-efficient IoT architecture, to address data delivery objectives according to the aforementioned characteristics of the IoT setting. The design objectives are to be met with respect to metrics such as delay-tolerance, cost and power-saving. Our proposed framework makes use of ubiquitous relays available in today's topologies to enhance connectivity and delivery rates between the components of the integrated topology. Our framework will as well provide delivery guarantees with respect to delay and connectivity over end-to-end links. Such guarantees will be carried out by dedicated components of our integrated architecture, in addition to other components incorporated within the wider IoT vision.

Our impact in this work comes in two-fold. First, presenting a routing approach customized for the heterogeneous IoT components. This is only possible with our second contribution, a pricing model which caters for the diverse requirements and conditions of nodes which are willing to relay IoT data packets without using the Internet backbone. This pricing model presented here joins measures of load balancing, delay, buffer space and link maintenance. An outline of the targeted routing problem and the dynamic constituents of the envisioned IoT is depicted in Figure 9.1.

The rest of this paper is organized as follows. Section 9.2 covers the background on IoT routing, and its enabling technologies. Then a rigorous definition of our proposed network model, manifesting the interactions of components in the IoT, and their governing constraints is proposed in Section 9.3. Next, we formally present our adaptive routing protocol in Section 9.4. Our proposed model is verified in Section 9.5 via use-cases and Markov-chain in Q-theory. Extensive results are performed and described in Section 9.6. Finally, our work is concluded in Section 9.7.

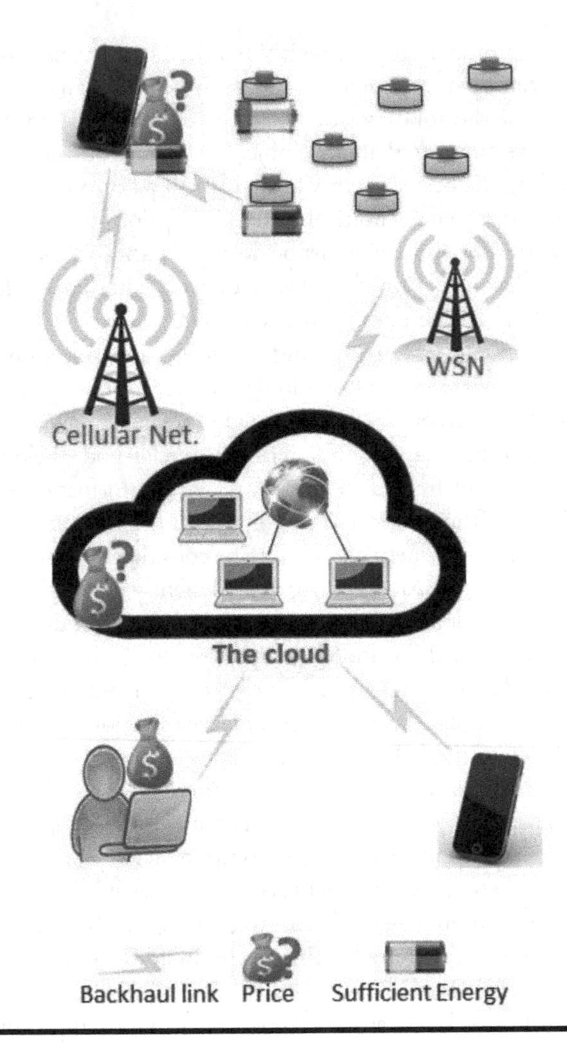

Figure 9.1 An outline of cross-network routing in the IoT and the pricing forced by the network heterogeneity.

9.2 Background

Nowadays, everywhere around us is surrounded with different types of networks. WiFi, LTE wireless communications, broadcasting, streaming etc. are quite common widely spread technologies. However, they bring their own limitations. These limitations can be in the form of cost or technology. Most often, it is about the cost of maintaining and placing an efficient network that can integrate all for what we call the IoT. Several attempts have been made for improvements and performance gains in the enablers of the IoT (especially WSNs and RFIDs). To present a perspective on these enablers, and the major domains of properties, Table 9.1 summarizes three main paradigms to the IoT. Accordingly, we emphasize two major driving forces. First, the lack of a distinctive routing approach that caters for dynamic IoT. The second drive lies in the tradeoff costs of routing over multiple entities, belonging to different service providers.

A major misconception was imposed by an inherent property of the IoT; namely being a descendent of the Internet. That is, as research on the IoT developed, it was expected that a significant pool of protocols previously developed for Internet services would migrate into the IoT. Nonetheless, as the IoT is set to encompass many stationary (static WSNs, RFID readers, etc.) and dynamic (laptops, PDAs, cell phones, etc.) components, we are challenged with multiple issues [3][28]. Most importantly, assuming that all components will inter-communicate via the Internet is insufficient and often degrading the intended dynamic paradigm performance.

A major obstacle would stem from the mounting number of messages that overload a network already handling millions of hosts. This is a noteworthy problem as recent endeavors are targeting higher levels of dynamic interaction between the IoT and its users, as in the Human Computer Interaction work presented by Kranz *et al* in [13]. As such, if a WSN needs to identify an object, with the aid of an

Table 9.1 IoT enablers and their properties

	Wireless Networks			
Property	*IoT*	*MANets*	*WSNs*	*RFIDs*
Topology	Dynamic	Dynamic[b]	Mostly static	Application dependent
Buffer size	Varies	High	Low	None
Medium contention	High	High	Medium	Low with singulation
Mobility	Frequent	Varies[b]	Limited	Frequent
Communication range	Varies	High	Medium (varies)	Reader dependent
Typical density	Very high	Small to medium	Medium to high	Medium to high
Computational power per node	Varies	High	Low	Low to none[a]
Inter-node communication	Heterogen-eous	Homogeneous		

[a] disregarding active-tags, as they equate many features of sensing nodes.
[b] since MANets encompass VANets as well.

RFID reader, direct communication between a sensing node (SN) and the reader would influence bottlenecks of communication and swarming the backhaul over the Internet with numerous packets. This is a prominent architecture, one that is strongly pushed for as a truly integrating IoT [14]. There is a need for establishing a cooperative scheme for routing in the IoT; one which includes all nodes with capabilities of relaying data. This includes those with only one access medium (e.g. WiFi routers) and others with multiple mediums (e.g. cell phones). Yet, due to obvious reasons of resource conservation, such entities would not participate in relaying data packets unless there is an incentive [15]. It is vital to note that some components only generate data (e.g., IDs), such as RFID tags.

Different incentives take part in the pricing model that dictates the choice of a group of candidates for relaying. Recent results in incentive based routing have been well studied. Zhong *et al* present an elaborate study on routing and forwarding in MANets, by emphasizing a scheme that ensures optimal gain for the individual nodes [16]. Auction pricing patterns [25] allocate resources to users through a bidding process conducted by the users. Auction pricing can accomplish equally resource allocation and service attributes. The scheme is based on profile bids where the seller computes an allocation to be given to the buyer. This sequence is repeated until all parts agree. We note that the lengthy negotiation process is ineffective particularly for mobile users while in high speed transit.

Dynamic priority pricing schemes [26][27], on the other hand, are applied on a wireless link shared by the subscribers, divided into different priority classes by the service provider. The mobile subscriber is allowed to select the preferable transmission rate, and its traffic allocation. In addition, the subscribers can split their traffic among several priority classes, and be charged accordingly. The efficiency of this scheme is in its simplicity, and its scalability. The provider's profit increases according to the user's satisfaction by the service. Priority schemes assume, however, that the network's capacity buffer is not exceeded and priority thresholds are kept under a maximum level. Other schemes have been presented to incorporate dynamic game theory models, for non-cooperative scenarios where local utility functions dictate the participation of nodes in relaying [17][18]. It is important to note as well that many of such factors are non-trivial to compute, and many nodes in the IoT would not possess the computational capacity to compute and execute local utility functions. Thus, it is intuitive to pursue a game theoretic approach for the IoT only if it caters for offloading the task of computing local utility functions to nearby high-end nodes.

Other problems stem from scalability issues in IoT, being an architecture that is envisioned to span continents and vast distances [19]. The major issue is being able to establish and maintain end-to-end links, and keeping track of nodes that are dynamically entering and exiting from the network. Remedies have been proposed by increasing the density of backhaul connections and multiple readers to enhance connectivity and capacity, respectively. However, recent studies highlighted the degrading effect of inter-reader and relay collisions [20].

9.3 IoT System Model

Many factors are intrinsically dominant in the operation of a routing protocol. More factors are further augmented as we devise a routing protocol for the IoT paradigm with dynamic topologies and heterogeneous data generating/sharing systems in place. In such a comprehensive paradigm, an incentive data sharing policy is required to motivate sensor owners to participate in the sensing process and to ensure that the provided data is fairly priced. And this in turn necessitates addressing IoT-specific challenges such as system's limitations in terms of lifetime, available capacity, reachability, and delay. In addition, a careful focus on quality management and assurance constraints is to be considered, as well.

Thus, it is the scope of this section to detail and elaborate upon the factors that are considered in IoT-specific routing protocols that tackles all the above mentioned concerns. No single protocol would achieve all objectives, as many objectives are inherently contradictory, thus routing belongs to the notorious NFL (no free lunch) class of algorithms.

Our system is presented in the remainder of this section and elaborated upon in four components. First we present the IoT network as a whole. In the following subsection, we discuss each of the resources pertaining to these nodes, and affecting the relaying scheme. The discussion is completed with a derivation for the utility functions that would govern the choice of nodes. In Table 9.2, a summary table of used notations is presented.

9.3.1 IoT Model

We assume a network of heterogeneous devices, those belonging to WSNs, MANets, RFIDs and stationary/mobile devices. Each communicating entity of these devices (i.e. wired/wirelessly enabled device) is considered as an active node in this design; hereon referred to as a node. Thus, given a set $\boldsymbol{N}$ covering all these devices, we represent each node as $\boldsymbol{n_i} \in \boldsymbol{N}$ where $\boldsymbol{i} = \{1, 2, \cdots, |\boldsymbol{N}|\}$. Thus the set N includes both nodes that are sole relays (access points, routers, WSN sinks, etc.) and other devices with relaying capabilities (communication and processing). We assume each $\boldsymbol{n_i}$ is connected to the network, as disconnected nodes would not take part in this scheme. *i.e.,* if there's no link from a node $\boldsymbol{n_j}$ to some other node $\boldsymbol{n_i} \in \boldsymbol{N}$ then $\boldsymbol{n_j} \notin \boldsymbol{N}$. It is important to note that the size of $\boldsymbol{N}$ varies over time as nodes enter, leave, and run out of energy.

Connectivity between nodes is assumed to take one of two modes. If nodes are in close proximity, then we advocate for direct communication between the nodes without re-routing through the Internet (via a backhaul). However, to sustain the important large scale aspect of the envisioned IoT, we dictate that packets travelling over a threshold of hops $\boldsymbol{\delta}$, would be routed through a backhaul as an intermediate stage, and then re-routed to the final destination from the closest backhaul to that destination. It is thus an important factor to cater for both short

Table 9.2 Summary of notations.

Notation	*Description*
N	Number of in-network devices.
n_i	A node/device $i \in N$.
δ	A threshold on number of hops per routed packet.
Ψ_i	A quintuple computed for each $n_i \in N$ based on residual energy, delay, trust and capacity per buffer.
u_i	Available storage capacity to compute and relay a message at n_i.
u_i'	Normalized buffer capacity per n_i.
π_i	Power consumption per n_i.
π_i'	Normalized power consumption per n_i.
E_i	Maximum charge per node n_i. It varies from ne node to another in a heterogeneous IoT.
D_k	K^{th} data packet size in the queue.
E_{ij}	Euclidian distance between a source node i and a destination j.
ω	A delay step; which is the distance a wireless signal would travel in one-time unit.
D_{single}	A single hop delay a packet will experience.
D_{total}	The total end-to-end delay a packet will experience.
T_{D_j}	A normalized value representing trust level of the exchanged packets between a node j and the destination D.
P_r	The probability to be connected within r communication range.
γ	Path loss exponent in a specific environment.
μ	A normally distributed random variable with zero mean and variance σ^2.
K_0	A constant value calculated based on the mean heights of the transmitter and receiver.
λ_d, λ_t	The arrival rates for data and trusted packets, respectively.
μ_d, μ_t	The departure rates for data and trusted packets, respectively.
μ_{cd}	The rate of departures caused by finding better price in the system.

(*Continued*)

Table 9.2 (*Continued*) Summary of notations

Notation	*Description*
MQL	Mean queue length in the system.
RT	Response time in of the system.
P_{ij}	The probability of being in an ij state shown in Figure 9.3b.
γ	The average percentage of transmitted packets that succeed in reaching the destination.
β	The inflection point of the randomly generated packets sequences.
$\in$	Represents the tolerance to variation in data quality expressed in a Sigmoid function according to Eq. 9.5.
α	A constant that determines the rate of decrease of the utility function in Eq. 9.5.

and long range communication between nodes, both directly or via the Internet backbone. We remark again the importance of obeying to approaches that utilize the Internet backbone only when necessary, and re-route spatially correlated data packets between neighboring nodes without loading the backbone.

9.3.2 IoT Node

Each node $\boldsymbol{n_i} \in \boldsymbol{N}$ takes part in relaying, as well as other tasks. Accordingly, each $\boldsymbol{n_i}$ encompasses a group of resources, with a minimum of communication and processing units. Moreover, in the case of cell phones, PDAs, WSN sinks and RFID readers, they would all encompass a larger pool of resources, not necessarily geared towards the routing task. Thus, it is important to consider how the load of performing these tasks could affect/hinder the relaying capabilities of such nodes. We note their existence but in this scope we account for their effect on residual energy and buffer capacity. And thus, main design aspects considered in the utility function of an IoT-specific node is depicted in Figure 9.2a. A quintuple Ψ_i is computed for each $\boldsymbol{n_i} \in \boldsymbol{N}$ aggregating the following parameters, for both their direct and implied effect on the routing scheme.

9.3.2.1 Residual Energy and Power Model

Each node operating on battery power would possess an energy reservoir, denoted by $\boldsymbol{e_i}$ where $0 \le e_i \le E_i$. Here we denote $\boldsymbol{E_i}$ as the maximum charge for n_i, since this varies across the different types of nodes. To normalize this representation across the heterogeneous nodes in this protocol, we define

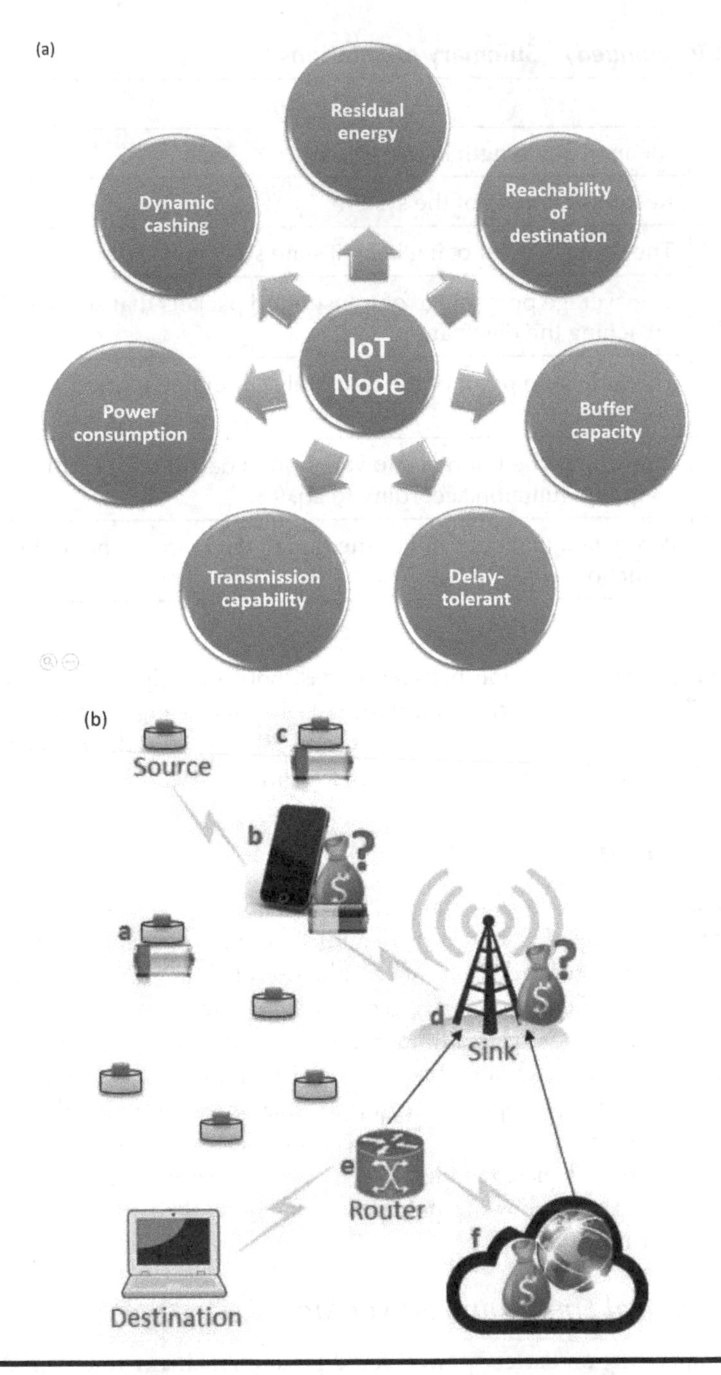

Figure 9.2 (a) The IoT-specific aspects incorporated in the pricing model for computing a utility function. (b) Use case demonstrating the routes taken by INs from the source to a destination (e.g., remote laptop).

$$e_i' = \frac{e_i}{E_i} \tag{9.1}$$

Knowing the size of data packet $\mathbf{D_k}$ to be forwarded, its distance to its next hop and the current load (u_i), each node would compute a value for the power consumption to be incurred by processing a given packet. The power consumption would be represented as $\boldsymbol{\pi}_i$. However, since this is a crude number dependent on the available resources at node n_i and their strength (of transceiver), this value is normalized by dividing by its maximal attainable load and transmission distance. This would favor high end nodes with longer transmission capabilities and more buffers. The normalized value is represented as $\boldsymbol{\pi}_i'$.

9.3.2.2 Load and Buffer Space

Since an intermediate node might be taking part in multiple tasks, each node will represent its available capacity to compute and relay a message as a utilization factor $\boldsymbol{u}_i$, which will be normalized by opposing it to its maximal capacity, thus yielding a normalized $\boldsymbol{u}_i'$. This is directly derived from memory and processing operations, and the yield of the node's MCU in handling different paths. Our delivery approach adopts a data delivery approach where a network intermediate node i has a limited capacity for the maximum amount of data that can be relayed over a specific time period. We define a normalized relaying capacity for the set of i's as:

$$\boldsymbol{u}_i' = \frac{\boldsymbol{u}_i}{max\_\boldsymbol{u}_i} \tag{9.2}$$

where $max\_\boldsymbol{u}_i$ is the maximum expected capacity.

9.3.2.3 Delay

We define a delay step ω which is the distance a wireless signal would travel in one-time unit. Let E_{ij} be the Euclidian distance between a source node i and a destination node j, then the discrete propagation delay over a single-hop link (i, j) would be $\left\lceil \frac{E_{ij}}{\omega} \right\rceil$. Hence, the discrete delay over a multi-hop path is the sum of the discrete delays of single-hop links that constitute that pathFor the sake of generality, Single-hop-delay (D_{single}) and Total-delay (D_{total}), can be defined in Eqs. 9.3 and 9.4, as follows:

$$D_{single} = \left\lceil \frac{E_{ij}}{\omega} + \psi \right\rceil \tag{9.3}$$

and

$$D_{total} = \sum_{total\ hops} D_{Single} \tag{9.4}$$

9.3.2.4 Trust

Trust parameter is a history-based function that is calculated at the network intermediate node per destination to represent a D_j fulfillment measure. A higher T_{D_j} indicates that previous data exchanges between $node_i$ and D_j have been fulfilled according to the predefined IoT characteristics (*e.g.*, capacity, delay, trust, etc.) promised by D_j. We remark also that this delay parameter can be defined alternatively according to the applied IoT application with varying weighting factors. For example, delay is a key parameter in risk management IoT applications. However, it can be more relaxed in other kind of entertainment applications.

9.3.3 Pricing Model

All the previous factors are pertaining to nodal resources and their operation levels, in contrast to the remaining energy each node could support. However, an important aspect to cater for, and possibly arbitrate upon, is the price the nodes are going to charge for relaying a given data packet. That is, since the heterogeneous nodes in the IoT system do not belong to the same network nor the same owner, it is imperative that a monetary cost would be associated with the forwarding action. This is an important aspect for integrating multiple heterogeneous nodes in the architecture and enhancing global scalability. The argument for utilizing current resources with a given cost/price is more dominant than claims of deploying enough resources to cater for all connectivity and coverage tasks of the envisioned IoT. We hereby adopt and build upon the former argument an IoT-specific pricing model.

Pricing schemes in heterogeneous networks such as the ones in IoT paradigm usually cover a wide range of factors to determine the value of a resource in usage. However, the most efficient schemes capitalize on the differential values of each of the heterogeneous components. We built our pricing strategy based on the original laws of supply and demand, the abundance of resources and their homogeneity decrease their value. And higher prices are usually assigned to nodes with rare services [29][30]. Moreover, an IoT-driven pricing model has to realize a level of service that aggregates data from several sources, including the network-context (e.g., 5G, WiFi, LiFi, etc.), the mobile apps' pool, and other sources, to produce better reliable readings. We remark also that an IoT-specific pricing model shall not assume a direct provider-client relation in determining their price mechanisms [31]. In contrast, in large economic systems such as the one we are targeting in this research, entities known as intermediate nodes (INs) are required for coordinating network managements tasks. These INs take care of the necessary authentication, billing and interfacing tasks to find the appropriate service provider within the heterogeneous crowd of resources in the IoT market. To this end, we define an IoT-specific setting by the following four main characteristics: 1) Node residual energy, 2) Load and buffer space, 3) Trust-level, and 4) Delay-tolerance. Those characteristics have been adopted into two different simulation environments;

MATLAB® and *Simulink*® in order to validate our price-based results. *Simulink*, a framework built on MATLAB, is used for validation purposes in order to obtain more realistic results by imitating the real multi-layered net-working process.

Hence, we provide a pricing framework for each node, encompassing a cost-efficient IoT architecture, to address data delivery and routing objectives according to the aforementioned characteristics of the IoT setting. We introduce γ_i which is a pricing factor for each node in the IoT. This is a factor that could be set as a flat rate per number of bytes transmitted, or computed based on the state of the current resources at node $\boldsymbol{n}_i$ represented by Ψ_i. In this work we adopt the latter, as a proof of concept to the monetary exchange for forwarding in the IoT under varying conditions. Thus, we denote the price charged by each node $\boldsymbol{n}_i$ as $\boldsymbol{p}_i$:

$$p_i = \gamma_i * \left[\frac{E_{Tx}\left(D_k, n_j\right) + E_{Rx}\left(D_k\right)}{e_i} + \pi_i' + \boldsymbol{u}_i' \right] \tag{9.5}$$

It is intuitive to note that owners of nodes in the vicinity of such a network, may choose to adaptively contribute or withdraw from the topology by varying the value assigned to γ_i. i.e. setting it to a relatively high value would diminish the chances of it being selected for relaying.

9.3.4 Communication Model

In practice, the signal level at distance d from a transmitter varies depending on the surrounding environment. These variations are captured through the so called log-normal shadowing model. According to this model, the signal level at distance d from a transmitter follows a log-normal distribution centered on the average power value at that point [23]. Mathematically, this can be written as

$$P_r = K_0 - 10P_l \log(d) - \mu d \tag{9.6}$$

where d is the Euclidian distance between the transmitter and receiver, P_l is the path loss exponent calculated based on experimental data, μ is a normally distributed random variable with zero mean and variance σ^2, i.e. $\mu \sim \mathcal{N}(0,\sigma^2)$, and K_0 is a constant calculated based on the mean heights of the transmitter and receiver.

9.4 ARA Routing Approach

The integrated architecture imposed by the heterogeneity of the IoT demands a scalable and inclusive routing protocol. The latter property refers to the exploitation of different relaying resources that are able to carry forward a data packet towards the destination. This section presents ARA protocol.

ARA is divided into two stages: forward and backward. The forward stage starts at the source node by broadcasting setup messages to its neighbors. A setup message includes the cost seen from the source to the current (intermediate/destination) node. A node that receives a setup message will forward it in the same manner to its neighbors after updating the cost based on the values computed in Ψ_i. All setup messages are assumed to contain a route record that includes all node's IDs used in establishing the path fragment from the source node to the current intermediate node. The destination collects arriving setup messages within a Route-Select (RS) period, which is a predefined user parameter.

The backward stage starts when an Acknowledgment (Ack) message is sent backward to the source along the best selected path (called **active** path) in terms of the parameters passed in Ψ_i. If a link on the selected path breaks (due to node movement or bad channel quality), the Ack at an intermediate node *i* is changed to setup message (called i_setup) and forwarded to neighbors of *i* which has discovered the error.

Once the source receives the i_setup, the active path between S and D is established. When no breaks are discovered, the source receives an Ack and it knows that the path has been established, and it starts transmission. If during the communication session (i.e., after selecting the active path) a break is detected, the intermediate node detecting the break will send data on an alternative route (if any) or it will buffer data and send an i_setup message to the destination to look for an alternative path.

In general, nodes can learn about their neighbors and update the Routing Table (RT) either by receiving a broadcasted setup message and accordingly it updates its neighborhood table, or by broadcasting a "*hello*" message periodically, if no messages have been exchanged. This hello message is sent only to the neighborhood of the node. A new neighbor, or failing to receive from a node for two consecutive hello periods, is an indication that the local connectivity has changed.

Algorithm 9.1: For Source node S.

1. **If** S has a new *data* msg and no route to D
2. **Then** forward a *setup* msg.
3. **If** S receives *D_Ack* or *i_setup* msg,
4. **Then** check local $\boldsymbol{p}_i$ and send the new *data* msg's if satisfied.
5. **If** S doesn't receive a response for a RD period,
6. **Then** go to line 2.
7. **If** no pkts are exchanged for *hello_interval* time units,
8. **Then** send a *hello* msg and update RT and $\boldsymbol{p}_i$.

A pseudo code describing the source node algorithm is shown in Algorithm 9.1. Lines 1–2 represents the beginning of the forward stage, where a request to establish an active path is initiated. Such that, if S has new packets to send and no route is known to targeted destination D, then a setup message is forwarded to all available neighbors of S. To do so, all INs node broadcast their identity at the deployment stage and each S node keeps a record of the next-hop towards some IN. Each source node n_i has a next node record which has the following: *ID field* to recognize the next relaying intermediate node ID, the *Geo_Loc* field to determine node geographical coordinates, and the *Number_of_hops field* which has the number of hops towards the destination D. Note that this process will construct a price-based tree for each S node; such that the tree of INs that is rooted at S and involves all price efficient INs towards the destination D will be identified at the initialization of the network. Lines 3–4 indicate that the path has been found.

Hence, active path between S and D is updated and source begins transmitting the new data packets. Lines 5–6 describe the case where a Route Discovery (RD) period is expired. Therefore, the source restarts the route discovery process by sending a new setup message. Finally, lines 7–8 indicate that S has not exchanged messages with neighbors for more than *hello_interval* time units. Thus a hello message is sent and RT is updated accordingly.

A pseudo code describing the intermediate node (IN) algorithm is shown below. Lines 1–2 handle the forward stage, such thatif an intermediate node *i* receives a setup message, it forwards this message to all its unvisited neighbors and records every visited node to establish a backward path. Contrarily, lines 3–7 handle the backward stage of the algorithm. If node *i* receives Ack from destination (called D_Ack), then it checks whether the neighbor towards S on the backward path is reachable or not (i.e. has a broken link). If reachable, it passes the D_Ack to this neighbor and records the necessary information to establish the active path. Otherwise, it initiates a new setup process between *i* and S, by sending i_setup message to *i*'s neighbors. Lines 8–9 keep forwarding this i_setup message until it reaches S to establish an active path between *i* and S instead of the broken one.

Algorithm 9.2: For an IN node *i*.

1. **If** *i* receives *setup* msg,
2. **Then** check thresholds and update/forward *setup* msg if satisfied. Also, the forwarded *setup* msg records visited nodes while traveling to D.
3. **If** *i* receives *D_ Ack*
4. **Then, If** a backward_neighbor is reachable,
5. **Then** forward the *D_Ack*
6. **If** backward_neighbor is not reachable,
7. **Then** send an *i_setup* msg and update RT and local $\boldsymbol{p}_i$.
8. **If** *i* receives *i_setup* msg
9. **Then** check thresholds and forward *i_setup* msg if satisfied. Also, the forwarded i_*setup* msg records visited nodes while traveling to destination.
10. **If** *i* receives *data* msg
11. **If** next hop is still reachable

12. **Then** send *data*
13. **If** a new active path was established
14. **Then** check the price, update RT and send *data* if satisfied.
15. **Else** buffer *data* **and** send *i_setup*

Similarly, lines 10–12 check for the availability of the next hop on the active path while data packets are transmitted through *i* towards the destination D. If next hop is not available, the intermediate node *i* checks for an alternative path. If a new path has been established, lines 13–14 detour the Data packets between S and D along this new partial route and update the active path. If no alternative path is found, line 15 buffers the data packets and initiate a new setup process. We remark that lines 2 and 9 will kill any setup message, if i is not willing to participate in routing.

Finally, a pseudo-code describing the algorithm at the destination node D is shown in Algorithm 9.3. Lines 1–10 handle the case when a setup process has been initiated by an intermediate node *i*. This also indicates link breakage at node *i* in active path between S and D. If there exist alternative path(s) passing through the node detecting link breakage (i.e., node *i*) or passing through the source S, lines 3–4 select the best-cost path and notify *i*. Otherwise, lines 5–10 initiate a new setup process and act as a source node in looking for a new path to S. Therefore, it sends to all D's neighbors and waits for an Ack from the source S (called S_Ack). Meanwhile, lines 11–14 represent the backward stage in response to the forward stage that has been initiated at S. The destination D keep receiving setup messages with the corresponding found paths between S and D for a Route Select (RS) interval. After RS time units, D acknowledges the source S that the active path has been established by sending a D_Ack message to it through the best-cost selected path.

Algorithm 9.3: For the Destination node D.

1. **If** *D* receives *i_setup*
2. **Then** remove paths containing broken links.
3. **If** there exist path(s) passing through *i* or S
4. **Then** select best-cost path and notify *i*.
5. **If** no paths found
6. **Then** send a *setup* msg
7. **If** *D* receives *S_Ack* or *i_setup*
8. **Then** select path indicated by received msg.
9. **If** *D* doesn't receive a response for a RD period,
10. **Then** go to line 5.
11. **If** *D* receives *setup* msg RS not expired
12. **Then** store the candidate path and cost.
13. **If** RS expired
14. **Then** select best-cost path and send *D_Ack* on it.

9.5 Use Case and Theoritical Analysis

To demonstrate the utility of the ARA protocol, we hereby adopt a use case that utilizes heterogeneous nodes in a sample IoT environment. The remainder of this use case will refer to Figure 9.2. A sensing node (the source) has obtained information to be sent to a destination computer. However, no direct link connects both devices, and intermediate devices belong to different networks. We assume that nodes *a*, *b*, *c*, *d*, *e* and *f* are all willing to relay, yet *a* and *c* are already depleted in energy. The sink, node *d*, is powered by electricity and acts as an intermediate node between the resourceful cell phone *b* and the router *e*. ARA will initiate a setup message sent to *a*, *b*, *c*, and its current neighbors. Since *a* and *c* have depleted batteries, they will terminate the flow of the setup request towards the destination. Since the cell phone *b* is in range of communication to the source, it will forward the message to its neighbors (not highlighted here as the pattern is clear).

Eventually the shortest path to the destination is established. The destination will receive two streams $\{S \rightarrow b \rightarrow d \rightarrow e \rightarrow D\}$ and $\{S \rightarrow b \rightarrow d \rightarrow f \rightarrow e \rightarrow D\}$. Since both f and e are resourceful entities, the arbitration of number of hops would manifest a preference for the former route, which will carry an Ack message back to the source node. It is important to note that an Internet link (both forward and backward), which would also incur a cost, takes part in the route options, as the setup message would also parse through it when it is beyond the preset threshold of hops dictated by the application and source request. Furthermore, the IoT network under study can be modelled using Queuing theory for steady state evaluation with an abstraction as illustrated in Figure 9.3a. The resulting continuous time Markov

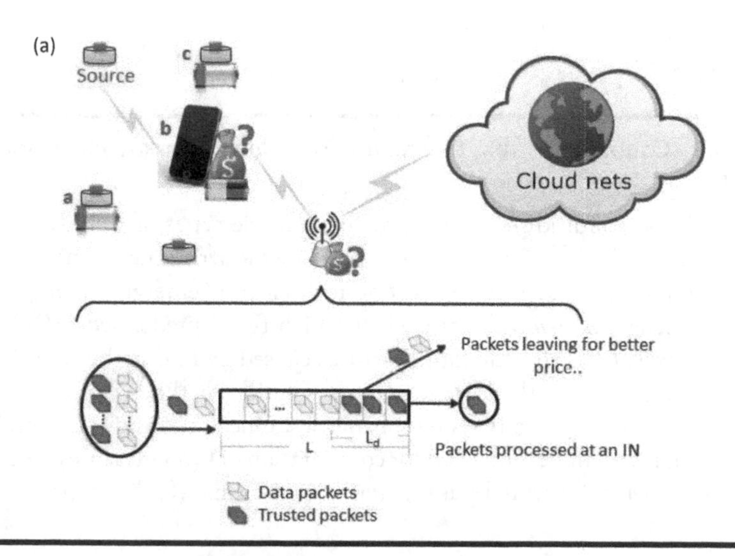

Figure 9.3 (a). The Q-model for the IoT networks.

(Continued)

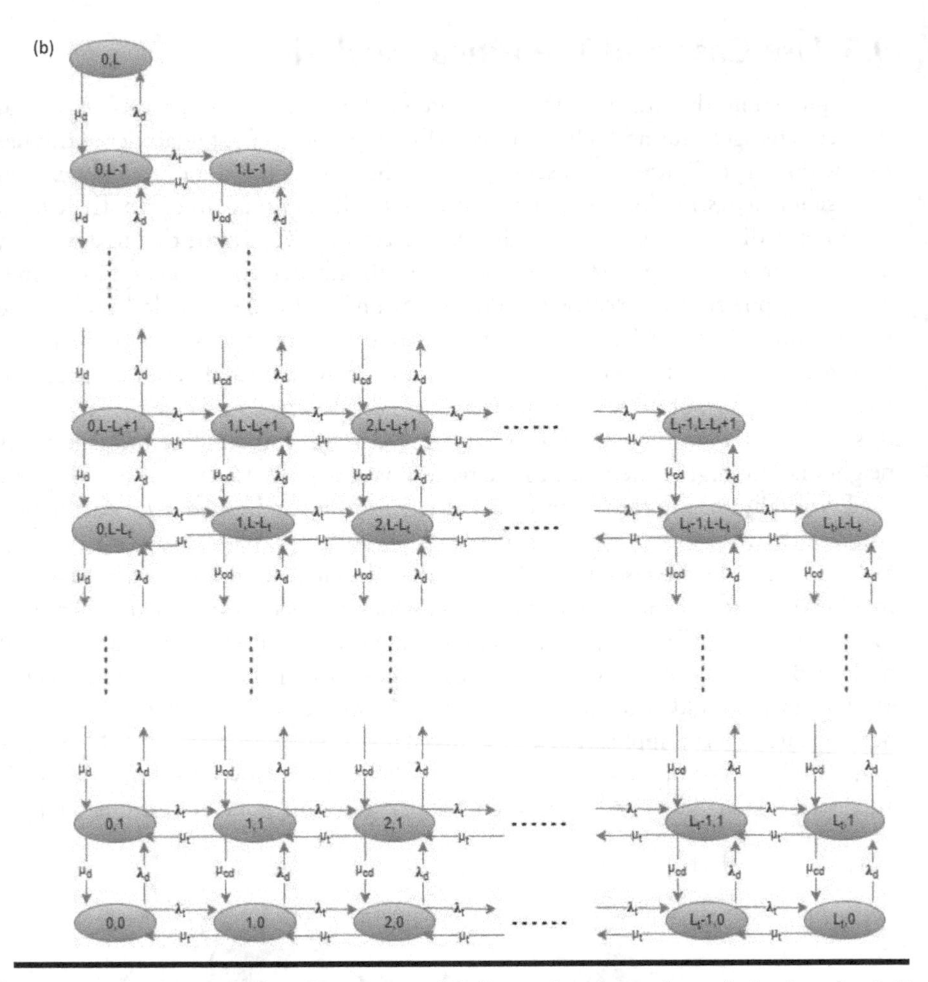

Figure 9.3 (Continued) (b). The multi-dimensional Markov-chain for the IoT networks.

chain would be a multidimensional one for multiple types of traffic similar to the one presented in Figure 9.3b. Where λ_d, and λ_t are the arrival rates, and μ_d, and μ_t are departure rates due to service completion for first time emerging *data* packets and previously exchanged *trusted* data packets which fulfill predefined IoT characteristics, respectively. μ_{cd} is the rate for departures caused by finding better price instead of service completion in the system. Various solution methods can be employed to solve such a system for steady state probabilities. Once the steady state probabilities are obtained, they can be employed for computation of QoS measures such as, mean queue length (MQL), throughput (ɣ), and response time (RT) as follows:

$$MQL_t = \sum_{i=0}^{L_t} \sum_{j=0}^{L} iP_{ij} \tag{9.7}$$

$$MQL_d = \sum_{i=0}^{L_d} \sum_{j=0}^{L} jP_{ij} \tag{9.8}$$

$$\gamma_t = \sum_{i=0}^{L_t} \sum_{j=0}^{L} \mu_t P_{ij} \tag{9.9}$$

$$\gamma_d = \sum_{i=0}^{L_d} \sum_{j=0}^{L} \mu_d P_{ij} \tag{9.10}$$

$$RT_t = \frac{MQL_t}{\gamma_t} \tag{9.11}$$

$$RT_d = \frac{MQL_d}{\gamma_d} \tag{9.12}$$

where P_{ij} is probability of being in a state *ij* in Figure 9.3b. Please note that, it is also possible to employ similarly the steady state probabilities in order to compute the expected value for energy consumption of the considered price-based IoT system.

9.6 Performance Evaluation

In this section, the effectiveness of ARA is validated while assuming a set of in-network heterogeneous nodes. Simulation results show the performance efficiency in terms of *average delay*, *price*, *idle time*, and *throughput* in comparison to key approaches in the literature. In addition, the quality of the data delivery approach is assessed under varying rates per number of bytes transmitted, average energy consumption, and several counts of the network nodes.

9.6.1 Simulation Setup and Baseline Approaches

Using MATLAB R2016a and Simulink 8.7, we simulate randomly generated heterogeneous networks. The generated networks are random in terms of their nodes' positions and densities. In order to route data in these randomly generated networks, we apply our ARA scheme. The output of the ARA scheme is compared to output of another four baseline approaches in the literature. These baseline approaches address the same problem tackled in this research; however, they use different routing strategies. The first approach forms a minimum spanning tree to find the most reliable route in a heterogeneous sensor network [21], and we call it Minimum Spanning Tree (MST) approach; the second is for solving a Steiner tree problem with minimum number of Steiner points [22], and we call it Steiner Tree (ST); the third for adaptive data delivery, and we call it Dynamic Routing Approach (DRA) [24], and forth one called LinGO [33]. In all these baselines we assume a packet size equal to 512 bits, which is typical size for the IoT communication protocols. Every S node has an initial energy of 50*J* and generates 150 pkts/round. A round is defined as the time span per which all S nodes have reported/requested a piece of data.

The MST opts to establish an MST through selected multi-hop paths. It first computes an MST for the given source and destination nodes and then forward messages over the minimum tree model; in which it finds the least count of hops to maintain the best path cost. ST first combines nodes that can directly reach each other into one connected graph. The algorithm then identifies for every three connected graphs a node x that is at most r (m) away. Then, these three connected graphs are merged into one. These steps are repeated until no such x could be identified (i.e., no isolated nodes). DRA takes into consideration the nodes' coordinates in order to limit the updates sent out by the moving node to a local area. LinGO, which is a Link quality and Geographical beaconless OR protocol, introduces a different progress calculation approach compared to the aforementioned ones. It takes into account both the progress of a given forwarding node towards the destination with respect to the last-hop, as well as the radio range. In this way, LinGO reduces the number of required hops. Both MST and ST routing strategies are used as a benchmark in this research due to their efficiency in finding the nearest next hop towards the destination while maintaining the minimum number of required nodes in the source-destination path. On the other hand, DRA is chosen due to its efficiency in adapting to any newly generated topology due to node mobility/heterogeneity. We remark that the original MST, ST and DRA approaches are not hierarchical. Thus, we employed the modified versions of them to make them suitable for our proposed hierarchical framework, where the modified versions take into consideration the in-network nodes heterogeneity and choose next hop based on types of the surrounding node types. For example, a Zigbee-based IoT node will scan for another Zigbee-based node to be considered as a candidate neighboring node for packet relaying/forwarding. Since a larger network size implies longer paths, and thus, higher probabilities for heterogeneity. We examined the four data delivery schemes while the size of the network increases in terms of the IoT-nodes' count. Knowing that larger node count in a data path raises the risk of node failure and, hence, dropped packets. Thus, choosing shorter peripheral paths is better for the overall quality/price gain.

The routing schemes: DRA, MST, ST, and ARA, are executed on 600 randomly generated wireless heterogeneous network topologies in order to get statistically stable results. The average results hold confidence intervals of no more than 2% of the average values at a 95% confidence level. We assume a predefined fixed time schedule for traffic generation at these networks. Data packets are delivered by applying these three approaches. Based on experimental measurements taken in a site of dense heterogeneous nodes [25], we set the communication model variables and other simulation parameters as shown in Table 9.3. We adopt the described signal propagation model in Section 9.3 where the utilized variables/parameters values, shown in Table 9.3, are set to be as follows: $P_l = 4.8$, $\delta = 10$, $P_r = -104$ (dB), and μ to be a random variable that follows a log-normal distribution function with mean 0 and variance of δ^2.

Table 9.3 Parameters of the simulated networks

Parameter	*Value*
τ	70%
n_c	110
ψ	0.001 (msec)
D_{max}	500 (msec)
ω	200,000 (km/s)
P_l	4.8
δ^2	10
P_r	−104(dB)
K_0	42.152
r	100 (m)

Moreover, we assume heterogeneous transceivers communication ranges, to validate our results in a typical IoT setups. For validation and verification purposes, we also used MATLAB with Simulink Framework [32]. *Simulink* can support wireless channel temporal variations, node mobility, and node failures. The simulations last for 2 hours and run with the lognormal shadowing path loss model [34]. In *Simulink*, we adopted also the same path loss and physical layer parameters shown in Table 9.3.

9.6.2 Performance Parameters and Metrics

To compare the performance of these three schemes, the following four performance metrics are used.

1. *Average Delay*: is measured in msec and is defined as the average amount of time required to deliver a data unit to the destination.
2. *Idle time*: this metric reflects the ratio of idle time every node spend while just waiting to forward a message. It is measured in μsec.
3. *Througput*: is set here as a quality measure. It is the average percentage of transmitted data packets that succeed in reaching the destination reflecting the effect of node heterogenuety and delay in IoT setups over the utilized data delivery approach.

4. *Average Price*: this metric is used to observe the influence of the utilized data delivery approach on the overall price to deliver a data unit from source to destination on average.

Meanwhile, the three data delivery performance is assessed using the following three parameters:

1. The size of the network in terms of total node count. This reflects the application's complexity and the scalability of the exploited routing scheme.
2. Average energy consumption rate per data unit (π_i') as an indicator of the network power saving.
3. Cost (γ_i) to observe the influence of the charged price rate over the utilized data delivery approach.

9.6.3 Simulation Results

For a varying number of heterogeneous nodes (between 50 and 350) and deployment space (= 1200 Km^2), Figure 9.4 compares ARA approach with DRA, MST and ST in terms of data delivery latency. It shows how ARA outperforms the other approaches under varying network size. Unlike the other approaches, exchanged messages using the ARA do not show a rapid increment in the end to end delay while the network size is growing. This is because of the utilized utility function in Eq. 9.5 that has been considered by ARA approach. Although DRA is the most adaptive approach, its delay is increasing rapidly while the network size is increasing. However, delay is slower in the monotonically increase while applying MST and ST approaches. It's worth mentioning also that when the network size is greater than or equal to 300, the ARA cannot improve anymore in terms of delay.

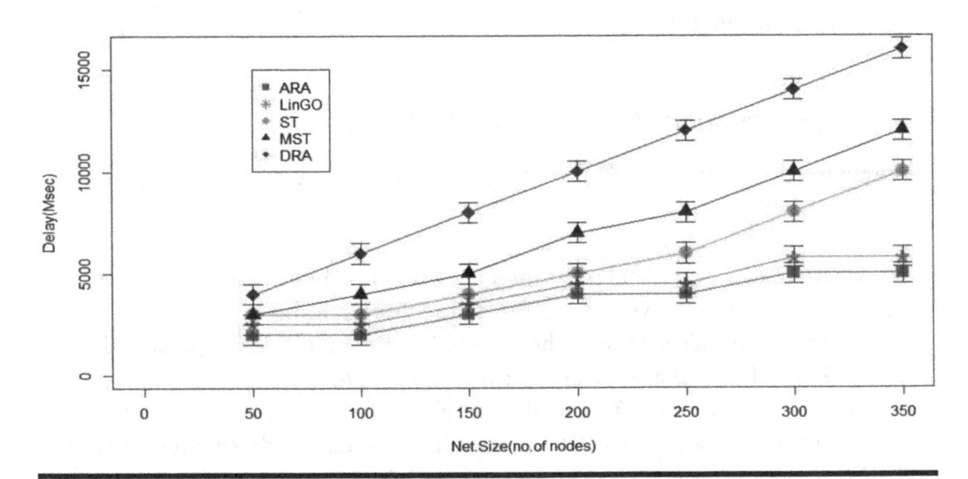

Figure 9.4 Latency vs. number of nodes.

However, it's obvious that ARA approach is achieving the lowest delay with respect to the varying network size.

Meanwhile, the overall network throughput levels achieved by the ARA outperform the levels achieved by other approaches due to considering the next hop status before forwarding the message to it, as depicted in Figure 9.5. In general, network throughput is increasing monotonically for all approaches while the network size is increasing. However, ST is the worst due to ignoring the current status of the node before message forwarding.

Furthermore, Figure 9.6 depicts the effects of the allowed average energy consumption level per node on the network throughput. It shows a monotonically

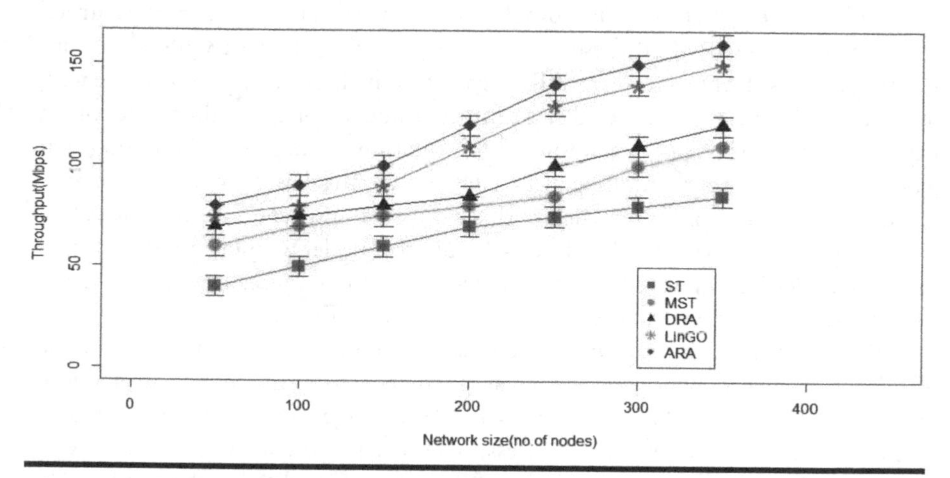

Figure 9.5 Throughput vs. the network size.

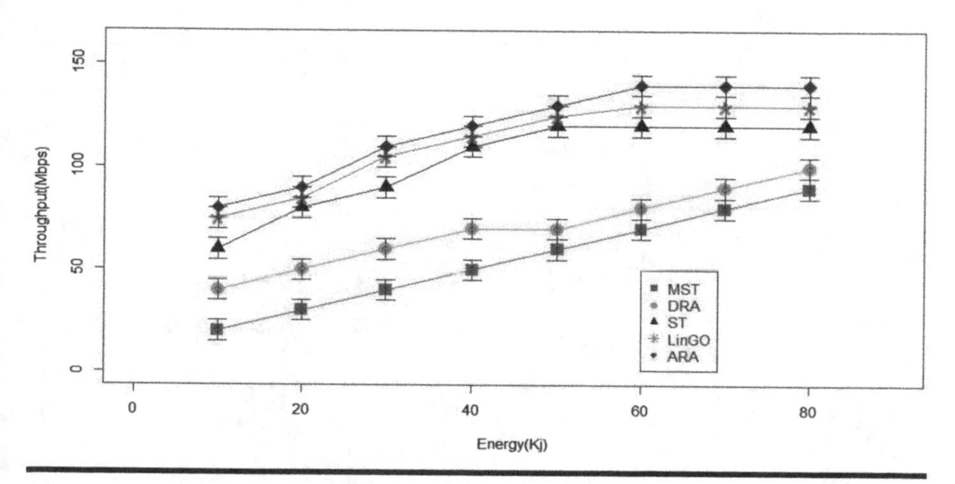

Figure 9.6 Throughput vs. the available energy.

increase in throughput for all approaches while varying the available energy budget. This comparison is performed while considering a fixed network size equal to 150 nodes, and average γ_i rate equal to 0.002 $/byte. Notably, more saving in terms of energy is achieved by applying the ARA and LinGO approaches. However, ARA approach achieve the highest throughput with respect to energy. When the energy budget is greater than or equal to 60 (KJoule), the network throughput is saturated due to other design factors such as γ_i and network size and capacity. Also, it worth remarking here that LinGO adds redundant packets in order to increase the packet delivery probability while experiencing link error periods. This leads to significant increment in the overall throughput.

In Figure 9.7, the network end-to-end delay is decreasing linearly for all methods while γ_i is less than or equal to 50% of the initial rate. This can be returned to the main objective of all these methods in providing the best QoS while considering the cost factor. However, ARA again has the best delay performance with respect to all other approaches due to direct influence of the utility function in Eq. 9.5 in choosing the next hop towards destination. Also, it is worth noting that when the γ_i is greater than or equal to 50%, all approaches cannot improve anymore in terms of the end-to-end delay. This has a great impact on the network QoS.

In Figure 9.8, average idle time is compared under varying total count of network nodes. All approaches are experiencing a monotonically increase in the average idle time while increasing the network size. This is expected due to the availability of several routing options/resources. In addition, we return the increase of idle time when network size is increased, to the dense distribution of network nodes within a fixed deployment space (=1200 km^2). Such a dense distribution provides idler resources as well. Nevertheless, ARA has the lowest average idle time in this comparison due to again the ability of the proposed utility function in Eq. 9.5, where better resource management and utilization is guaranteed.

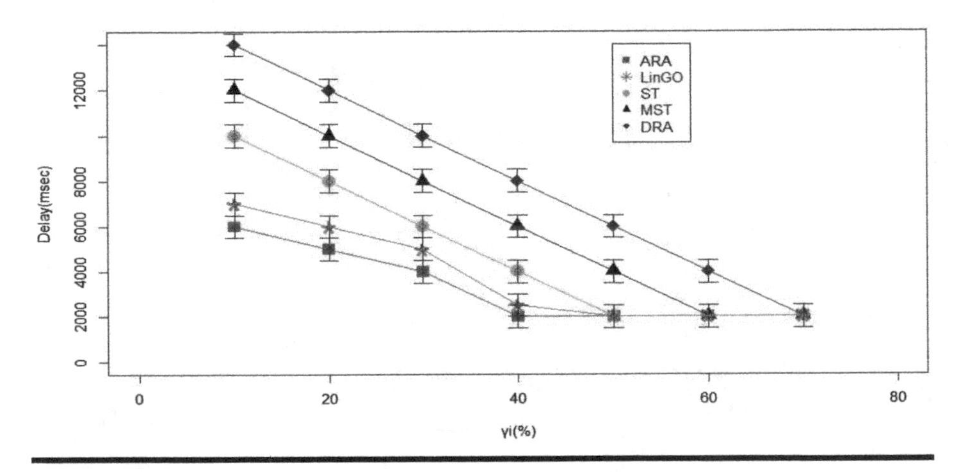

Figure 9.7 Delay vs. the average γi rate percentage.

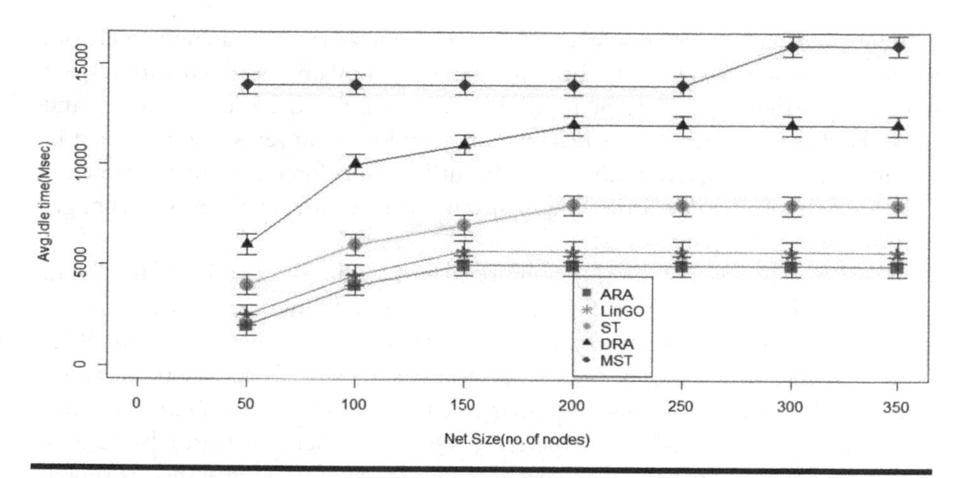

Figure 9.8 Avg. idle time vs. the network size.

In Figure 9.9, ARA consistently outperforms the MST, ST, LinGO and DRA while increasing the nodes' counts. ST and DRA are the worst in terms of average price and that can be returned to their complexity in locating the next hop. Unlike ST and DRA, MST and LinGO approaches are very close to ARA. However, ARA is still outperforming them. The reason is that LinGO and MST adds redundant packets in order to increase the packet delivery probability while experiencing link error periods. This leads to significant increment in the overall price. In general, ARA is better because of the computed price factor based on the state of the current resources at every node $\boldsymbol{n}_i$ represented by Ψ_{i}.

It is noted also that the average price is increasing while the network size is increasing. Again, this has been accomplished due to considering longer routes

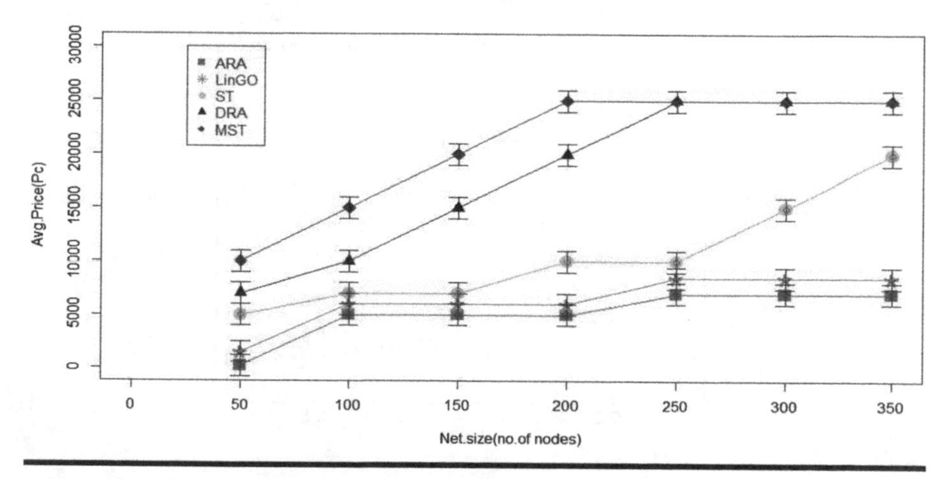

Figure 9.9 Avg. price vs. the network size.

while expanding the network size. We remark that as the nodes count increases the total price achieved by ARA becomes more identical and minimal with respect to other approaches. This can be returned to the excess in the available nodes, and thus, the effect of better choices becomes observable and prevents any increment in the price. In general, the function in Eq. 9.5 utilizes the aforementioned parameters after normalization, in a manner that maps the expected user experience to changes in individual utility parameters.

To show the impact of the aforementioned parameters on our utility function in Eq. 9.5, we present the plots in Figures 9.10–9.12 for each of the utility parameters *Delay*, *Network Quality* (i.e., throughput) and Trust T, respectively. In Figure 9.10, *Delay* is plotted with a constant α that determines the rate of decrease of the exponential utility in Eq. 9.5. This particular function was chosen for *Delay* to reflect the rate of loss in the Quality of Experience (QoE) [1] as delay increases. By varying the value of α, it is possible to achieve different levels of delay-tolerance as shown in Figure 9.10, where we chose $\alpha = 0.5$ for delay-tolerant data and $\alpha = 0.1$ for more delay-sensitive data. We note that, for a delay-sensitive data request, a very low delay has to be achieved in order to provide a high delay utility component.

The Quality parameter is plotted in Figure 9.11. We note that we adopt a Sigmoid function according to Eq. 9.5 where the tolerance to variation in data quality is expressed by fixing the value of ϵ (set here to 10) and varying the inflection point denoted by the value of β. Thus, if the requested data is quality-sensitive (*e.g.*, VoIP is sensitive to low transmission rate) the function will require a higher value before the utility increases (as depicted in the lower plot with $\beta = 0.8$ in Figure 9.11). In contrast, lower constraints on quality require a utility that increases rapidly at a lower value of *Quality*, which can be achieved with an early inflection point ($\beta = 0.5$ in Figure 9.11).

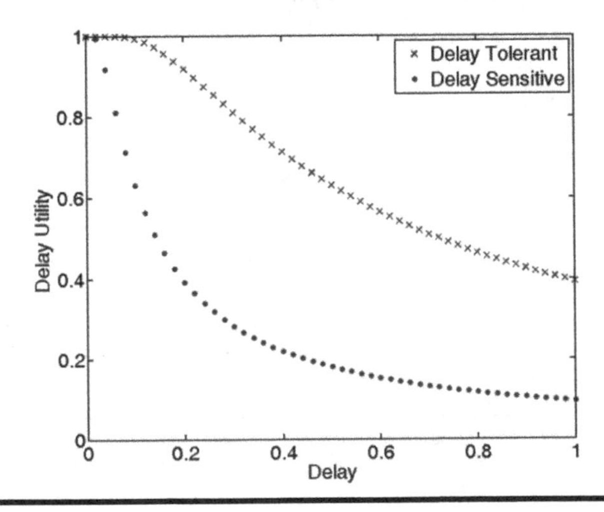

Figure 9.10 Delay function plot.

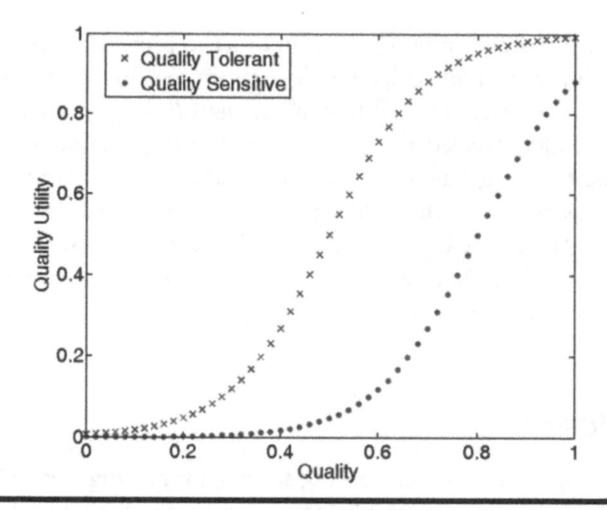

Figure 9.11 Quality function plot.

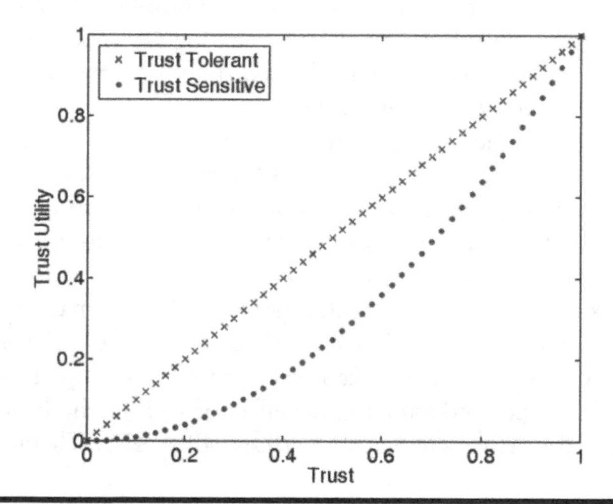

Figure 9.12 Trust function plot.

The value of Quality in Figure 9.11 ranges from 0 to 1, where 1 indicates the best level of quality attainable depending on the quality metric.

Lastly, Figure 9.12 shows the plot of the Trust function where $T \in [0,1]$. Note that an intermediate node can give more emphasis to this parameter through the factor σ to particularly penalize sinks with bad service accounts. This is shown in the lower plot of Figure 9.12 where $\sigma = 2$, whereas the upper (better T) plot is a result of $\sigma = 1$.

We note here that we multiply the Trust component by the rest of the utility function in Eq. 9.5 in order to balance the effect of "deceiving" intermediate nodes that may offer attractively low P_i in order to pass false quality promises. Moreover, we divide the utility function by $P_{intermediate}$ in order to protect the client from situations where two or more intermediate nodes happen to achieve almost equal utility scores while charging prices that, although less than P_i, largely vary.

Evaluation results revealed that price based routing scheme is better in most cases. Price based routing scheme has a longer path length in contrast to shortest routing scheme as expected. But it has got better results in terms of cost, latency, idle time, and scalability. ARA avoids using busy, centric nodes, thus data in ARA deliver over more nodes. Data reach to destination over available nodes instead of waiting for queue in busy nodes.

9.7 Conclusion

This paper proposed a price based routing scheme for heterogeneous IoT networks called the ARA. ARA aims at establishing a cap on the inter-nodal routing price to dynamically utilize the Internet backbone if the source to destination distance surpasses a preset (case optimized) threshold. Promising simulation results are achieved by altering significant parameters including network size and the available energy budgets, and seeing how it affects certain metrics in the IoT paradigm such as latency, cost and resource utilization.

These results show the efficiency of our framework when compared to three prominent Ad-hoc data delivery protocols. Our simulation results show that the ARA exhibits superior performance for different network sizes, lifetime, end-to-end delays, quality, and prices. It is strongly recommended with huge size networks as it is the most cost-effective one on the long run.

Future work would investigate utilizing IoT nodes in smart city settings as mobile data collectors with semi-deterministic mobility trajectories. Also, of practical interest is the application of localization methods among sensors operating on different technologies and studying the effect of such methods on the system's performance and the delivery rate of the corresponding ARA scheme.

References

[1] F. Al-Turjman, A. Alfagih, and H. Hassanein, "A Novel Cost-Effective Architecture and Deployment Strategy for Integrated RFID and WSN Systems", *In Proc. of the IEEE Int.Conf. on Computing, Networking and Communications (ICNC)*, Maui, Hawaii, 2012, pp. 835–839.

[2] L. Boloni and D. Turgut, "Should I send now or send later? A decision-theoretic approach to transmission scheduling in sensor networks with mobile sinks", *Wiley's Wireless Communications and Mobile Computing Journal (WCMC)*, vol. 8, no. 3, pp. 385–403, 2008.

[3] F. Al-Turjman, "Impact of User's Habits on Smartphones' Sensors: An Overview", *HONET-ICT International IEEE Symposium*, 2016.

[4] G. Wang, D. Turgut, L. Boloni, Y. Ji, and D. Marinescu, "Improving routing performance through m-limited forwarding in power-constrained wireless networks", Journal of Parallel and Distributed Computing (JPDC), vol. 68, no. 4, pp. 501–514, 2008.
[5] Atzori, L., Iera, A. and Morabito, G. "The internet of things: A survey Computer Networks", *Computer Networks, Elsevier*, vol. 54, pp. 2787–2805, 2010.
[6] G. Singh, and F. Al-Turjman, "Learning Data Delivery Paths in QoI-Aware Information-Centric Sensor Networks", *IEEE Internet of Things Journal*, vol. 3, no. 4, pp. 572–580, 2016.
[7] G. Solmaz, M.I. Akbas and D. Turgut, "A Mobility Model of Theme Park Visitors", *IEEE Transactions on Mobile Computing (TMC)*, vol. 14, no. 12, pp. 2406–2418, 2015.
[8] D. Turgut and L. Boloni, "Heuristic approaches for transmission scheduling in sensor networks with multiple mobile sinks", The Computer Journal, vol. 54, no. 3, pp. 332–344, 2011.
[9] Q. Zhu, et al. "IOT Gateway: BridgingWireless Sensor Networks into Internet of Things", In *Proc. of the IEEE/IFIP Int. Conf. on Embedded and Ubiquitous Computing (EUC)*, Hong Kong, China, 2010, pp. 347–352.
[10] F. Al-Turjman, "Hybrid Approach for Mobile Couriers Election in Smart-cities", In *Proc. of the IEEE Local Computer Networks (LCN)*, Dubai, UAE, pp. 507–510, 2016.
[11] E. Welbourne, et al. "Building the Internet of Things Using RFID: The RFID Ecosystem Experience", *Internet Computing, IEEE*, vol. 13, pp. 48–55, 2009.
[12] I. Fajjari, et al., "New Virtual Network Static Embedding Strategy within the Cloud's Private Backbone Network", *Elsevier Computer Networks*, April 2014.
[13] M. Kranz, et al., "Embedded interaction: Interacting with the internet of things" *Internet Computing, IEEE*, vol. 14, no. 12, pp. 46–53, 2010.
[14] Sarma, A. and GirÃo, J. "Identities in the Future Internet of Things" *Wireless Personal Communications*, Springer Netherlands, vol. 49, pp. 353–363, 2009.
[15] M. Afergan, "Using repeated games to design incentive-based routing systems" In *Proc. of the IEEE Int. Conf. on Computer Communication, (INFOCOM)*, Barcelona, Spain, 2006, pp. 1–13.
[16] S. Zhong, et al., "On designing incentive-compatible routing and forwarding protocols in wireless ad-hoc networks", *Wireless networks*, vol. 13, no. 9, pp. 799–816, 2007.
[17] W. Saad, Z. Han, M. Debbah, A. Hjorungnes, T. Basar, "Coalitional game theory for communication networks", *Signal Processing Magazine, IEEE*, vol 26, pp. 77–97, 2009.
[18] G. Singh, and F. Al-Turjman, "A Data Delivery Framework for Cognitive Information-Centric Sensor Networks in Smart Outdoor Monitoring", *Elsevier Computer Communications*, vol. 74, no. 1, pp. 38–51, 2016.
[19] H. Sundmaeker, P. Guillemin, P. Friess, S. Woelfflé, "Vision and challenges for realising the Internet of Things", *CERP-IoT, European Commission*, Luxembourg, 2010.
[20] W. Heinzelman, A. Chandrakasan, and H. Balakrishnan, "Energy-efficient communication protocols for wireless microsensor networks," In *Proc. of Hawaiian Int. Conf. on Systems Sci, Big Island*, Hawai, 2000, pp. 2–10.
[21] F. Al-Turjman, H. Hassanein, and M. Ibnkahla, "Towards prolonged lifetime for deployed WSNs in outdoor environment monitoring", *Elsevier Ad Hoc Networks Journal*, vol. 24, no. A, pp. 172–185, Jan., 2015.

[22] F. Senel and M. Younis, "Relay node placement in structurally damaged wireless sensor networks via triangular steiner tree approximation" *Comput. Commun.* vol. 34, no. 16, pp. 1932–1941, Oct. 2011.
[23] M. Z. Hasan, et al. "A Survey on Multipath Routing Protocols for QoS Assurances in Real-Time Multimedia Wireless Sensor Networks", *IEEE Communications Surveys and Tutorials*, 2017. DOI: 10.1109/COMST.2017.2661201.
[24] R. Rahmatizadeh, S. Khan, A.P. Jayasumana, D. Turgut and L. Boloni, "Routing towards a mobile sink using virtual coordinates in a wireless sensor network", In *proc. of the IEEE ICC*, pp. 12–17, June 2014.
[25] X. Wang, H. Schulzrinne, "Pricing network resources for adaptive applications," *IEEE/ACM Trans. Netw.*, vol. 14, no. 3, pp. 506–519, Jun. 2006.
[26] G. V. Ozianyi, N. Ventura, and E. Golovins, "A novel pricing approach to support QoS in 3G networks," *Computer Networks*, vol. 52, no. 7, pp. 1433–1450, May 2008.
[27] F. Al-Turjman, "Cognition in Information-Centric Sensor Networks for IoT Applications: An Overview", *Springer Annals of Telecommunications*, pp. 1–16, 2016. DOI: 10.1007/s12243-016-0533-8.
[28] F. Al-Turjman and H. Hassanein, "Towards augmented connectivity with delay constraints in WSN federation", *International Journal of Ad Hoc and Ubiquitous Computing*, vol. 11, no. 2, pp. 97–108, 2012.
[29] A. DaSilva, "Pricing for QoS-enabled Networks: A Survey," *IEEE Commun. Surveys and Tutorials*, vol. 3, no. 2, pp. 2–8, 2000.
[30] D. Turgut, L. Bölöni, "IVE: Improving the value of information in energy-constrained intruder tracking sensor networks", In *proc. of the IEEE ICC*, pp. 6360–6364, 2013.
[31] F. Al-Turjman, "Cognitive Routing Protocol for Disaster-inspired Internet of Things", *Elsevier Future Generation Computer Systems*, 2017. DOI: 10.1016/j.future.2017.03.014.
[32] https://www.mathworks.com/products/simulink/. SIMUTOOLS2010.8727. doi: http://dx.doi.org/10.4108/ICST.SIMUTOOLS2010.8727.
[33] D. Rosário, Z. Zhao, A. Santos, T. Braun, E. Cerqueira, "A beaconless Opportunistic Routing based on a cross-layer approach for efficient video dissemination in mobile multimedia IoT applications", *Elsevier Computer Communications*, vol. 45, pp. 21–31, 2014.
[34] M. Z. Hasan, et al., "Optimized Multi-Constrained Quality-of-Service Multipath Routing Approach for Multimedia Sensor Networks", *IEEE Sensors Journal*, vol. 17, no. 7, pp. 2298–2309, 2017.

Chapter 10

Planting & Farming in Cloud-Based IoT[1]

Fadi Al-Turjman
Antalya Bilim University

Contents

[1] **F. Al-Turjman**, "The Road Towards Plant Phenotyping via WSNs: An Overview", *Elsevier Computers & Electronics in Agriculture*, 2019. DOI: 10.1016/j.compag.2018.09.018.

10.1 Introduction

Plant phenotyping (PP) is the identification process of the genetic code differences and the environment effects on the phenotype (or plant appearance/behavior). Phenotyping is a significant research direction on plant biological processes, and is used in both forward and reverse genetic approaches to obtain fundamental insights or advance crop improvement [1]. Phenotyping has been vital for many years where small scale farming, herding, and fishing were the principal means of existence for eras before the finding of oil and gas. Although the relative significance of these activities has decayed in recent years, governments are attempting to revive the animal and plant agriculture to provide a reasonable degree of self-support in food production. Moreover, they have recently issued a number of phenotyping projects such as medical phenotyping [1], Mediterranean fever in Turkish population [2], familial Mediterranean fever [3], and Alpha-Thalassemia Mutations [4]. Moreover, food supplies imports in Turkey for example are recently decreased by 7.1% [5] while it can be much more effective when the Wireless Sensor Network (WSN) technology is utilized [6]. Worth mentioning also is that the agricultural raw materials trading shortage is about $5 million [7]. In Qatar for example, they import over 90% of its food supplies, and the agricultural trading shortage is about US $1.2 billion [8]. Moreover, Qatar and many other similar countries are still providing less than 8% of its poultry, 4% of its cattle, 7% of its sheep, and 6% of its liquid fresh milk from domestic sources [8]. These numbers are expected to increase in the coming few years due to increasing population growth.

In general, the agriculture development worldwide is facing major obstacles due to the scarcity of the problem of irrigation, use of fertilizers, seed breeding, machinery, pesticides, soil treatment, and soil analysis complications [7]. Therefore, reviving the agricultural sector requires the application of a novel farming model that has high economic efficiency, optimal use of scarce resources, minimum impact on the environment, and should be sustainable. One key strategic solution is to deploy advanced farming management systems (AFMS) to observe, measure and respond to inter and intra-field variability in the farms. Plant Phenotyping (PP) arises as one of the key state-of-the-art technologies that can be adopted to build AFMS, which can introduce a significant increase to the crops, poultry and livestock productivity by effectively managing the available resources and providing the optimum quantitates in terms of water, food, temperature, humidity, fertilizers, etc. The integration of PP will allow the farmers to establish a comprehensive and informative records of their farms and crops, improve the decision-making process and enable better management of their farms. Consequently, using PP can be one of the pivotal solutions for the aforementioned challenges. However, the process of taking phenotypic measurements had to be performed in laboratories under costly and time consuming conditions. And thus,

In this survey article, we overview the existing attempts in WSNs towards realizing smart agriculture applications in general and PP in specific. We discuss key design factors in deploying WSNs under harsh operational conditions in outdoor environments. Moreover, we assess and evaluate the existing solutions and key enabling technologies that can realize the PP in the near future in practice.

The rest of this article is organized as follows. In Section 10.2, we discuss the PP project in details in addition to providing some real examples from the field in Section 10.3. Next we introduce and discuss key design aspects related to PP testbeds in Section 10.4. In Section 10.5, we discuss the possibility of prototyping and implementing a custom WSN for PP in practice using the existing enabling technologies. Finally, we conclude this work with few recommendations and possible future work in Section 10.6.

10.2 Background

The concept of PP, which was alternatively called site-specific management, is not new. The application of PP started more than 20 years ago [9]. The PP is generally associated with the application of crop production input elements based on the assessment of the variability of need for a particular input. The nature of the variability can be either temporal or spatial. The inputs are usually the regular daily requirements of the farms such as seeds, fertilizers, insecticides, drainage tiles and subsoiling. The input can vary by rate (quantity) or type (e.g., two kinds of herbicide). PP suggests that the application of the input does not have to be uniform across the field, but the analyses of variability are essential to enable such an application. Based on this definition of PP, a grower who spot-sprays a field for weed control, or alters fertilizer quantities for a sandy knoll in his fields, is applying PP. Scheduling of irrigation, fertigation, and banding of fertilizers can be classified as PP practices as well. More recently, PP has progressed to include the use of advanced technologies such global positioning systems (GPS), yield monitors, field mapping and record archiving, variable rate application and planting equipment. It is worth noting that PP should not be understood as the incorporation of advanced automated equipment in agriculture, but the gathering and effective use of information obtained from the field. Information collection and exploitation is the core of PP. Therefore, PP will not replace humans, but it will increase the capability and requirement for more highly trained farmers and engineers. Relatively speaking, very few groups of farmers presently have such proficiency in the world. The PP can also be applied to livestock; however, it is commonly called Precision Livestock Farming (PLF). The PLF enables the recognition of each individual animal. Using PLF has enabled farmers to record several aspects of each animal, such as age, pedigree, production, growth, health status and feed conversion. The result is significantly higher reproduction out- comes, high quality food and general safety, animal farming that is highly efficient and sustain- able, healthy animals, and a low impact of livestock production to the environment [10].

10.3 Plant Phenotyping (PP) Examples

In [11], authors summarize current progress in plant disease phenotyping and suggest future directions that will accelerate the development of resistant crop varieties. In Figure 10.1A, Pseudomonas syringae infection on Arabidopsis thaliana with gray water-soaked lesions surrounded by chlorosis. Figure 10.1B, Early-stage Xanthomonas euvesicatoria infection on pepper with small water-soaked lesions. Figure 10.1C, Xanthomonas oryzae pv. oryzae infection on rice with grayish green water-soaked lesions coalescing into yellow streaks. Figure 10.1D, Xanthomonas axonopodis pv. manihotis infection on cassava with dark water-soaked lesions that are spreading and leading to leaf wilt. In [12], authors summarize as an example of the integrative automated high-throughput phenotyping platform, grow chamber-based phenotyping (see Figure 10.2). This reference focuses on recent advances towards development of integrative automated platforms for high-throughput plant phenotyping that employ multiple sensors for simultaneous analysis of plant shoots. In both basic and applied science, the recently

Figure 10.1 Examples of disease symptoms caused by bacterial plant pathogens discovered by phenotyping [11]. (A–D) Please see text for descriptions.

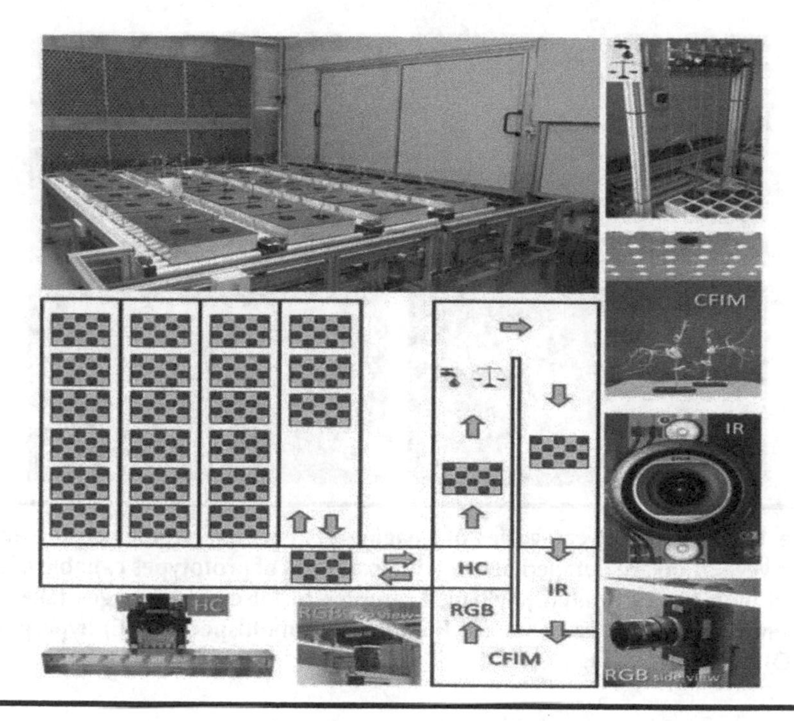

Figure 10.2 Scheme of grow chamber-based automated high-throughput phenotyping platform [12].

emerging approaches have found importance as tools in unravelling complex questions of plant growth, development, responses to environment, as well as selection of appropriate genotypes in molecular breeding strategies. The development of effective field-based High-Throughput Phenotyping Platforms (HTPPs) remains a bottleneck for future breeding advances. However, progress in sensors, aeronautics, and high-performance computing are paving the way. In [13], authors review recent advances in field HTPPs, which should combine at an affordable cost, high capacity for data recording, scoring and processing, and non-invasive remote sensing methods, together with automated environmental data collection (see Figure 10.3). Laboratory analyses of key plant parts may complement direct phenotyping under field conditions. Improvements in user-friendly data management together with a more powerful interpretation of results should increase the use of field HTPPs, therefore increasing the efficiency of crop genetic improvement to meet the needs of future generations. In [14], authors have assessed a range of different wavelength imaging techniques in plant phenotyping (see Figure 10.4). For the imaging sensors applied to plant phenotyping, physical properties, depth knowledge, robust software, and image analysis pipelines are prerequisites to enable the collection of phenotype data. This have been achieved a WSN. Visible imaging for the estimation of shoot biomass and growth patterns in 2D (individual leaves to

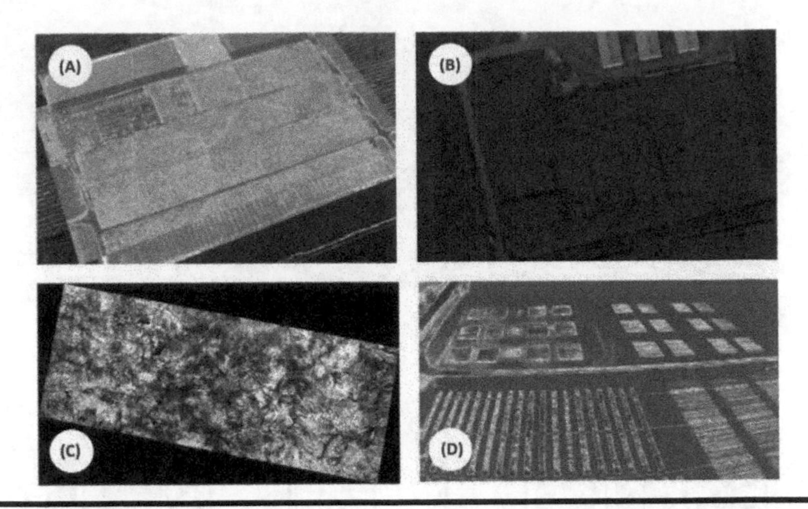

Figure 10.3 Different categories of imaging systems for remote sensing evaluation of vegetation are detailed below with examples of prototypes capable of being carried by UAPs of limited payload. Examples of false-color images taken with different categories of cameras: (A) RGB/CIR, (B) multispectral, (C) hypespectral, and (D) thermal imaging [13].

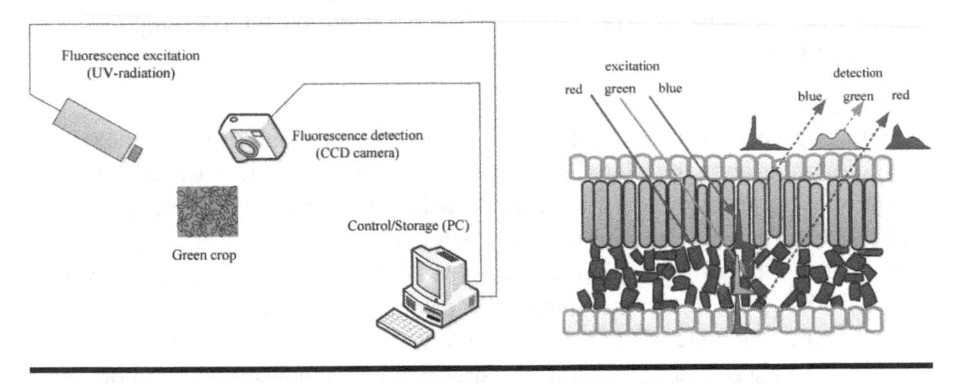

Figure 10.4 A scheme for the multi-color fluorescence imaging system and the chlorophyll fluorescence emission of green leaves as induced blue, red and green excitation light [14].

canopies) has been used reliably for crops in breeding. Fluorescence imaging was primarily used for foliar disease detection and thermal imaging for plant water status detection. A 3D surface reconstruction requires calibration for biomass estimation. Imaging spectroscopy requires standard procedures for the extraction of spectral features to reduce raw data in plant phenotyping. There is a large difference in the reliability of imaging methods between controlled environments and the field. This reliability must be considered to understand the measurement principle for each experimental design, proper sensor calibration, and regular calibration of the imaging-based systems.

10.4 WSN Design Aspects in Phenotyping

The main objective of this paper is to investigate design and implementation aspects of a wireless sensor network in phenotyping. The WSN shall handle remote monitoring and control of a large number of sensing and monitoring devices under variable density and mobility conditions. The network should operate under the constraints of low energy consumption and low power transmission, using reliable and cost effective networking techniques. The design should prolong the network lifetime and overcome various unpredictable elements at both network deployment and operation stages. Specifically aiming at PP applications, our interest is to enable features such as connectivity, coverage, reliability, survivability and lifetime, in an energy efficient manner. In the following we propose a detailed discussion for the key design aspects to be considered in any PP application that involve a WSN as enabling technology.

10.4.1 Deployment

Deployment of nodes in the field is an important design aspect that is usually overlooked, or not thoroughly investigated, by designers of WSNs. Node placement can have a profound effect on connectivity, coverage and reliability [14][15][16]. Therefore, deployment can be a significant factor in improving the cost efficiency of the entire systems and the overall network performance. In the literature, most of the existing research on deployment tackles the problem from a 2D perspective [17][18]. However, in PP applications the coverage area can change significantly over time in the 3D space, causing an optimized 2D deployment strategy to fail. For example, animal migration over seasons, plant growth in different directions, different concentration of gases, are factors that can cause a significant change in channel propagation conditions. 3D deployment strategies can provide higher degrees of freedom for node placement thus leading to better connectivity and longer network lifetime. The main challenge of 3D deployment is that the search space is very large, thus optimization of node placement becomes computationally overwhelming. To address this issue, we have proposed a 3D grid-based deployment strategy [19], which limits the search space and simplifies the placement optimization problem. In this work, the monitored environment is divided into a virtual 3D grid, and the positions of sensor nodes (SN) and relay nodes (RN) are limited to the intersection points. We formulated an optimization problem that maximizes connectivity between the SNs and the base-station (BS), with constraints on the required lifetime of the network, and the number of RNs available. Our proposed deployment strategy achieved outstanding performance results in terms of cost and network lifetime as compared to the 2D models. However, the work [19] was performed for low density static WSN where the nodes are placed in a deterministic manner, which is mainly suitable for small scale PP applications. In a previous work, we considered the 3D deployment problem taking into consideration the variable density, randomness and mobility of the nodes [19]. This work optimizes the network connectivity, while guarantying specific network lifetime and cost. To evaluate the performance of WSN more accurately in PP environment, a new metric shall be developed and utilized to measure the WSN lifetime. The revised definition for the network lifetime shall be developed by considering node redundancy and heterogeneity, in addition to addressing the problem of the huge search space and the network connectivity representation. The effectiveness of the deployment strategy shall be validated through extensive simulations and comparisons assuming practical considerations of signal propagation and connectivity under varying probabilities of node/link failures. Significant improvements are expected after considering a revised WSN lifetime definition in terms of the network connectivity and lifetime under harsh operational conditions where the probabilities of node/link failures are high.

It is worth noting that the deployment of the nodes shall be performed based on the recommendation of an expert (consultant) in the field of agriculture to optimize the nodes deployment process. For example, the number of nodes per unit area, and the variable sampling frequency has to be determined for targeted environment in particular. If such process is not optimized, it can negatively impact the efficiency of the system.

10.4.2 Localization

In the more general PP models, the node density will be variable and thus varying Probability of Link Failures (PLF) might apply according to their location. Moreover, some of the nodes might be mounted on the farming machines, which also can change its location with respect to the other nodes. Mobility of the nodes can create variable density clusters of nodes and variable link quality. Detecting the location of the nodes can be used to maintain their connectivity to the network [20]. A few attempts in localization have focused on the development of efficient techniques to enhance the localization using wireless networks as well as satellite positioning systems [21][22][23]. For example, in [23] authors presented novel technique for asynchronous WCDMA multipath delay estimation by de-convolving the received signal with a specific pulse shape, followed by Teager-Kaiser operator. The deconvolution process is applied to reduce the impact of the pulse shaping process utilized for band limiting purposes. In [20], authors considered quality-of-service (QoS) position-based routing for ultra wide band (UWB) mobile ad hoc networks. The considered technique applies call admission control and temporary bandwidth reservation for the discovered routes through the network, taking into account the interactions of the medium access control (MAC) protocol. Using cross-layer design, it exploits the advantages of UWB at the network layer by exploiting the location information for optimizing the routing and bandwidth allocation, and by enabling the multi-rate feature.

10.4.3 Medium Access Control (MAC) & Security

Energy waste at the link layer mainly comes from idle listening and overhearing [24]. In idle listening, a mobile node (MN) listens to an idle channel for possible incoming data traffic, while in overhearing the MN receives packets that are intended for other nodes due to the broadcast nature of radio transmission. The common approach for energy efficiency in sensor networks is to put the MN into a sleep mode during its idle listening and/or overhearing time, i.e., to let the MN alternate its operation mode between active and sleep periods [25][26]. Extending such an approach to highly dynamic data traffic and node mobility in both temporal and spatial domains requires further studies, in order to satisfy QoS requirements. Specifically, providing statistical delay guarantee for data delivery and satisfying requirements for packet dropping probability and throughput, while achieving energy saving, have been overlooked in literature. The situation is even more challenging in a multi-hop scenario, where it is necessary to avoid packet forwarding interruptions due to the next hop MN being in a sleep state. Focusing on distributed contention-based channel access, we recommend to develop new methodology to effectively schedule the active and sleep periods so as to maximize energy efficiency while ensuring service quality in the presence of node mobility and data traffic dynamics. Intuitively, a longer sleep period for an MN saves more energy. However, it can jeopardize a desired upper bound on packet delivery delay at the destination. Due to the random nature of packet arrivals, wireless fading channel, and node mobility, a statistical delay guarantee can be more efficient than a deterministic guarantee. On the other hand, energy consumption can be reduced by decreasing MAC overhead and transmission collisions among nodes [27]. One approach is to let a node contends only once to transmit a batch of packets within required delay bound, followed by assigned contention-free channel time.

The sleep duration of an MN should be limited by its required throughput and the service demands of other MNs in the neighborhood. With distributed MAC, each node needs to contend for the channel according to incomplete information of its neighboring nodes, with possible asynchronous active/sleep schedules, via limited information exchange. The energy efficiency depends on MN active/sleep schedules, propagation environment, node locations, and traffic load. The source MN transmission shall be modeled as an M/G/1/K queue, given a Poisson data traffic arrival process, with a finite buffer size K to capture the delay bound requirement. Also establishing an analytical relation between the MN sleep duration and packet dropping probability with buffer size K, using the M/G/1/K queuing model [28] is recommended. The relation will help to capture the trade-off between energy efficiency and system throughput. Extension to a multi-hop scenario involves studying a G/G/1/K queuing system, which is technically very complex. Based on the single-hop analysis, we aim to develop some approximations for the energy and throughput relation.

Most existing works on energy efficient MAC protocols rely on time synchronization among MNs. However, it is not always practical to assume that MNs can satisfy the required time synchronization condition [29]. We recommend developing a distributed asynchronous MAC protocol for a multi-hop multi-source and multi-destination network, based on cooperative information sharing, to minimize energy consumption at each MN while satisfying required delay bound and throughput in PP.

The type of exchanged information in the WSN has different constraints and urgency in accordance with the content of the communicated packets. Thus, the way security protocols are applied must match with the confidentiality required for that specific packet [30][31]. This creates a need to classify different levels of communication before even relaying/broadcasting them at the WSN used in PP. The most crucial part in any WSN application nowadays is ensuring that the network supports an end-to-end encryption and authentication. Critical key points to be considered in small cells applications for guaranteed security and privacy protection are as follow: 1) personal data collection, which if limited to certain extent can significantly help in mitigating several security issues, 2) data and traffic analysis for WSN based applications requires information sharing, therefore service providers and the technology partners should come into an agreement for a secure data handling methods which assures the mobile user privacy protections by considering the de-identification concept for example, 3) reliability of the WSN itself, encryptions and digital signature per user are also important aspects to be considered in this domain alike with the management protocols and physical security of the WSN, 4) human errors, which can elevate security risks and breaches, and thus, customized policies and procedures are required to mitigate the oversight issues, 5) lastly, transparency of the WSN usage/configuration assures the integrity of such systems in wireless technology domains and necessitates accountably clear policies with respect to the offloaded data security and privacy.

10.4.4 Routing

Many new routing strategies have been proposed to solve routing problems in traditional WSNs [32][33]. A Particle Swarm based routing approach for fault-tolerant optimization in heterogeneous WSNs which uses a hybrid routing scheme to calculate and maintain k-disjoint paths from source to Sink is proposed as well in [34]. It presents a model to solve the fault-tolerant optimization routing for sensor nodes that are densely distributed in a heterogeneous wireless environment. It employs an intelligent swarm algorithm that provides a faster way to recover the k-disjoint multipath from failure. It is a cooperative algorithm, which defines each particle as an antibody to generate a new population in the space search. Cooperative routing algorithms provide accuracy in finding the optimal solution by jumping out to the local optima, while minimizing the energy consumption, and shorting the end-to-end delay in the packet delivery process [35]-[37].

Meanwhile, authors in [38] have used a simple form of fault tolerance mechanism, which relies on the directed connected graph concept. The construction of topological heterogeneous WSNs consists of two types of sensor equipment, arranged in two layers. The lower layer is formed by traditional sensor nodes with restricted resources, which respond for any task, such as processing, transmission, and sensing data. The second layer consists of macro nodes with more capabilities in energy, processing, and storage and have the responsibility of decision routing. The design of routing protocols for PP, however, is still an open research area. In addition to the major issues of designing routing protocols in WSNs, there are new characteristics and constraints due to the nature of the energy-hungry multimedia exchange that must be handled over the network such that routing protocols for WSNs are not applicable to PP. Energy waste occurs mainly due to signaling overhead and unsuccessful transmissions during route discovery and repair. Routing protocols can be broadly classified into table-driven and on-demand protocols [32]. Table-driven routing is proactive, requires periodic advertisement of routing information, and is not considered to be energy efficient due to energy consumption associated with the routing overhead. On-demand routing is reactive, in which a transmission path is created only when needed. We will focus on on-demand routing, taking account of node mobility. There exist three main approaches to improve energy efficiency: i) minimum energy routing which selects the most energy efficient path between the source and destination [33][39], ii) data aggregation to concentrate the data traffic on some paths while switching off the MNs at the light loaded paths [40], and iii) energy balancing which aims at balancing the remaining battery energy at the MNs in establishing a route from the source to the destination [41][42]. However, the first two approaches can exhaust the batteries of MNs along the selected paths. It can result in network partitioning which will eventually lead to a network failure. The third approach does not minimize energy consumption. We suggest developing energy efficient routing solutions that can balance energy saving with network lifetime maximization while providing a stable queuing behavior. Also, we recommend more studies on cross-layer optimization for MAC and routing protocols. In a packet switching network, packets are often queued in transmitters for statistical multiplexing over a radio channel. While data aggregation routing can achieve energy saving, the queuing behavior of the network can be unstable if many paths are switched off and all data traffic is concentrated on only a few paths, due to high traffic intensity in the active paths. The queuing instability can have a severe detrimental impact on network performance and has not been studied. We will investigate the performance trade-off between the amount of energy saving (via data aggregation and path on/off switching) and the network stability. The alternating active/sleep modes at the link layer pose significant technical challenges in the queuing instability analysis. It is strongly recommended to address these challenges by using the statistical link layer model to be developed for PP, and by exploring conditions on the inter-arrival and service times of the queuing network [43]. This will shed some light on how many paths can be switched off and how to perform data aggregation so as to balance the amount of energy saving with the network stability. While data aggregation routing can reduce energy consumption, it can also reduce the overall network lifetime. Network lifetime can be the time to the first node failure, the time to the first network partition, or the time to unavailability of some application functionality [44]. Concentrating data routing on the minimum energy paths or data centric paths can exhaust the batteries of the MNs along the paths, which in turn reduces the network lifetime. Hence, an energy efficient routing protocol should balance energy saving with network lifetime maximization. Thus, we suggest first establishing an analytical model of the network lifetime as a function of MN residual energy, data traffic characteristics, and QoS requirements, based on MN energy models [45][20] and graph theory [21]; Second, to examine the energy consumption amounts for the minimum energy paths and for the data traffic centric paths, and evaluate performance trade-off between energy efficiency and network lifetime for the routing strategies; Then, a routing protocol to balance energy efficiency with network lifetime maximization can be developed. The problem can be formulated as a multi-objective optimization, subject to QoS requirements and queuing stability constraints. We recommend using a weighting parameter to achieve desired flexibility in the routing protocol, such as balancing energy saving and network lifetime or favoring one over the other. When each MN alternates its state between the active and sleep modes with the duty cycle adaptive to data traffic load and radio channel condition, the link layer exhibits a dynamic and random on/off behavior. A key issue to be addressed also is how to achieve a reliable end-to-end path under the link dynamics. One approach is to explore group routing, to route data packets through MN groups instead of employing a single forwarding MN [22]. If an MN along the most energy efficient path is in a sleep state, the MN's active neighbor(s) group can create an alternate suboptimal energy efficient path towards the destination. Further, MN transmit power is a control variable in the cross-layer design. By exploiting wireless broadcast nature in packet forwarding, we will develop a cooperative multi-hop relaying solution, which adaptively re-organizes the end-to-end path according to the instantaneous active/sleep modes of MNs, channel conditions, and MN locations to minimize the end-to-end outage probability while reducing the associated route establishment and maintenance cost.

10.4.5 Node-to-Node Energy Efficient Distributed Resource Allocation

The most substantial obstacle facing the deployment of small long-life sensor nodes is the need for major reductions in energy consumption to maximize the energy saving of the system. An energy-aware design that highlights the elegant scalability of energy consumption with factors such as available resources and their significance, event frequency of occurrence, and desired output quality, at all levels of the system hierarchy. The design for energy-aware sensing nodes emphasizes the association between different layers of the sensing nodes stack to provide energy-quality trade-offs given that the hardware is designed for scalable energy consumption. The quality of service (QoS) in PP applications is mainly determined by the duty cycle of the data collection and the reliability of the collected information. In PP applications, the collected data will be varying in unpredictable manner. For example, a sensor for the soil moisture will not expect major changes over night in the absence of rain fall or irrigation. On the contrary, a rain fall or starting the irrigation system will cause a drastic change that has to be reported to the data processing center in real time to enable the irrigation control system to stop the irrigation process at certain times. Such behavior creates a variable data traffic for the network and variable duty cycle for the sensing nodes. Therefore, an event triggered transmission is necessary to maximize the energy efficiency of the nodes since the duty cycle for particular nodes can be as low as 5%.

One of the major limitations of the event-triggered WSN is that large number of sensing nodes might be triggered simultaneously, which can cause network congestion, and it may overwhelm the data processing center with a massive amount of data to process, analyze, store and interpret. However, a large part of the collected data could be correlated temporally and specially, which results in transmitting a large amount of redundant information. Consequently, an optimum energy-aware WSN should consider the data correlation before even sending this data to the processing center. Towards this goal, it is necessary to enable node-to-node (n2n) communications so that the group of nodes in a given cluster are aware of the status of each other and hence they can select a small subgroup of nodes to transmit their data. At the data processing center, data fusion techniques can be utilized to create the desired field map. One possible approach is to borrow video processing techniques where particular nodes can be configured as base nodes and other nodes can be configured as enhancement nodes. The enhancement nodes can be classified with different enhancement levels. Then, we can use differential information measurements to select the frequency and level of nodes that will transmit. A simple example of a node cluster that consists of 20 nodes is given in Figure 10.5. The base nodes' layer has four (black) nodes, the first enhancement layer (dark gray) has eight nodes and finally the second enhancement layers (light gray) consists of 8 nodes as well. This structure has some similarity with the hierarchal modulation used in video transmission.

In the literature, the n2n communication is commonly referred to as device-to-device (D2D) communications. The D2D has received increased attention for wireless cellular networks because it can noticeably off-load the network traffic and reduce the transmission energy since the communicating devices are in close proximity of each other. Bluetooth is a simple example of D2D communications. However, the n2n approach and adopting the hierarchal topology requires mainly

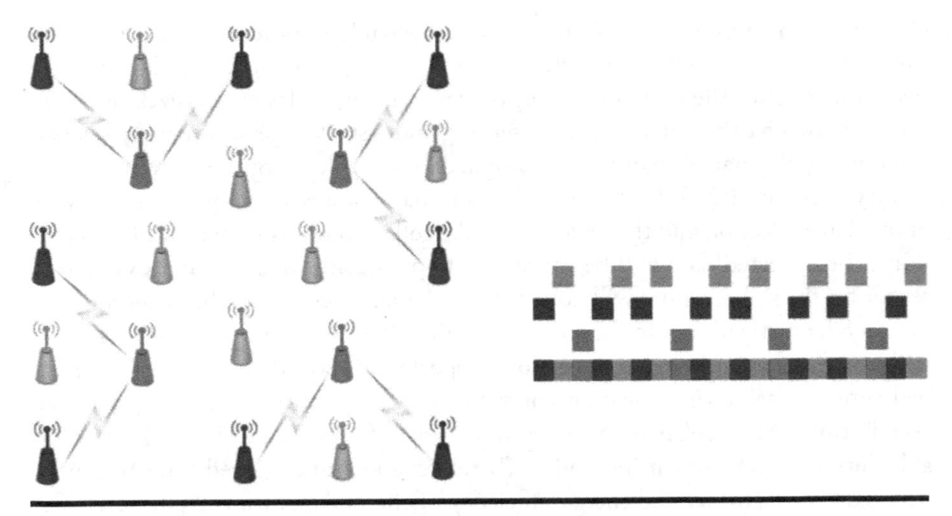

Figure 10.5 Example for a hierarchical node configuration with three levels. The black nodes are the base nodes. The dark gray are the enhancement nodes level-1 and the light gray nodes are the enhancement nodes level-2.

a specifically designed protocols because with n2n configuration the network becomes heterogeneous in the sense that it has distributed and centralized processing components. In [46][47], authors considered the problem of distributed throughput optimization in wireless ZigBee networks as well as the problem of resource allocation in distributed heterogeneous wireless cellular networks. The results obtained revealed that the distributed topologies can be of significant impact on the network performance. However, the works in [46] and [47] were not designed under the same assumptions and conditions we are considering for PP applications. Hence, extending these algorithms to WSN for PP applications is not straightforward, and it will be one of the aims of any WSN applied in PP in the near future.

10.5 WSNs Prototyping and Implementation in PP

A typical wireless sensor node, depicted in Figure 10.6, consists of a power unit, processing unit, memory unit, sensing unit, and a communication unit. Each of these units has a strong impact on the overall performance of the WSN. The processing unit is the core component that collects the signals captured by the sensors, processes the gathered information, and transmits the information to the network via the communication unit. The processing unit has also access to a memory unit where the information can be stored if required. The communication unit allows communication with other nodes in the WSN using a wireless communication channel via a radio, infrared or laser based link.

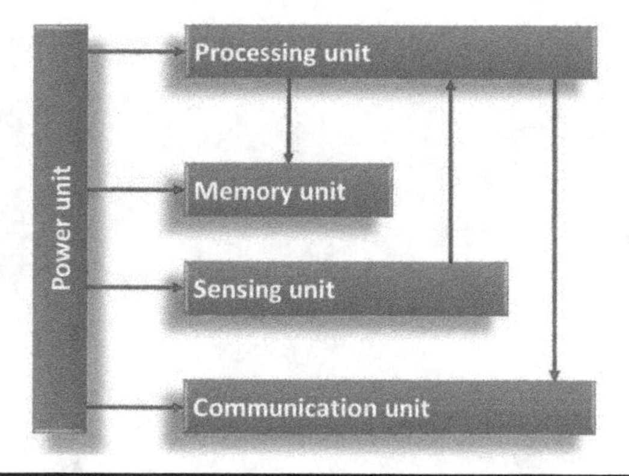

Figure 10.6 A high level block diagram of a sensing node.

The power unit provides the energy supply for the node and is composed of batteries, which may be rechargeable. The sensor unit is composed of multiple sensors and their choice depends on the application domain of WSNs. Typical sensors used for PP applications include soil moisture, relative humidity, temperature, and gas (carbon dioxide (CO2), methane (CO4), carbon monoxide (CO)) detection sensors [48]. Besides the above mentioned hardware modules, the most essential component of a WSN node is a dedicated micro-kernel. The operating systems used for other hand-held devices cannot be used for WSN nodes because of their small size and rugged environments, and thus special attention has to be paid to problems like localization, filtering, energy consumption and security. The most important implementation aspect of a WSN node is the choice of the processing unit, which greatly affects the energy consumption and the computational ability of a WSN node.

The typical choices are to use an off-the-shelf WSN node with a built-in processing unit, a microcontroller or a FPGA unit as the processing unit. Some of the commonly used off-the-shelf WSN nodes include Mica2, Mica2Dot, MicaZ, Fleck 3node and TinyNode mote. These WSN nodes contain a microcontroller, which is used as the processing unit, along with most of the other blocks depicted in Figure 10.7 on the same board. The WSN can be setup by programming the microcontrollers and thus this choice provides the best time-to-market but on the downside it does not provide the flexibility to change components. The other two choices, i.e., microcontroller and FPGA based processing units overcome this problem and are discussed in detail below.

10.5.1 Microcontrollers

Nowadays microcontrollers include their own memory (both volatile and non-volatile) units and various typically used standard modules like Analog-to-Digital converters, UART, timers and counters. Thus, a WSN node can be constructed using a

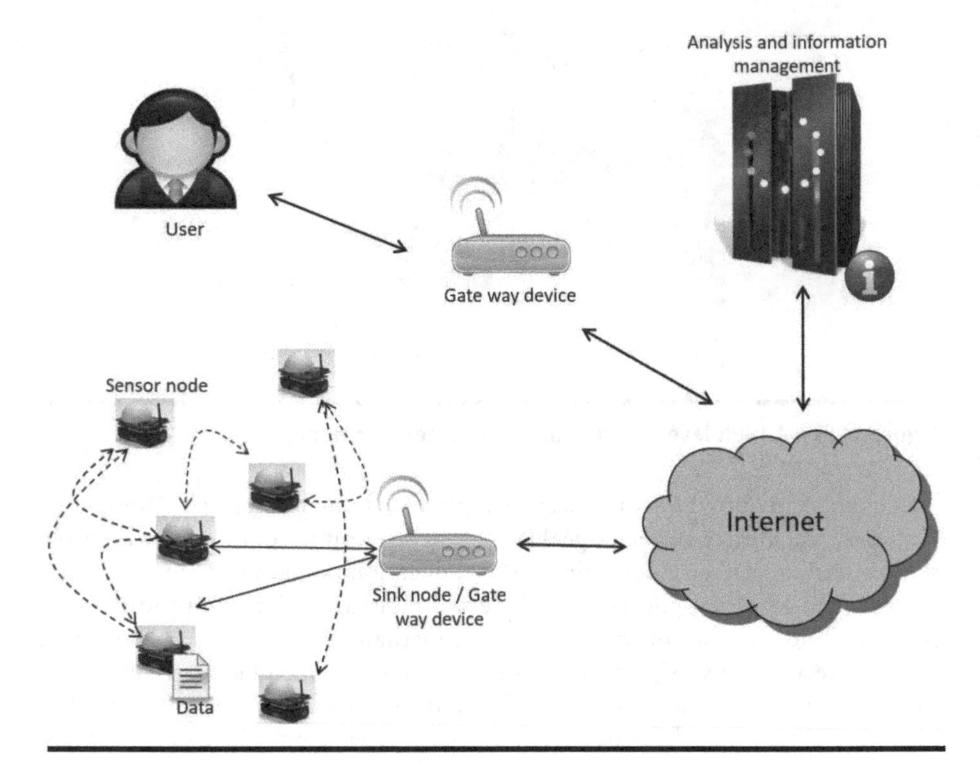

Figure 10.7 WSN architecture for PP applications.

microcontroller by integrating such a microcontroller with a communicating unit and sensors. A wide range of microcontrollers dedicated for WSN applications are available [49] and can have 4 to 32 data bits, 512 Bytes to 128KB RAM, 1 to 6 timers, power dissipation as low as 1.9mW and the appropriate choice can be made based on the requirements.

10.5.2 FPGA

The FPGA technology provides the ability to develop customized WSN nodes and a dedicated single-purpose processor can be developed in HDL along with the required timers, counters, and communication protocols such as UART and USB. Moreover, the soft-core of an already developed general purpose or an application processor can also be tweaked to meet the design requirements. The major challenge in this regard is the comparatively high power dissipation. For this purpose, ultra-low-power FPGA units need to be used for WSN node development. The main benefit for using FPGA compared to microcontrollers for implementing WSN nodes is their performance as has been demonstrated in [50]. The high performance feature of FPGAs can be leveraged upon to build on-chip image-processing sensors [51] for WSN nodes.

10.5.3 WSN Prototype Implementation

In general, we recommend to use Flash based FPGA, such as ACTEL Fusion FPGA module, for developing the desired sensor nodes in WSNs designed for PP. The availability of Flash would allow using low power sleep modes instead of the wake-up reconfiguration that is required in the other types of FPGAs. We also recommend to use the ARM Cortex M1 core on our FPGA and this decision is mainly motivated by the high performance and small size of this core, which is desirable for PP applications. This processor also provides a high and a low speed buses. For the analog components, like ADC and multiplexer, numerous off-the-shelf components can be placed on the same board as the FPGA. The required filters can be implemented on the FPGA. These FPGA-based sensor nodes will be remotely accessed by researchers and/or public users via the internet connection as shown in Figure 10.7. Moreover, advances in cloud computing will empower such kind of light-weight networks, i.e., the WSNs, by offloading the heavy analysis and processing part up to the High Performance Computing (HPC) machines to filter and process raw data coming directly from the field. Accordingly, farmers and researchers can collect comprehensive and precise yield data without significant efforts and inexpensively as shown in Figure 10.8. Yield maps can be generated in real-time subsequent to data collection to identify yield general patterns within fields. These maps allow recognizing within-field spatial variability for variable-rate applications, empowering farmers to estimate the economic revenues of different farming management plans. Furthermore, they are essential for field-level developments such as leveling the land, timing of irrigation systems, drainage, building fences, and for off the field data usage.

The information provided in Plant Phenotyping (PP) represents a valuable resource because it enables real-time decision making with regard to critical issues, such as

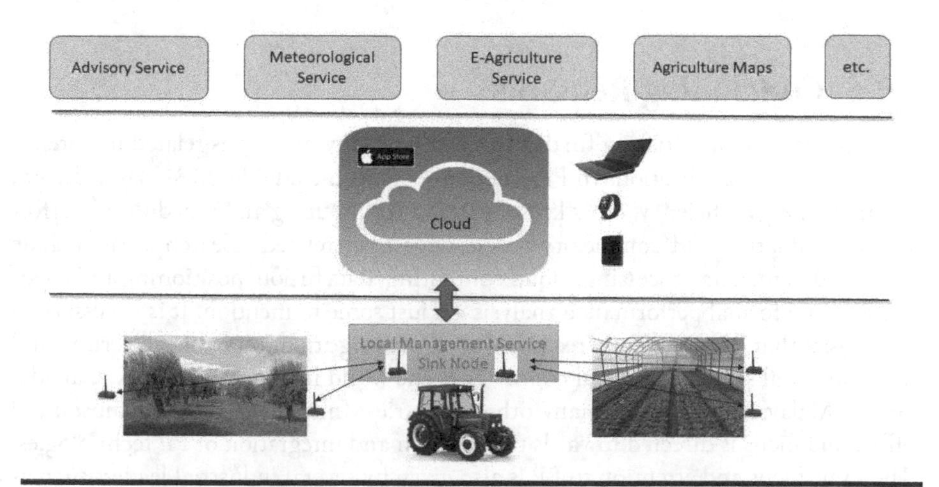

Figure 10.8 A block diagram of the wireless sensor network nodes in the Cloud.

establishing water saving policies while providing adequate irrigation and choosing the right time to farming activities such as planting, harvesting, specifying the fruit maturity etc. In order to provide services for relatively large areas of coverage, it may be necessary to employ large numbers of WSN nodes, which use applications and communication protocols tailored mainly to provide higher energy efficiency. A popular approach in energy efficient routing is to use cooperative game theory. Recently, this theory is widely used in studies associated with wireless networks [52]. Since energy is highly wasted in such networks due to extensive cooperation in data forwarding. Therefore, the game theoretic mechanism which applies the concept of gains and losses has been implemented in [53] for WSNs. Game theory can also be used to analyze the gains a WSN nodes can make and can certainly lead towards the achievement of equilibria for all the concerned nodes. Moreover, game theory has been used in various articles for selfish nodes management. Incentive based mechanism has been successfully applied in many approaches for load balancing, energy efficiency etc. in WSNs. Such mechanisms are categorized into credit-based and reputation based [54]. The reputation-based mechanism relies on the evaluation of nodes' behavior. In this type of approaches, different reputation stages are made to determine the nodes' cooperation level [55]. The message forwarding by the WSN intermediate nodes are made according to a reputation value at the source [56][57]. In the credit-based approach, a WSN node can gain credit-scores by offering these relaying services.

It's worth pointing out that the initial deployment and routing based decisions for the WSN nodes and the configuration of wireless communications related components can make significant differences for the evaluation process in PP. Starting the evaluation process by employing the prototype straight from the beginning and checking various strategies can be quite costly especially in terms of time and efforts. Therefore, similar to the existing studies in the literature, we believe that it can be ideal to use simulation tools prior to deployment in an attempt to have a degree of optimization beforehand.

10.6 Concluding Remarks

This article aims at initiating further research in many directions related to wireless sensor networks applications in PP. These directions include the MAC layer design, security, energy efficiency, cross layer optimization, routing and scheduling, performance evaluation and complexity reduction. Other related research areas include sensors' design, data processing, cloud computing, data fusion, positioning and localization, and formal performance analysis are just some to mention. It is necessary to emphasize that PP is currently receiving an increasing attention by the governmental and industrial sectors in several countries in the world including the USA, Canada, Brazil, Malaysia, India, and many other countries. In such countries, a substantial effort and focus is directed towards the research and integration of PP technologies. The same focus and attention to PP is also given by several industrial leaders such as IBM [58], AG Leader [59], and Precision Planting [60]. The immediate next steps towards involving WSNs in large scale PP deployments, is the incorporation of the state-of-the-art technology such as IoT and Cloud computing.

As a future work, we foresee the immediate next steps to include field studies with large-scale deployments, data collection and mining, in addition to incorporating the state-of-the-art information technology such as the Internet of things and cloud/edge computing in order to further improve the performance. These kind of technologies are expected to have a significant impact on the field of PP in particular, and wireless networks in general. Therefore, the output of any research in this direction will initiate further advances in many directions related to the PP applications. Furthermore, the measurement criteria for the success of any PP underlying infrastructure will be based on the energy efficiency of the custom nodes, and MAC/Routing algorithms, as well as other well-known quality of service (QoS) related measures which are relevant to agriculture applications such as delay, consistency and reliability.

References

[1] F. Yalçınkaya, N. Çakar, M. Mısırlıoğlu, N. Tümer, N. Akar, M. Tekin, H. Taştan, H. Koçak, N. Özkaya, A. H. Elhan, "Genotype–phenotype correlation in a large group of Turkish patients with familial Mediterranean fever: evidence for mutation-independent amyloidosis", *Rheumatology*, vol. 39, no. 1, 2000, pp. 67–72.
[2] http://www.ncbi.nlm.nih.gov/pubmed/18353061.
[3] http://www.ncbi.nlm.nih.gov/pubmed/10662876.
[4 http://link.springer.com/article/10.1007%2Fs12288-014-0406-0#page-1.
[5] http://www.tuik.gov.tr/PreHaberBultenleri.do?id=18566.
[6] M. F. Domingues, and A. Radwan, "Optical Fiber Sensors for IoT and Smart Devices," Springer Publishing Company, Feb 2017.
[7] http://www.cografya.gen.tr/egitim/ekonomik/turkiye-de-tarim.htm.
[8] Qatar National Food Security Program, available at: www.qnfsp.gov.qa.
[9] R. Heuvel, "The promise of precision agriculture", Journal of Soil and Water Conservation, vol. 51.1, p. 38, 1996.
[10] D. Berckmans, "Automatic on-line monitoring of animals by precision livestock'" Proceedings of the International Society for Animal Hygiene, Saint-Malo, 2004, pp. 27–30.
[11] Mutka and R. Bart, "Image-based phenotyping of plant disease symptoms", Plant Sc., 2014.
[12] J. Humplik, D. Lazar, A. Husickova and L. Spichal, "Automated phenotyping of plant shoots using imaging methods for analysis of plant stress responses – a review", Plant Methods, vol. 11, no. 29, pp. 1–10, April 2015.
[13] J. Araus and J. Cairn, "Field high-throughput phenotyping: the new crop breeding frontier", Trends in Plant Science, vol. 19, no. 1, pp 52–61, January 2014.
[14] L. Li, Q. Zhang and D. Huang, "A Review of Imaging Techniques for Plant Phenotyping", Sensors, pp. 20078–20111, 2014.
[15] F. Al-Turjman, H. Hassanein and M. Ibnkahla, "Towards prolonged lifetime for deployed WSNs in outdoor environment monitoring", Elsevier Ad Hoc Networks Journal, vol. 24, no. A, pp. 172–185, Jan 2015.

[16] F. Al-Turjman, "Price-based Data Delivery Framework for Dynamic and Pervasive IoT", Elsevier Pervasive and Mobile Computing Journal, vol. 42, pp. 299–316, 2017.
[17] M. Ishizuka, M. Aida, "Performance study of node placement in sensor networks," Proc. 24th Int. Conference on Distributed Computing Systems Workshops (Icdcsw), vol. 7, Mar. 2004.
[18] F. Al-Turjman, L. J. Poncha, S. Alturjman and L. Mostarda, "Enhanced Deployment Strategy for the 5G Drone-BS Using Artificial Intelligence", IEEE Access, vol. 7, no. 1, pp. 75999–76008, 2019.
[19] F. Al-Turjman, H. Hassanein and M. Ibnkahla, "Efficient deployment of wireless sensor networks targeting environment monitoring applications", Elsevier: Computer Communications Journal, vol. 36, no. 2, pp. 135–148, Jan 2013.
[20] M. Jongerden and B.R. Haverkort, "Battery modeling," Technical Report TR-CTIT-08-01, 2008.
[21] J. Aldous and R. Wilson, Graphs and Applications: An Introductory Approach. Springer-Verlag, 2000.
[22] Y. Chang and C. Hsu, "Routing in wireless/mobile ad-hoc networks via dynamic group construction," Mobile Networks and Applications, vol. 5, pp. 27–37, 2000.
[23] A. Abdrabou and W. Zhuang, "A position-based QoS routing scheme for UWB ad hoc networks," IEEE Journal on Selected Areas of Communications, vol. 24, no. 4, April 2006.
[24] N. Ray and A. Turuk, "A review on energy efficient MAC protocols for Wireless LANs," Proc. 4th Int. Conf. Industrial and Information Systems, pp. 137–142, Dec. 2009.
[25] S. Choudhury and F. Al-Turjman, "Dominating Set Algorithms for Wireless Sensor Networks Survivability" IEEE Access Journal, vol. 6, no. 1, pp. 17527–17532, 2018.
[26] J. Zhang, G. Zhou, C. Huang, S. H. Son, J. A. Stankovic, "TMMAC: an energy efficient multi-channel MAC protocol," Proc. IEEE ICC, 2007.
[27] F. Al-Turjman and M. Abujubbeh, "IoT-enabled Smart Grid via SM: An Overview", Elsevier Future Generation Computer Systems, vol. 96, no. 1, pp. 579–590, 2019.
[28] F. Al-Turjman, "Modelling Green Femtocells in Smart-grids", Springer Mobile Networks and Applications, vol. 23, no. 4, pp. 940–955, 2018.
[29] F. Al-Turjman, and S. Alturjman, "Context-sensitive Access in Industrial Internet of Things (IIoT) Healthcare Applications", IEEE Transactions on Industrial Informatics, vol. 14, no. 6, pp. 2736–2744, 2018.
[30] V. Sucasas et al. "An OAuth2-based protocol with strong user privacy preservation for smart city mobile e-Health apps," ICC, 2016.
[31] F. Al-Turjman, and S. Alturjman, "Confidential Smart-Sensing Framework in the IoT Era", The Springer Journal of Supercomputing, 2018. DOI. 10.1007/s11227-018-2524-1.
[32] A. Singh, H. Tiwari, A. Vajpayee, and S. Prakash, "A survey of energy efficient routing protocols for mobile ad-hoc networks," Int. J. Computer Science & Engineering, vol. 2, no. 99, pp. 3111–3119, 2010.
[33] Y. Wang, X. Li, W. Song, M. Huang, and T. Dahlberg, "Energy-efficient localized routing in random multihop wireless networks," IEEE Trans. Parallel and Distributed Systems, vol. 22, no. 8, pp. 1249–1257, Aug. 2011.

[34] M. Z. Hasan, and F. Al-Turjman, "SWARM-based data delivery in Social Internet of Things", Elsevier Future Generation Computer Systems, 2017. DOI: 10.1016/j.future.2017.10.032.
[35] F. B. Saghezchi, et al, "Coalition Formation game toward green mobile terminals in heterogeneous wireless networks," IEEE Wireless Communications, 20(5), 85–91, Oct. 2013.
[36] F. B. Saghezchi, et al, "Energy-aware relay selection in cooperative wireless networks: An assignment game approach," Ad Hoc Netowrks, vol. 56, Mar. 2017.
[37] F. Al-Turjman and H. Zahmatkesh, "An Overview of Security and Privacy in Smart Cities' IoT Communications", Wiley Transactions on Emerging Telecommunications Technologies, 2019. DOI. 10.1002/ett.3677.
[38] Y. Hu, Y. Ding, and K. Hao, "An immune cooperative particle swarm optimization algorithm for fault-tolerant routing optimization in heterogeneous wireless sensor networks," J. Math. Problems Eng., vol. 2012, pp. 19, Aug. 2011.
[39] M. Dehghan, M. Ghaderi, and D. Goeckel, "Minimum-energy cooperative routing in wireless networks with channel variations," IEEE Trans. Wireless Communications, vol. 10, no. 11, pp. 3813–3823, Nov. 2011.
[40] Y. Kim, E. Lee, and H. Park, "Ant colony optimization based energy saving routing for energy-efficient networks," IEEE Communications Lett., vol. 15, no. 7, pp. 779–781, July 2011.
[41] C. Ma and Y. Yang, "A battery-aware scheme for routing in wireless ad hoc networks," IEEE Trans. Vehicular Technology, vol. 60, no. 8, pp. 3919–3932, Oct. 2011.
[42] F. Ren, J. Zhang, T. He, C. Lin, and S. K. Das, "EBRP: Energy-balanced routing protocol for data gathering in wireless sensor networks," IEEE Trans. Parallel and Distributed Systems, vol. 22, no. 12, pp. 2108–2125, Dec. 2011.
[43] M. Bramson, Stability of Queuing Networks. Springer, May 2008.
[44] M. Z. Hasan, and F. Al-Turjman, "Analysis of Cross-layer Design of Quality-of-Service Forward Geographic Wireless Sensor Network Routing Strategies in Green Internet of Things", IEEE Access Journal, vol. 6, no. 1, pp. 20371–20389, 2018.
[45] D. Chen, H. Ji, and X. Li, "An energy-efficient distributed relay selection and power allocation optimization scheme over wireless cooperative networks," IEEE ICC, pp. 1–5, June 2011.
[46] F. Al-Turjman, and S. Alturjman, "5G/IoT-Enabled UAVs for Multimedia Delivery in Industry-oriented Applications", Springer's Multimedia Tools and Applications Journal, 2018.
[47] M. Ismail and W. Zhuang, "A distributed multi-service resource allocation algorithm in heterogeneous wireless access medium," IEEE Journal on Selected Areas of Communications, vol. 30, no. 2, pp. 425–432, Feb. 2012.
[48] F. Al-Turjman, "A Novel Approach for Drones positioning in Mission Critical Applications", Wiley Transactions on Emerging Telecommunications Technologies, 2019. DOI. 10.1002/ett.3603.
[49] M. Lakhzouri, Simona Lohan, Ridha Hamila, and Markku Renfors, "Extended Kalman filter for LOS estimation in WCDMA mobile positioning," in EURASIP Journal on Applied Signal Processing, vol. 13, pp. 1268–1278, Dec. 2003.
[50] R. Hamila, Markku Renfors, Gudni Gunnarsson, and Marko Alanen, "Data processing for mobile phone positioning," in Proc. IEEE 50th Vehicular Technology Conference, VTC'99, pp. 446–449, Sep. 1999.

[51] Yoo, Seong-eun, et al. "A2S: Automated agriculture system based on WSN," IEEE International Symposium on Consumer Electronics (ISCE), pp. 1–5, 2007.
[52] F. Al-Turjman, "Cognitive Routing Protocol for Disaster-inspired Internet of Things", Elsevier Future Generation Computer Systems, vol. 92, 1103–1115, 2019.
[53] T. AlSkaif, M. Guerrero Zapata, and B. Bellalta, Game theory for energy efficiency in Wireless Sensor Networks: Latest trends, Journal of Network and Computer Applications, vol. 54, pp. 33–61, Aug. 2015.
[54] B. Arisian and K. Eshghi, A Game Theory Approach for Optimal Routing: In Wireless Sensor Networks," International Conference on Computational Intelligence and Software Engineering, Sep. 2010.
[55] M. Umar; S. Khan; R. Ahmad; D. Singh "Game Theoretic Reward Based Adaptive Data Communication in Wireless Sensor Networks", IEEE Access, 2018. DOI: 10.1109/ACCESS.2018.2833468.
[56] Z. Chu, H. X. Nguyen, T. A. Le, M. Karamanoglu, D. To, E. Ever, F. Al-Turjman and A. Yazici, "Game Theory Based Secure Wireless Powered D2D Communications with Cooperative Jamming", IEEE Wireless Days conference, Porto, Portugal, pp. 95–98, 2017.
[57] C. Zhu, Y. Wang, G. Han, J.J.P.C. Rodrigues, J. Lloret, LPTA: location predictive and time adaptive dat a gathering scheme with mobile sink for wireless sensor networks, The Scientific World Journal 2014.
[58] www.research.ibm.com/articles/precision_agriculture.shtml.
[59] http://www.agleader.com.
[60] www.precisionplanting.com.

Index

D

N

O